Analysing Design Activity

Analysing Design Activity

Edited by

Nigel Cross,
Henri Christiaans,
Kees Dorst,
Delft University of Technology, The Netherlands

John Wiley & Sons
Chichester · New York · Brisbane · Toronto · Singapore

Other Wiley Editorial Offices

John Wiley & Sons, Inc., 605 Third Avenue,
New York, NY 10158-0012, USA

Jacaranda Wiley Ltd, 33 Park Road, Milton,
Queensland 4064, Australia

John Wiley & Sons (Canada) Ltd, 22 Worcester Road,
Rexdale, Ontario M9W 1L1, Canada

John Wiley & Sons (SEA) Pte Ltd, 2 Clementi Loop #02-01,
Jin Xing Distripark, Singapore 0512

Britrish Library Cataloguing in Publication Data

A catalogue record for this book is available from the British Library

ISBN 0 471 96060 8

Typeset in 10/12pt Palatino by Acorn Bookwork, Salisbury

Contents

Introduction: The Delft Protocols Workshop

Nigel Cross, Henri Christiaans and Kees Dorst
Delft University of Technology, The Netherlands

Design activity encompasses some of the highest cognitive abilities of human beings, including creativity, synthesis and problem solving. Every normal person is capable of exercising such abilities, but it is in design activity that they are most frequently stretched to their limits, and the most able designers clearly exercise some exceptional levels of ability. The study and analysis of design activity therefore offers significant intellectual challenges. Nonetheless, this is a growing research field, stimulated perhaps both by the challenge and by the increasingly wide recognition of the value of design ability.

A substantial and varied range of research methods has been developed and adopted for the analysis of design activity. The range extends from philosophical reflection to empirical investigation, and includes study of both the natural and the artificial intelligence of design. This variety was well reviewed in the first Delft Workshop on 'Research in Design Thinking' in 1991[1].

This book contains the proceedings of the second Delft Workshop, 'Research in Design Thinking II – Analysing Design Activity', in 1994, which focused on one particular research method, that of protocol analysis. Of all the empirical, observational research methods for the analysis of design activity, protocol analysis is the one that has received the most use and attention in recent years. It has become regarded as the most likely method (perhaps the only method) to bring out into the open the somewhat mysterious cognitive abilities of designers.

In essence, protocol analysis relies on the verbal accounts given by subjects of their own cognitive activities. It is difficult to imagine how else we might examine what is going on inside people's heads, other than by asking them to tell us what they are thinking. Of course, this is fraught with difficulties. People do not necessarily know what is going on inside their own heads, let alone have the ability to verbalize it.

Nevertheless, people do normally find it relatively straightforward to give a verbal account of what they believe they are thinking, or what they were thinking recently. Retrospective verbal accounts (i.e. recalling what one was thinking recently) offer one means of getting at cognitive activity which is frequently used not only in research but also in everyday interchanges: 'What were you thinking when you were doing that?'. Concurrent verbal accounts (i.e. 'thinking aloud') offer the researcher the hope that they really do externalize – or allow insight into – at least some of the subjects' cognitive activities. As Ericsson and Simon claim, in their standard work on protocol analysis, 'There is a dramatic increase in the amount of behaviour that can be observed when a subject is performing a task while thinking aloud compared to the same subject working under silent conditions. A brief instruction to think aloud usually suffices to bring about this major change in observable behaviour'.[2]

However, there are some significant disadvantages of think-aloud protocols, including the following. Firstly, there may well be side-effects of the verbalization, such that it actually changes the subject's behaviour and their cognitive performance. Secondly, what the subject reports may well be incomplete accounts of what their cognitive activity actually is. Thirdly, the subject may, quite unintentionally, give irrelevant accounts, reporting parallel but independent thoughts to those that are actually being employed in the task. All of these disadvantages weigh particularly heavily on the validity of protocol analysis in design, where 'non-verbal thinking' is believed to be a significant feature of the relevant cognitive activities, and where the use of sketches and similar externalization of thought processes seems to be fundamental. (See particularly the paper by Lloyd, Lawson and Scott, Chapter 20, in this volume.)

1 History of Protocol Analysis

The analysis of think-aloud protocols emerged as a method of psychological research in the 1920s. From the beginning it was a method of seeking insight into problem solving. The early studies were limited, however, by the researcher's ability to take accurate notes of the subject's verbalizations. It was not until after 1945, when tape recorders became available, that more accurate methods of verbal data collection allowed more precise, and less selective, studies to be undertaken. However, it was not until the 1960s that the first major protocol studies were made. Notably, thinking-aloud protocol analyses were used by de Groot[3] in his studies of chess playing, and by Newell and Simon[4] in their studies of cryptarithmetic and logical problem solving. In the 1970s, the availability of video-recording added a new dimension, in which the non-verbal behaviour of subjects could be studied alongside their verbal reports.

The number and variety of protocol studies of design activity have grown significantly in recent years. The first report of such a study was that by Eastman[5], who studied architects. Architectural design has continued to be a rich subject area for such studies[6-11]. As well as the conventional, single-subject, think-aloud protocol studies, the use of dialogue exchanges between two or more collaborating subjects has also been employed[12].

It was not until the late 1980s that protocol studies of engineering design began to appear[13-15], but the studies in that domain have increased rapidly since then[16-19]. A significant feature of several of these engineering design studies has been their further extension of the method of analysis to team design activity[20-22]. The classical protocol study relies on an individual subject thinking aloud; this is not possible in teamwork, of course, but the verbal exchanges of members of a team engaged in a joint task do seem to provide data indicative of the cognitive activities that are being undertaken by the team members.

The engineering design studies have been undertaken mainly in the mechanical engineering domain, although some work in electronic engineering has also been undertaken[23]. A rapidly-growing field of study which has also begun to use protocol analysis is that of software design[24-26], where teamwork has also begun to figure significantly, particularly related to the design of computer-based support systems (both software and hardware design) for cooperative or collaborative work[27]. Most researchers using protocol analysis have studied design activity within one domain, but there are a few who have attempted to make comparisons across several domains[28,29]. The industrial (product) design domain has been studied relatively little through protocol analysis until very recently[30,31]. Industrial design is the domain in which we, the originators of the Delft Workshop, are situated, and which we therefore particularly wished to focus upon.

The last decade has seen a major growth in the use of protocol studies as a method of analysing design activity. However, it has not been easy to draw general comparisons, nor even to agree general procedures or standards for such studies, because of their scattered and independent nature. It was therefore most appropriate that the 1994 Delft Protocols Workshop was designed precisely to force a concentration on the research methodology itself, and to encourage more coordinated progress in analysing design activity in the near future.

2 Workshop on Analysing Design Activity

The idea for this Workshop originated with Kees Dorst, in conversation at the ASME92 Conference with David Radcliffe, who brought the XeroxPARC and Stanford collaborators into the project. The aim of the Workshop was to bring together a distinguished group of design researchers (all well versed in protocol analysis) to compare analyses of the same data and to discuss the state of the art in protocol research.

Although the number of protocol studies in design has been growing steadily, this has not resulted in intensifying and broadening the discussion around protocol analysis. Most of the studies have been relatively isolated projects of research groups trying out the method for themselves, and exploring the design universe. The buildup of a 'community of knowledge' out of protocol research has been limited; discussions vital for the dialectic of the scientific inquiry have been few and far between.

One of the factors that keeps the dialectic low in the field of protocol analysis is the large number of variables that have to be set before the actual protocol study can get under way. The experimental layout, the choice and number of subjects, the domain and the assignment given to the subjects all influence the outcome of the protocol studies in hitherto unexplored ways. The different settings of all these variables in the studies make it very hard to compare and learn from each other's work. And protocol analysis happens to be so labour-intensive that single research projects just cannot yield statistically significant stand-alone data.

To foster discussion of protocol analysis we eliminated many of the above variables from the research equation. This was done by providing the Workshop participants with a set of 'standard' data, to be analysed by all of them. They were asked to perform the analysis in any form they saw fit. The common set of data should make it easier to compare and criticize each other's work. The aims of this whole Workshop undertaking were, therefore:

- to get an overview of the accumulated knowledge on design of these researchers
- to seek a common language in discussing protocol analysis and detailed design processes
- to 'validate' in this forum protocol analysis as a research technique
- to get a discussion going on the properties and limitations of protocol analysis research in design
- to discuss possible ways of using protocol analysis in the future, alone or in combination with other research techniques
- to form the basis of an international research network, a platform for discussion on these matters.

The data prepared for analysis at this Workshop were recordings of an individual designer and a three-person team of designers working for two hours designing a typical industrial design product, *a 'fastening device' that should allow a given backpack to be fastened onto a mountain bike*. The video tapes and transcribed protocols were sent to the participants some six months before the Workshop. The 20 participants in the Workshop were selected from the many applications received in response to a call for participation, with the aim of putting together a group of experienced researchers with a great diversity in goals and approaches to the analysis of design activities.

3 Experimental Procedure

The recordings were made in January 1994 at the Xerox Palo Alto Research Center (XeroxPARC), California, USA, and in collaboration with the Design Research Center of Stanford University. The experiment subjects were invited by either Xerox or Stanford to participate in a research project into design thinking. They were told that the research would involve them doing a short (two-hour) design exercise, within their familiar design domain. All subjects volunteered, and were not paid. The experiments were conducted in a design room equipped with video and audio recording facilities, and supplied with table and chairs, drawing pad, pens and pencils and whiteboard. Also in the room were a mountain bicycle and the backpack to be used for the design assignment. An experimenter was present in the room, who had a file of information and data for supply to the designers if they were to ask for it.

Two microphones and four video cameras recorded the activity in the design room. The four cameras were installed to capture different views:

- a general overview of the room
- a closer view of the subjects' faces when sitting at the table
- the whiteboard
- overhead view of the drawing pad/tabletop.

All four camera views were recorded separately on Hi-8 (NTSC) videocassettes. At the same time the images were mixed into one 4-PIP (four pictures in picture) combined view, recorded on SVHS (NTSC) videocassette. Time codes were added later to all tapes.

The experiment used the same setting as in earlier experiments in Delft by Christiaans and Dorst[31,32]. One experimenter (EXP1) is in the design room; a second experimenter (EXP2) is in a separate room, connected to EXP1 with a telephone headset and able to view and hear the activity in the design room. A third experimenter (EXP3) controls the video recording, logs the sequence of information cards which are asked for by the subjects, and keeps a summary of the procedures.

During the experiment the actions of the experimenter EXP1 are limited to:

- giving information, either by supplying the information on cards or by answering questions directly (sometimes drawing on advice from EXP2)
- reminding the individuals to think aloud, or speak louder, if necessary
- in some cases asking the designers to work in positions where they can be adequately recorded by the video cameras
- reminding the subjects of the time when they have half an hour and a quarter of an hour to go, and to stop them when time is up.

At the start, the experimenter in the room (EXP1) read aloud the general experiment instructions. There were slightly different instructions for individual designers and teams of designers – an individual subject was asked to 'think aloud' continuously throughout the experiment, whereas a team was given no such instruction, and an individual subject was asked to perform a short, initial 'think aloud' training exercise in order to become accustomed to thinking aloud. After reading the instructions, the experimenter asked the subjects if there were any questions about the procedure. The subjects were then given the design assignment, which they read for themselves, and the session began.

The instructions read to the subjects at the commencement of each session were as follows:

Thank you for agreeing to participate in this research project. We are interested in the ways that designers work on conceptual design projects. This session is being videotaped for later analysis by a number of researchers. I am going to ask you to undertake a short design project. Let me assure you that no commercial advantage will be taken of any design work or ideas that you produce.

(To an individual subject:
   Because we want to understand what you are thinking during the task, we want you to think aloud continually during the session. This would be as if you are talking to yourself but loud enough for me to hear. In order for you to get used to thinking aloud, I will give you a small problem which has nothing to do with the design task. Try to solve this problem while thinking aloud. This is only a training exercise, which lasts 10 minutes: so please do not worry if you do not solve the problem. Here is the problem I want you to work on.

Written instructions:
Three missionaries and three cannibals are together on one side of a river. They have one rowing-boat, which can hold up to two people. They all know how to row. How can they all reach the other side of the river, given that – for obvious reasons – there must never be more cannibals than missionaries on either river-bank?)

[10 minutes allowed for this exercise, in which the experimenter reminds the subject to keep thinking aloud if there are silences of more than 30 seconds.]
I am now going to ask you to undertake the design project. The time available is limited to two hours. In a moment I will give you a short written design brief for the project.
   Additional information which you may require during the project is available in an information file which I keep

---

and have access to. So, if you think that you need additional information during the project, please ask me for what you need to know. Please be specific about what you ask for. The information available in the file includes both technical and client information. Although access to the file is only available through me, please do not feel constrained about asking for any information that you feel you need during the project. I also have an outside helper who can give me additional information if it is not available in the file. Apart from providing you with information when you ask for it I will not interact with you in any other way. You should try to ignore my presence in the room.

Drawing materials – paper, pens and markers – are available for your use. There is also a whiteboard which you may use if you wish. The particular design project which we are asking you to undertake is concerned with a new product related to mountain bikes and backpacking. Therefore we also have in the room a mountain bike and a backpack, which you may use or refer to as you wish. Please work as you normally would on such a design assignment as this.

(To an individual subject:
   But remember to keep thinking aloud.)

Before I give you the design brief, are there any questions about the procedure?

If there are no further questions we will start the session now. I remind you that the maximum time available to you is two hours. I will remind you of the time at 30 minutes and 15 minutes before the end of the session, if you are still working.

(After the subjects have read the assignment the experimenter says:
The bicycle here in the room is not the Batavus Buster, but the backpack provided is the HiStar backpack.)

At the end of the allotted two-hour period for the experiment session, the subjects were debriefed in a short interview. The debriefing questions were:

1. What did you think of this assignment?
2. Are you satisfied with the result you managed to achieve?

3. What were the key decisions which influenced your final design?
4. Which particular items of information were especially helpful?
5. What did you find particularly difficult in the project?

4 The Experiment Assignment

The design assignment given to the subjects for the experiment was as follows:

HiAdventure Inc. is a fairly large US firm (some 2000 employees) making backpacks and other hiking gear. They have been very successful over the last ten years, and are well known nationwide for making some of the best external-frame backpacks around. Their best selling backpack, the midrange HiStar, is also sold in Europe. In the last one and a half years, this European activity has suffered some setbacks in the market; in Europe internal-frame backpacks are gaining a larger and larger market share.

As a response, HiAdventure has hired a marketing firm to look for new trends and opportunities for the European market. On the basis of this marketing report, HiAdventure has decided to develop an accessory for the HiStar: a special carrying/fastening device that would enable you to fasten and carry the backpack on mountain bikes.

The device would have to fit on most touring and mountain bikes, and should fold down, or at any rate be stacked away easily. A quick survey has shown that there is nothing like this on the European market.

This idea is particularly interesting for HiAdventure, because the director, Mr Christiansen, has a long-standing private association with one of the chief product managers at the Batavus bicycle company (one of the larger bicycle manufacturers in northern Europe, based in Holland). Mr Christiansen sees this as a good opportunity to strike up a cooperation and to profit from the European marketing capabilities of Batavus.

The Batavus product manager, Mr Lemmens, is very enthusiastic about putting a combination product on the market, a mountain bike and a backpack that can be fastened to it. The idea is to base the combination product on the Batavus Buster (a midrange mountain bike), and to sell it under the name Batavus HikeStar.

The design department at Batavus has made a preliminary design for the carrying/fastening device, but both
Mr Christiansen and Mr Lemmens are not very content with
it. The user's test performed on a prototype also showed
some serious shortcomings.

That is why they have hired you as a consultant to make a
different proposal. Tomorrow there is going to be a meeting
between Mr Christiansen and Mr Lemmens, scheduled as
the last one before presenting the idea to the board of
Batavus. Before then, they need to have a clearer idea of the
kind of product it is going to be, its feasibility and price.

You are hired by HiAdventure to make a concept design
for the device, determining the layout of the product, concentrating on

- ease of use
- a sporty, appealing form
- demonstrating the technical feasibility of the design
- ensuring that the product stays within a reasonable price
  range.

You are asked to produce annotated sketches explaining
your concept design.

Good luck!

See the first paper in this volume, 'The design problem and its
structure' by Dorst, for an analysis of this problem and the
solution possibilities open to the designers.

Most of the information that designers could be expected to
need to complete this assignment was available on information
cards kept by EXP1. The designer or designers could ask the
experimenter for anything they wanted. Information regularly
asked for included:

1. Marketing research
   – this included a profile (age, habits) and some estimations
   of the size of the target group.
2. Use of the backpack
   – some details about the use of the backpack (such as the
   position of the bedroll).
3. Users' trials/evaluation

> – vital information on the evaluation of the Batavus design. Stability is an issue, freedom of leg movements, and the dangling straps.

4. Batavus design (three cards)
   > – an exploded view and some side-views of the preliminary design. It is a simple tube-frame product, fastened to the bike at the rear axle and under the saddle. The backpack is in an upright position, and the frame can fold down for storage.
5. Underlay drawing of a bike (22 inch frame)
6. Technical drawings of the 'Buster' (two cards)
7. 'HiStar' product information
   > – giving the weight and dimensions of the backpack
8. Technical drawing of the 'HiStar'
9. Comparable products – Blackburn (three cards)
   > – some copies from the Blackburn catalogue, featuring all the basic forms of bike racks and panniers.

On the whole, the designers asked for only a small portion of the information that was available to them. Things not asked for (but prepared by us) included in-depth interviews with the director of the backpacking firm and the bicycle manufacturer. Occasionally, a designer would ask for some information included in these interviews, without asking for the complete interview. In that case the information (if available) was given verbally by the experimenter in the room.

The designers also had access to the backpack featuring in the assignment, and to a mountain bike. These were used very intensively all through the project. (See the paper by Harrison and Minneman, Chapter 19, in this volume.) Two technical reference books which were available were not used at all, except to fill up and increase the weight of the backpack.

5 Data Analysis

In all, three teams of three designers and two individuals were recorded at XeroxPARC:

- Trial experiment: team (Stanford students) (code: DPW 94.1.12.1)
- Second experiment: individual (code: DPW 94.1.13.2)
- Third experiment: individual (code: DPW 94.1.13.3)
- Fourth experiment: team (code: DPW 94.1.14.4)
- Fifth experiment: team (code: DPW 94.1.14.5)

Of these five experiments two were selected for analysis (DPW 94.1.13.2 and DPW 94.1.14.5). The main criteria for the selection of the tapes were as follows:

1. We wanted to include both individual and group work, because some of the Workshop participants had been working with either groups or individuals; they should at least have one tape that they can process in their own way, though they might or might not be particularly interested in the other one.
2. In a discussion including the Delft Workshop organizers and the Xerox team (six people of very different backgrounds in design and design research), the two selected tapes were identified unanimously as being very interesting for their content. We deemed this informal agreement adequate for the purpose of preselecting the tapes. If the agreement had been less strong, we would have had to resort to throwing dice.
3. Recording quality (not a major factor, but the audio recording of the team in DPW 94.1.14.4 was slightly inferior to that of DPW 94.1.14.5).

The main reason for not sending tapes of all five experiments to the participants of the Workshop was that we wanted the participants to focus on the same data, to ensure that conclusions could be compared. It simply would be no use sending people more data than they could possibly process.

A copy of the 4-PIP video recording, a transcript and copies of the designers' drawings, whiteboard notes, etc. were sent to all the Workshop participants, for each of the two selected experiments. (For cost reasons, only the 4-PIP recordings were sent, but any of the separate recorded views was also available on request.) In the transcribed protocols, the subjects were identified by fictional names – 'Dan' (DPW 94.1.13.2) and 'Ivan', 'Kerry' and 'John' (DPW 94.1.14.5). 'Dan' is an engineering designer with more than 20 years of design experience in mechanical and electro-mechanical engineering. 'Ivan', 'Kerry' and 'John' all work for the same leading product design consultancy firm, and all have approximately 5–8 years design experience in engineering product design.

Participants were asked to use the data to make their own analyses, and write research papers for the Workshop. We did not specify how the analyses should be made, which features from the experimental sessions should be emphasized, what encoding systems should be used, or whether to concentrate on either one or both of the recordings. The given goal was to perform research aimed at understanding the design process; different participants were expected to have different interests and to use different methods to achieve that goal.

6 Results

The Workshop meeting was held in Delft in September 1994. Draft papers were presented and discussed, and final papers were then prepared by the participants. It is these papers that form the contents of this book.

Together the participants produced an outstanding set of papers. Even a book like this cannot be more than a shadow of what happened at the Workshop itself. The variety of papers and the arguments brought up in the discussions were such that it is very hard to summarize them and draw a finite number of conclusions. Protocol analysis is clearly not a universal cure for design research problems. Nevertheless, we feel that we can say that protocol analysis as a research technique for design has been 'validated' with some qualifications, as follows.

1. Protocol analysis has severe limitations in capturing the non-verbal thought processes going on in design work. 'Completeness' of protocol data is an illusion. Protocol analysis is a very specific research technique, capturing a few aspects of design activity in great detail.
2. Even within these restrictions we have to be very careful. The experimental setup heavily influences the protocol data, and the amount of interpretation needed to wrench conclusions from protocol data is also comparatively large. It is impossible to claim to have all of these factors under control. In the end, conclusions and generalizations drawn on the basis of protocol research will only be valid if and when we have a coherent picture of the influence of the experimental technique and situation. In the meantime, we can do little more than identify the sources of bias in protocol research, and be very explicit on the way the experiment was set up, the background of the researchers, the aim of the research project, etc.
3. The adoption of protocol analysis as a research technique for design is an effort on the part of design methodologists to find a rigorous form for their empirical research. Protocol analysis is somewhere in the middle ground between the 'hard' experimental methods of the natural sciences and the 'weaker',

purely observational methods of the social sciences. The whole of empirical design research can be seen balancing between these, perhaps trying to lean both ways. The general feeling in the discussions at the Workshop was that the balance has tipped too much to the side of rigour and 'safe' research techniques, at the expense of the 'relevance' of results for design practice and education.

The emerging picture for empirical design research in the near future is that of larger research projects, in which multi-disciplinary teams of researchers work together on analysing more complex and realistic design tasks. Techniques like partici-pant observation and interviews can be used for analysing the long-term processes. Protocol analysis can take its place among the more detailed techniques for studying short-term processes or some especially significant turns in the design saga. The great potential of protocol analysis can then be used to the fullest extent.

The Workshop format has proved to be very valuable in tying together the many specialisms and disciplines within design research. We hope this event in Delft will spark off other initia-tives. Cooperation in research projects and regular discussion platforms like these are a good way to make progress in tackling the baffling complexities of design activity.

Acknowledgments

We wish to thank the Faculty of Industrial Design Engineering of the Delft University of Technology for its generous (financial and moral) support of the Workshop. Thanks are also due to Steve Harrison and Scott Minneman of XeroxPARC for setting up and working with us on the taping of the experiments, and to Stan-ford University's Design Research Center for helping to find the subjects; and of course we are especially indebted to the anon-ymous subjects themselves. We are grateful to Anita Clayburn Cross for transcribing the recordings and to the Workshop par-ticipants for preparing their papers.

References

1   Cross, N., C. Dorst, and N. Roozenburg (Eds), *Research in Design Thinking*, Delft University Press, Delft (1992)
2   Ericsson, K.A. and H.A. Simon, *Protocol Analysis: Verbal Reports as Data*, MIT Press, Cambridge, MA, (1993) (Rev. Edn)

3   de Groot, A.D., *Thought and Choice in Chess*, Mouton, The Hague (1965)

4   Newell, A. and H.A. Simon, *Human Problem Solving*, Prentice Hall, Englewood Cliffs, NJ (1972)

5   Eastman, C.M., On the analysis of intuitive design processes, in G.T. Moore (Ed.), *Emerging Methods in Environmental Design and Planning*, MIT Press, Cambridge, MA, (1970)

6   Foz, A., Observations on designer behaviour in the *Parti*, *DMG-DRS Journal: Design Research and Methods*, **7**(4) (1973) 320–323

7   Akin, Ö., How do architects design?, in J.-C. Latombe (Ed.), *Artificial Intelligence and Pattern Recognition in Computer Aided Design*, IFIP/ North-Holland, New York (1978)

8   Chan, C.-S., Cognitive processes in architectural design problem solving, *Design Studies*, **11**(2) (1990) 60–80

9   Schön, D.A., Designing: rules, types and worlds, *Design Studies*, **9**(3) (1988) 181–190

10   Klein, M. and S.C.-Y. Lu, Conflict resolution in cooperative design, *Artificial Intelligence in Engineering*, **4**(4) (1989) 168–180

11   Goldschmidt, G., Linkography: assessing design productivity, *Proc. Tenth European Meeting on Cybernetics and Systems Research*, Vienna (1990)

12   Schön, D.A., Problems, frames and perspectives on designing, *Design Studies*, **5**(3) (1984) 132–136

13   Ullman, D.G., T.G. Dietterich, and L.A. Stauffer, A model of the mechanical design process based on empirical data, *Artificial Intelligence in Engineering Design and Manufacturing*, **2**(1) (1988) 33–52

14   Adelson, B., Cognitive research: uncovering how designers design, *Research in Engineering Design*, **1**(1) (1989) 35–42

15   Whitefield, A. and C. Warren, A blackboard framework for modelling designers' behaviour, *Design Studies*, **10**(3) (1989) 179–187

16   Ennis, C.W. and S.W. Gyeszly, Protocol analysis of the engineering design process, *Research in Engineering Design*, **3**(1) (1991) 15–22

17   Kuffner, T.A. and D. Ullman, The information requests of mechanical design engineers, *Design Studies*, **12**(1) (1991) 42–50

18   Ehrlenspiel, K. and N. Dylla, Experimental investigation of designers' thinking methods and design procedures, *Journal of Engineering Design*, **4**(3) (1993) 201–212

19   Lloyd, P. and P. Scott, Discovering the design problem, *Design Studies*, **15**(2) (1994) 125–140

20   Tang, J.C., Findings from observational studies of collaborative work, *International Journal of Man–Machine Studies*, **34** (1991) 143–160

21   Minneman, S. and L. Leifer, Group engineering design practice: the

social construction of a technical reality, *Proc. Int. Conf. on Engineering Design ICED93* (Ed. N. Roozenburg), The Hague: Heurista, Zürich (1993)

22   Visser, W., Collective design: a cognitive analysis of cooperation in practice, *Proc. Int. Conf. on Engineering Design ICED93* (Ed. N. Roozenburg), The Hague: Heurista, Zürich (1993)

23   Colgan, L. and R. Spence, Cognitive modelling of electronic design, in J. Gero (Ed.), *AI in Design 91*, Butterworth-Heinemann, Oxford (1991)

24   Jeffries, R., *et al.*, The processes involved in designing software, in J.R. Anderson (Ed.), *Cognitive Skills and their Acquisition*, Lawrence Erlbaum Associates, Hillsdale, NJ (1981)

25   Guindon, R., Knowledge exploited by experts during software system design, *International Journal of Man–Machine Studies*, **33** (1990) 279–304

26   Davies, S. and A. Castell, Contextualizing design: narratives and rationalization in empirical studies of software design, *Design Studies*, **13**(4) (1992) 379–392

27   Olson, G.M., *et al.*, Small group design meetings: an analysis of collaboration, *Human–Computer Interaction*, **7** (1992) 347–374

28   Thomas, J.C. and J.M. Carroll, The psychological study of design, *Design Studies*, **1**(1) (1979) 5–11

29   Goel, V. and P. Pirolli, The structure of design problem spaces, *Cognitive Science*, **16** (1992) 395–429

30   Ballay, J., An experimental view of the design process, in W.B. Rouse and K.B. Roff (Eds), *System Design: Behavioral Perspectives on Designers, Tools and Organisations*, North-Holland, New York (1987)

31   Christiaans, H. and C. Dorst, Cognitive models in industrial design engineering: a protocol study, in D.L. Taylor and D.A. Stauffer (Eds), *Design Theory and Methodology – DTM92*, ASME, New York (1992)

32 Christiaans, H., *Creativity in Design: The Role of Domain Knowledge in Designing*, Lemma, Utrecht (1992)

1 The Design Problem and its Structure

Kees Dorst

Delft University of Technology, The Netherlands

The choice and precise setting of the design problem is a crucial part of protocol analysis-type research. Researchers in this line of work have to spend a lot of time, care and effort on producing a design assignment that will be suitable for their research purposes. But there is very little by way of theory, or even descriptions of the considerations that led researchers towards setting a certain task. (Ericsson and Simon[1] also observed this: '... if the purpose of obtaining verbal reports is mainly to generate hypotheses and ideas, investigators need not concern themselves with the methodological questions about data collection. As a result, there is little published literature on such issues ..'.)

This paper is meant to aid the buildup of knowledge on this point, by exposing our choices as experimenters in setting this design task and the reasons behind them to the public eye. It should thus be seen as a primer, informally written and largely without the comfortable backup of established theory or informal consensus in the field. Most of the considerations to be put on display here are based on our own experience in setting tasks like this, and will be of a commonsense nature.

1 On the Design Task

The design task (a carrying/fastening device) was close to one used in earlier research at the Delft Faculty of Industrial Design Engineering, reported upon on a number of occasions by Christiaans and Dorst[2-4]. This new problem was given a trial run some two months before the recordings were made, with three groups each of three design teachers as subjects. We also spent some considerable time going through the problem ourselves mapping the whole solution space, and getting an overview of issues and tradeoffs that could possibly occur during the design processes.

The design task was meant to be

1. challenging,
2. realistic,
3. appropriate for these subjects,
4. not too large,
5. feasible in the time available, and
6. within the sphere of knowledge of the researchers.

1. *Challenge.* The designers were challenged to work on a totally new product. They were given the basic idea in rather abstract terms: something that can be used to fasten a backpack to a mountain bike. The word 'fastening device' was used not to bias them towards making something luggage rack-like right away. In the comments on a preliminary design (available in the information system) they were prodded not to make a 'universal' luggage rack, but a product that should be dedicated to the carrying of backpacks. Thus, they were lured into uncharted territory. The target price was set high enough so they did not really need to worry too much about that in this conceptual stage. The task was meant to give the impression of a brand new design problem with the opportunity to make something 'nice' and surprising.

2. *Realism.* Realism in this case means that the product idea and all the information came from real life, and that the subjects were put in a realistic situation. What was offered to the designers was a company problem, to be interpreted and turned into a design problem. The precise problem statement and the performance specification had to be derived from the assignment and the information that was available on request. The designer was cast as being in a design firm, with only a few hours to spend on getting his/her act together for a meeting. There was some realistic – for this stage in a product development process – vagueness and ambiguity in the information, in particular the standpoints of the different stakeholders.

3. *Appropriateness.* This is the kind of problem that would be pretty regular stuff in a small design consultancy. The time pressure (a few hours to come up with some concept to start

talking about) is also typical for that sector. The subjects were chosen for their all-round design abilities, and their experience with these kinds of challenges. They were basically asked to do something in which they had displayed competence. The product idea introduced in this problem was novel enough to make sure that they probably would not have experience on something quite like this. Otherwise, that could lead to routine design work, or could give some a head start because of their level of background knowledge. That would spoil the research by the implicitness of routine reasoning, and by making the protocols less comparable.

4. *Size.* The problem was limited in size by making it one that is easy to relate to for the designers – they themselves are also in the target group. The information load was limited by the conciseness of the information cards, providing the bike and backpack, and by providing information on a comparable product (the preliminary design). This preliminary design, and its evaluation, actually characterize the companies and their view of the project a lot better than the statements from the assignment itself. The evaluation also hints which factors are important in the design, and which problems could be expected. It serves as the much-needed link between the rather abstract description of the companies and the product idea, and the very concrete possibilities and problems to be encountered in designing such a product.

This preliminary design was introduced on the basis of problems encountered in the trial run of the assignment. It seemed that designers had too much trouble getting from the abstract problem description to a workable (albeit implicit) design problem statement. Some even started rejecting the assignment as plainly impossible. Having a clear 'nearest product' could be vital for design work; designers in practice tend to look for one, because it makes picturing the problem and evaluating the solution much easier.

An earlier study by this author has shown designers to spend up to 25% of their time (in a research setup comparable to this one) working in close reference to the 'current product'. The role of the 'closest comparable product' or 'current product of the company' is important enough to be recommended for further study by design methodologists.

5. *Feasibility.* The last section of the assignment sets the scene for the designer's activities; he or she is asked to produce some

concept on a level detailed enough to serve as the basis for a
worthwhile discussion. This is a trick sentence, allowing people
who get stuck in the problem analysis to see this analysis itself as
a good result of the design session. When these were asked if they
were content, they would say things like 'at least I exposed the
basic problems involved', 'I would need some company feedback
on this before I would go on making a concept'. This wording of
the assignment evokes a fairly concrete image of what they are
expected to produce, but also reassures that a basic (satisfying)
level of ideas and problem analysis can easily be reached within
the two hour limit.

6. *Experience.* The author of this paper has been involved, as a
designer, in the development of a bicycle seat for children. Thus
there was a clear prior overview of the issues involved, the kinds
of solutions that would probably be generated, etc, etc.

2 On the Design Assignment

The design assignment will now be introduced and discussed
section-by-section. The most probable interpretations and conclusions will be discussed briefly.

Design Assignment

HiAdventure Inc. is a fairly large US firm (some 2000
employees) making backpacks and other hiking gear. They
have been very successful over the last ten years, and are
well known nationwide for making some of the best external-frame backpacks around. Their best selling backpack,
the midrange HiStar, is also sold in Europe. In the last one
and a half years, this European activity has suffered some
setbacks in the market; in Europe internal-frame backpacks
are gaining a larger and larger market share.

• The company, HiAdventure, is large enough to have some
reserves, and to be able to spend some money on new product
development. They have a quality product, so technical
knowhow will be available, and production facilities will presumably be under control.
• They are not in grave financial trouble, but the European
market (where they are not very active) poses a 'local' problem.

The problem of being small and 'local' can be an advantage; it means that there is probably no undue pressure on company time and resources. But it could also be a negative point: smallness could lead to a low priority being given to this design problem ('also' sold in Europe). One would need further information to find out.

- Their external-frame backpack is not going to make it in Europe in the long run, but they want a temporary solution to crank up sales a bit. So, investment and interest will be low, unless the designer comes up with something special, appealing, and also suitable for other uses or (US) markets...

> As a response, HiAdventure has hired a marketing firm to look for new trends and opportunities for the European market. On the basis of this marketing report, HiAdventure has decided to develop an accessory for the HiStar: A special carrying/fastening device that would enable you to fasten and carry the backpack on mountain bikes.
>
> The device would have to fit on most touring and mountain bikes, and should fold down, or at any rate be stacked away easily. A quick survey has shown that there is nothing like this on the European market.

- The idea for the new product is introduced, in rather abstract functional terms. This could mean that the company doesn't have a clear idea of what it should look like. At any rate, it would like the designer to start at this level of abstraction.
- The new idea is introduced as having been decided upon. There are no problems with company acceptance of the new product idea.
- There is market information available.
- But the product hinted at (something like a bike-rack) is quite different from the products the company normally designs and makes. That might influence the technical support the designer can get from the company, and could severely limit the choice of materials and production methods to be used.

  The initial associations reading about the 'carrying/fastening device' are: luggage racks, clamps, panniers, straps, bungee cords, etc.

> This idea is particularly interesting for HiAdventure,
> because the director, Mr Christiansen, has a long-standing
> private association with one of the chief product managers
> at the Batavus bicycle company (one of the larger bicycle
> manufacturers in northern Europe, based in Holland). Mr
> Christiansen sees this as a good opportunity to strike up a
> cooperation and to profit from the European marketing
> capabilities of Batavus.
>
> The Batavus product manager, Mr Lemmens, is very
> enthusiastic about putting a combination-product on the
> market, a mountain bike and a backpack that can be fas-
> tened to it. The idea is to base the combination product on
> the Batavus Buster (a midrange mountain bike), and to sell it
> under the name Batavus HikeStar.

- A second company is introduced, that is going to be an associate in the project. This confuses the assignment on two
  points: for whom is the designer really working, HiAdventure
  or Batavus, and is this going to be a product that is tailor-made
  for the Batavus bike? The product is going to be made by
  Batavus, but it is clear from the assignment that HiAdventure
  would profit more from the whole project if the fastening
  device could also be used on other bikes. But in doing this, and
  selling it as a separate product, they would risk losing the
  goodwill and marketing support of Batavus. Still, it would be a
  decided advantage if the designed product could fit on more
  bikes; HiAdventure could then always decide to enter the
  market independently or not. If it could be 'universal' without
  any extra costs, that would be perfect.
- The series in which the combination product is going to be
  made is fairly small; this leaves little room for injection
  moulding and other high-investment production techniques. If
  the product is going to be marketed more widely, these would
  be a good choice. It would probably be good to design the
  product in such a way that it can be produced in small series at
  little investment (maybe using standard parts), and also in
  larger series, when the issue of Batavus participation has been
  sorted out and the idea of the product has caught on.
- The information that the product is going to be made by
  Batavus is greeted with a sigh of relief: bike companies are
  pretty competent in making sturdy framelike structures. And

they know about carrying luggage on bicycles, probably making luggage carriers themselves. Steel or aluminium are now obvious choices of material for the series product, whatever it is going to be like.

- The combination product (bike–fastener–backpack) is going to be sold as a Batavus product. This suggests that Batavus could be the stronger partner in the cooperation.

> The design department at Batavus has made a preliminary design for the carrying/fastening device, but both Mr Christiansen and Mr Lemmens are not very content with it. The user's test performed on a prototype also showed some serious shortcomings.

- This is absolutely vital information; the assignment is very vague, the role and outlook of the companies is unspecified and the idea of the product is dim (see Figure 1.1). All of this can be cleared up by knowing what they think of the preliminary design. (This defines what is 'quality' in this situation.)
- Also, the product to be designed has to compete for attention and resources with this other design. The preliminary design defines the minimal quality to be reached. As a designer, I would ask for this information first, use this as a basis and fill in possible knowledge gaps by asking some small additional questions.

> That is why they have hired you as a consultant to make a different proposal. Tomorrow there is going to be a meeting between Mr Christiansen and Mr Lemmens, scheduled as the last one before presenting the idea to the board of Batavus. Before then, they need to have a clearer idea of the kind of product it is going to be, its feasibility and price.
>
> You are hired by HiAdventure to make a concept design for the device, determining the layout of the product, concentrating on
>
> - ease of use
> - a sporty, appealing form
> - demonstrating the technical feasibility of the design
> - ensuring that the product stays within a reasonable price range.

You are asked to produce annotated sketches explaining your concept design.

Good luck!

• The designer is asked to produce an all-round concept, to be used as a starter for the project. This concept should be reasonably good, though, because the brighter the idea, the more chance there is of the project going through. And concepts like this often get a life of their own, determining everyone's view of what the product should be like. The company could easily start concentrating on the embodiment of this idea, instead of just considering it as a primer. It had better be good, new and appealing as well as feasible.

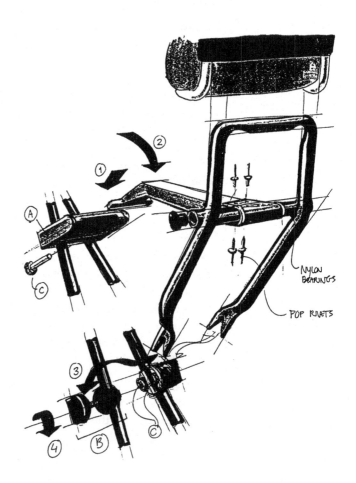

Figure 1.1
The preliminary
design (not included
in the assignment
but available on
request)

The assignment is really too much for the two hours available, but something acceptable can be achieved in that time. The least thing you should get is a clear overview of the design issues, and an idea of what the product could look like. Building up knowledge and making a concept are goals of equal importance in a situation like this. Just try to get as far as possible in the time available.

3 On the Problem Structure: Routes to a Solution

3.1 Introduction: on the Methodology of Problem Structures

Much has been said in the design methodology literature about the ill-structuredness or 'wickedness' of design problems[5,6], and closely linked to that the theoretically endless number of possible solutions. These papers have had an enormous impact on the view that design methodology has taken of design problems. The structure of design problems has all but been given up as a worthwhile subject for methodological reflection. However, observations from practice do not justify such a stance. Designers do not approach design problems as if they were mystical clouds of unique and baffling issues. They seem to have some implicit taxonomy of different kinds of design problems, and are very competent in tackling them without much ado.

The number of solution possibilities is logically virtually endless when one looks at a design problem as such. But of these 'solutions' only a few will be good enough to solve the problem within the real-life contraints. A morphological chart will give thousands of possibilities, but the selection process will reveal most of them as being obviously impossible or impractical nonsense. The number of solutions is only great when you artificially refrain from evaluating them. And as design generally is a process of continuous alternation between generation and evaluation, the situation of solution-overload simply does not occur.

The real number of solutions to be considered within a design process is very limited. When a designer starts analysing the problem, and designing possible solutions, the problem often becomes so constrained and tangled in structure that there is no 'total' solution at all. Satisficing behaviour, widely observed in designers, comes in because of this lack of all-round good solutions.

In the case of the fastening device, as in maybe 95% of the cases one encounters in practice, the design problem turns out to be fairly well structured, and the number of solutions very small, when you start thinking about it. (Note that the notion of problem structures used here is dynamic in character: it has to do with the importance of design issues, but also with the order in which they are tackled. The part of the design problem to be solved continually changes shape in the light of the decisions taken[7]. Provided that no single issue dominantly commands attention, there are several possible structures of the design problem and routes towards solving it. A structure is then imposed by the designer.)

This design assignment looks quite open-ended, but there is a 'natural' sequence in which the problems should be tackled. This 'natural' sequence is determined by the amount of influence different issues have on the end result, and the degree of independence of these issues. The mechanism at work is that at any one moment, the designer determines what the 'main' (most influential) issue is, and seeks to solve that, keeping track of the influence his or her decisions have on other items in the problem or solution.

In this case, the first two or three main issues are pretty clear, as will be demonstrated below. And a thorough analysis/synthesis session on these first two (or three) main issues leaves only one or a few options for the layout of the product. Every competent designer will end up with more or less the same kind of concept, in more or less the same way. This may seem very surprising for an assignment that is wide open (there are only a few constraints given, and the product idea is totally new) compared to what we normally find in practice.

After three issues, the structure becomes more clouded: there is a number of different routes from there, the issues are less independent, and the design decisions are less clear-cut, too. The solutions designers come up with will differ on these points.

This paper was not meant to grow into a theoretical exposé on design problem structures. The stance taken above will now be illustrated by revealing and discussing the structure of the fastening device problem. The impression I would like to leave in doing this is that much can be said about the structure of design problems, and about the possible solutions. I would recommend these issues for renewed methodological attention.

3.2 The Case of the Fastening Device

The structure of this design problem will be discussed at some length now. This will be done by tackling the design issues, and listing the most probable design decisons.

To set the scene: the designer has read the assignment, and just finished the first round of information collection, clarifying just a few points. There is a clear but fairly general view of what has to be developed.

The first issue: location

The location of the backpack on the bike determines the where, what and how of the fastening device, so it is no use starting with anything else. The location of the backpack is not a creative design problem, it is just a puzzle to be solved by generating all possible solutions and commenting on them, choosing one or two (see Figure 1.2).

Figure 1.2
A possible solution
(DPW 94.1.14.2-23)

All designers will probably come up with the same possible positions. With the bike and the backpack available, the easiest way to go about solving this problem is to take up the pack, and to hold it in different positions relative to the bike. This also gives a feel for the size of the two objects, and for the weight distribution. The gesture of holding the backpack above the bike also happens to mimic the gestures prospective users are going to have to make, in putting the pack on the bike.

Because the preliminary Batavus design is very clumsy, and positions the backpack (upright) above the rear wheel of the bike, that position will initially be avoided by the designers. They will try to innovate on the fundamental level of product layout. In the end, they will have to accept that above the rear wheel is the only feasible position. The orientation of the backpack is more of a question; lying down or standing upright are both possible from a stability standpoint. Standing upright has the disadvantage of making it harder to mount the bike (to swing your leg over it), so lying down seems to be the best option. (This is not necessarily so: see the subsection below on radically different designs.)

In the process of trying these different positions, and commenting upon them, the designers generate a lot of knowledge about bike stability, bike use, etc. They act as users and bodily mimic different ways of handling the bike and fastening device. They thus construct the argument supporting their choice for the most obvious solution.

The only 'mistake' the author has ever seen being made in this phase is that the designer starts to circumvent the stability problem (putting the backpack in the rear puts your centre of gravity perilously far to the rear, too) by distributing the weight. To put a little bag on the handlebars is good stability thinking, but it goes against the basic principle of the product. That was based on the idea that one should be able to hike with the backpack only, not having to bother with other bags (and not having to leave things on the unattended bicycle). If one chooses a distributing-the-weight option like this, the additional bag(s) should be easy to strap to the backpack. (See the subsection on radically different designs.)

The second issue: kind of product

The second issue is what kind of product the fastening device should be (see Figure 1.3). This is what Schön would call the 'framing' of the product[8,9]: is it a clamp, or a bike rack, a pannier? Could it be just a few snaps put on the bike in the right position, using the frame of the backpack itself as the frame of the fastening device? (It is good engineering practice not to use more material than you need, and unnecessarily putting frames upon frames is considered not done.)

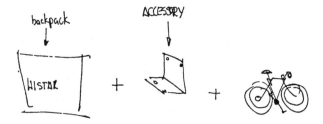

Figure 1.3
What kind of product
(DPW 94.1.14.2-21)

The kind of product is in the end determined by the fact that the frame of the backpack isn't very stable, so it needs to be supported at several points. This leads to a bike rack-like layout, after all. The distribution of the supports also makes for greater stability, and a reasonably low weight. Fancy forms and materials are out, because the series are small to start with, the producing company probably couldnt handle them, and the time for the assignment is far too short to go into those.

The third and fourth issues: attaching to the bike, or attaching to the backpack

Now there are two issues of almost equal priority: how to attach the fastening device to the bike, and how to attach the backpack to the fastening device. The designers will generally start with the first problem, because it puts some contraints on the kind of framelike structure to be made, and on the choice of materials. The three-point construction (see Figure 1.4) gives the most stability for the amount of material used.

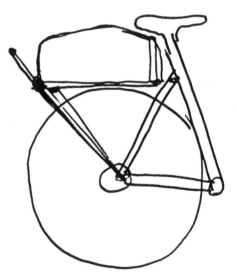

Figure 1.4
Side view of the threepoint construction (DPW 94.1.13.1-7)

For fixing, there are two brazed-on fixation points on the rear stays of the bike, under the saddle, that are almost too good not to be used. But the use of these would make the fastening device less adaptable to different kinds of bikes – to use them or not depends on the view the designer has of the cooperation between HiAdventure and Batavus. For the lower fixation points, everyone is bound to use the standard fixing holes near the rear axle. These choices could give problems because of the different frame sizes the bike comes in; most designers will probably overlook or try to ignore this. For the actual fixing to the bike, some quick-release fixing mechanism, either a snap or a latch, will be chosen.

This has been left until now, because the frame of the backpack can be gripped almost anywhere – that choice hardly influences the basic shape of the fastening device. The ergonomic aspect, the effort and motion of putting the backpack on the bike, has already played a part in solving the first issue, and possibly in the second. Some designers will already have decided at that stage what the loading action is going to be; others have at least developed an idea of what it could be like. The backpack can be attached in a number of ways, without an obvious favourite. The designers will differ in their choices (see Figure 1.5).

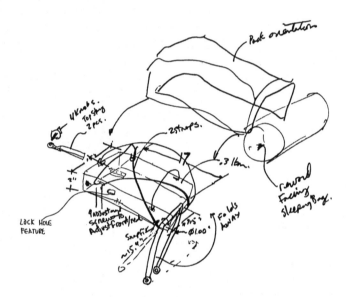

Figure 1.5
Strapping the
backpack on the
bike
(DPW 94.1.14.2-24)

The basic frame, with the stays coming up on the back, is now sketched a few times, and the tubes are brought parallel with the frame of the backpack. This is a slight puzzle, trying to do this as elegantly as possible without using too much tube or introducing too many extra parts. For the actual fixing, they prefer a quick-release latch, or a snap with some other kind of pressure-lock.

Minor conceptual issues   There are some minor issues that tend to be solved at the end of the conceptual stage, or in the embodiment of the basic design:

1. The product should fold down, or at any rate be stowed away easily when not in use. The little frame the designers have at this stage is looked at for folding possibilities that are, in most

cases, not very hard to find or build in. The ergonomics of this are simple, and not very critical as the product will be handled like this only once or twice a year.

2. The product should protect the backpack against mud and gravel that can be thrown up by the rear wheel. At least the part of the pack that touches your back in hiking use should be kept clean. This can change the character of the solution a bit, because it calls for something tray-like to protect the pack for at least the width of the tyre. Designers working on a pure-tube solution until now will have to start to rethink part of their product in sheet metal, vacuum-moulded or injection-moulded plastics. They can circumvent this problem by introducing a sack to cover the backpack. (This is also of some use for keeping out the rain, of course.)

3. The straps of the backpack should be kept away from the spinning wheels, chain and gears of the bike. A fastening device should preferably incorporate this function automatically. This favours laying down the backpack horizontally, with straps up – but that increases the distance of the frame of the bike to that of the backpack. Most designers won't opt for this, but patch up the design to help solve this problem. Using Velcro to hold down the straps is a good and imaginative solution.

Non-conceptual issues

1. Solving the aesthetic problems of the form of the fastening device has taken a back seat until now. It has had a minor influence on the main-issue decisions. The form of the fixing points is left to a later moment by most designers. They will work out some details of the construction, making sure they arrive at some provisional solution that is also reasonably pleasing aesthetically.

2. Most designers will not arrive at a precise choice of materials. They will, however, have seen to it that the decisions on the main issues have landed them with parts that can be made, and be strong and stiff enough, with the kinds of materials they have in mind.

3. The same goes for production methods. They will have avoided a design that is critical on these points. This is partly brought about by the time constraints. Choosing novel or high-tech production methods would increase the risk of ending up with a concept that is not feasible.

4. Safety issues will also be discussed in a passive, checking manner at this point in the design project.

Radically different
designs

The long line of reasoning set up above seems to lead inevitably to one (kind of) solution. There are some other possibilities, though, arising from different views of the design problem.

1. A hidden assumption in the treatment of the first issue above (location) is that the prospective user will approach the bike with his or her backpack in hand, attach the backpack to the bike, and then mount the bike. (This precluded the upright position of the backpack, when it is situated behind the saddle.) But one might think of a *different scenario*: a user could mount the bike with the backpack on his or her back, and then let it slide down behind the saddle, where it is gripped by the fastening device. This would make the fastening device much smaller, clamp-like, and fixed to the bike below the saddle, perhaps with some connections to the rear stays – a totally different product.

2. One could also seek to solve possible stability problems of the bike-and-pack combination by (as discussed in the section on issue 1) distributing the weight of the luggage. This is, in fact, choosing a *different system boundary* of the object under consideration. The easiest way to do this is to locate the backpack on the rear of the bike, and design a small bag (for a sleeping bag, for instance) that is suspended from the handlebars. This small bag should have some snaps or straps that would make it easy to attach to the backpack.

But the problem structure and design path spelled out above is still the most probable: the level-headed way towards a design solution, in this case. The alternatives chosen on each of the issues are the reasonable ones, the most feasible ones, the ones that evade unnecessary risk. But there might be other, maverick solutions that could even be radically better. We will never be quite sure.

4 Conclusions

At the end of this paper, it will be clear that very much can be said and known about the structure of design problems like this. The static structure of a design problem (such as Alexander uses in his *Notes on the synthesis of form*[10]) might quickly become a tangled web of dependencies between relationships. But the dynamic structure, as laid down above, can be very simple indeed.

Thoughts like these could lead to a taxonomy of design problems, taking things like the number of main issues, the existence of a dominant issue and the number of dependencies between issues as a starting point. This would lead to a division of design problems into categories that would closely mirror the ways they are (to be) solved by designers. That would be one of the best services design methodology could do to design practice and education.

Most empirical research in design methodology has worked on the assumption that the reasoning patterns found are those of an active designer imposing his or her decisions upon the design concept. But most of these decisions logically follow from the kind of 'creative analysis' outlined above. One should be careful, in the analysis of protocol data like these, not to presume a very great degree of freedom of the designer (the view of design as a creative, almost mystical process could lead to such assumptions). A thorough study of the structure of the design problem is indispensable for finding the patches where this freedom really lies.

The posing of design problems for protocol studies and also the describing of problem structures has never been discussed in (written) design methodology. The author would like to invite criticism on the design assignment used in this project, and on the way its structure has been described above, in the hopes of starting up new lines of discussion and enquiry in design methodology. Please write. We need these perspectives to help us build up an image of the design activity. Design is a maddeningly difficult subject to capture in empirical research.

Acknowledgments   I would like to thank Karen Wijkhuijs for her contributions to the making of the design assignment and information system, and Judith Dijkhuis and Nigel Cross for their constructive comments on this paper.

References

1   Ericsson, K.A. and H.A. Simon, *Protocol Analysis*, MIT Press, Cambridge MA (1993)
2   Christiaans, H.H.C.M. and C.H. Dorst, Cognitive models in industrial design engineering: a protocol study, *Design Theory and Methodology–DTM 92*, ASME, New York (1992)
3   Dorst, C.H, The structuring of industrial design problems, *Proc.Int.*

Conf. on Engineering Design ICED93 (Ed. N. Roozenburg), The Hague: Heurista, Zürich (1993)

4   Cross, N.G., H.H.C.M. Christiaans and C.H. Dorst, Design expertise amongst student designers, *Journal of Art & Design Education*, **13**(1) (1994)

5   Rittel, H.W.J. and M.W. Webber, Planning problems are wicked problems, in N. Cross (Ed.), *Developments in Design Methodology*, Wiley, Chichester (1984)

6   Simon, H.A., The structure of ill-structured problems, in N. Cross (Ed.), *Developments in Design Methodology*, Wiley, Chichester (1984)

7   McGinnis, B.D., and D.G. Ullman, The evolution of commitments in the design of a component, *Journal of Mechanical Design*, **144** (March) (1992)

8   Schön, D.A., *The Reflective Practitioner*, Harper Collins, New York (1983)

9   Schön, D.A. and G. Wiggins, Kinds of seeing and their functions in designing, *Design Studies*, **13**(2) (1992)

10 Alexander, C., *Notes on the Synthesis of Form*, Harvard University Press, Cambridge, MA (1964)

2 Design Protocol Data and Novel Design Decisions

Ömer Akin and **Chengtah Lin**
Carnegie-Mellon University, Pittsburgh, PA, USA

Inspired by developments in management science, cognitive psychology and computer science, Eastman[1] conducted the first published protocol study of architectural design. In this study, experienced designers were asked to redesign the interior of a given residential bathroom based on orthogonal drawings and user evaluations of the current situation.

This work was followed by a number of studies which resemble either methodologically, substantively or both. Krauss and Myer[2] tracked the design actions of a team of architects designing a school building over a period of 18 months. Here the collective design behaviour in a professional office was observed in the form of an unabridged design delivery process. In 1973, Adel Foz[3] completed a study of the process of designing an architectural *parti* for a small institutional building, by both experts and novice designers. These early studies of the design process were concerned with characterizing it in its most general form: identifying the operations and representations which were responsible for the development of designs, calibrating the human cognitive system, and describing design in the context of a general taxonomy of tasks.

Subsequently, these researchers and others entering the field from related areas, such as engineering design, built upon this early foundation. Research agendas in the area of design thinking were diversified. Some dealt with the internal and external representations of designed objects[4], others with the issues of design generation[5-7], the knowledge base of design thinking[8,9], the formulation of design problems[10], and the thought processes that apply to learning[11], and yet others with refining the general descriptions of the design process offered by the initial group of studies[12,13].

1 Data in Design Protocols

One of the ubiquitous aspects of design protocol studies has been the complementary relationship of two forms of data: verbal–conceptual and visual–graphic. The distinction we are drawing here is akin to the distinction provided by Pylyshyn[14]. Almost exclusively all of the design protocol studies we cite here have recorded verbalizations, usually through audio taping, and have kept a record of drawings produced by the subjects, in some systematic way. These appear to constitute the minimum data set required for design protocol analysis. While there is considerable attention paid to the contents of these, there is little written about their role and relative importance for design. It is clear that experimenters require think-aloud protocols to access the mental processes of designers, but the role that verbalizations play in the design process is not clearly determined. We know that drawings play a key role in design. They are used as a matter of course. Otherwise experimenters would have to remind designers to keep on drawing. Instead, drawing appears to be a fundamental aspect of design[15] and one would be hard pressed to prevent subjects from drawing.

What then is the role of verbalizations in this context? Since they are induced by the experimental setting, how essential are they? If they are not essential for design, then why are they still used for the analysis of the design process? Similarly, why is drawing so essential in design?

1.1 Dual Mode Model

A justification for the coexistence of *verbal* and *visual* data in design protocols is one that is motivated by cognitive science. A debate has been raging among cognitive psychologists since the early 1970s about the nature of the mental encodings that handle visual information. There is little doubt about the necessity of symbolic encodings. There is ample evidence which supports the idea that humans process information whether it is in the form of a language or other conceptual representations. The debate is about the sufficiency of this for visual data processing. Can one encode and process graphic data without encoding mechanisms that form analogues to visual entities? In this respect, Shepard and Metzler's work on mental rotation (not to mention the work of Kosslyn[16] and Loftus[17]) comes to mind. We find it neither useful nor possible to engage in this debate here. Instead, we envision a situation for the designer, in which his medium of work is represented through a pair of modalities: speech and graphics.

Visual aspects of design are explored in the graphic mode. This is reflected in the drawings that are produced. When this is the primary activity of the designer, any verbal information generated is usually its reflection, almost like an echo. In fact, in think-aloud protocol experiments, this is explicitly required by experimenters. Conversely, there can be periods in a design session during which the verbal–conceptual mode is primary. The designer may be interested in solving a symbolic problem, such as one involving a mathematical calculation. In this case, the designer uses external media solely as a reflection of the symbolic processing activity. This, in a design protocol, can be seen as the drawn and written echo of symbolic processing (Figure 2.1).

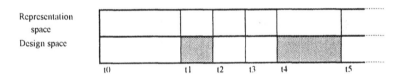

Figure 2.1
Dual-Mode Model

Using this model we see the protocol as a *dual mode* process in which the verbal–conceptual and the visual–graphic data are dealt with, alternatively. The two rows of Figure 2.1 represent this along the time dimension, *t*. The dark shaded segments in each row indicate the location of the foci of the designer's attention, or the primary design activity, while the lighter segment parallel to it, in the other row, represents its echo. Applying this model in the analysis of a protocol, one should be able to see which mode is particularly useful in capturing the essentials of the design activity. As a result one could assess the relative importance of each in the overall design process.

The difficulty of this approach is that neither the verbal (transcriptions) nor the visual (drawings) data alone can explain the design process adequately. Looking at these data, we are never sure if the designer is doing purely visual or conceptual processing. Also, one of these modalities, i.e. verbalizations which constitute 50% of the entire model, is by and large experiment induced. This further complicates the instrumentality of this model for our purposes.

1.2 Activity Based Model

In order to overcome these problems, we augmented the *dual mode* model with an *activity based* model of the design process. We categorized all activities observed in a set of protocols, one of which will be analysed in detail in this paper, into six: drawing, thinking, examining, talking, writing, and listening. We were able to sort the data entirely into these categories using a consistent set of criteria. Three of these correspond directly to the components of the *dual mode* model: drawing, writing and talking. In addition, there are those activities (thinking, examining, and listening) which are present in the video recordings of design protocols. These do not result in any particular data products, like drawings and transcriptions, but nevertheless play an important role in the resulting design.

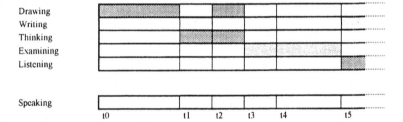

Figure 2.2
Activity Based
Model

Given this breakdown of six activities (Figure 2.2), we envision the designer performing the task at hand using one or more of these activity modes, at any one time. For example, he or she may be examining a document relevant to the design problem, then writing down a few key words that capture the gist of the document, then drawing a diagram based on the document, then pausing and thinking about what to do next, then asking the experimenter a question or two, listening to the answer given and starting a new sketch. We propose that all of these activities are essential to design and provide much richer data regarding what is responsible for the development of successful designs.

Verbalization of the designer in this scheme is treated differently from all the others, since it seems to function as self-annotation of the designer's behaviour. Thus we include speech in a separate category at the bottom, as an echo of all that goes on in the other five categories. While this activity is not a direct outcome of the design process, it serves the important function of reflecting the contents of the other activities. For example, looking at specific documents, if the designers state what they consider important, this can reveal something about the other design activities, such as thinking or examining. Thus speaking becomes the means through which the other activities can be better understood.

The result of speaking corresponds to the 'transcription' part of the *dual mode* model. The other component of the *dual mode* model, 'drawings', is represented by two activities in the *activity based* model: drawing and writing. While there is this correspondence between the two models these components are not identical. In the *dual mode* model the breakdown is based on the products of the protocol experiment while in the *activity based* model it is based on the behaviour of subjects. Consequently, the *activity based* model is not only more comprehensive but also a better representation of the design process.

1.3 Correlating Decisions with Activity Based Design Modes

Given such modes of activities in a protocol, we are interested in the question of how these activities influence the emergence of design decisions, particularly those that represent important features of the design. Is it the spatial–graphic processing, such as the drawing activity, that produces the most significant features of a design; or are there verbal–conceptual activities that influence the emergence of designs more prominently?

In order to accomplish this more had to be done to find correspondence between these activities and the cognitive equivalents of visual and verbal data. Drawing, for example, falls in the visual–graphic category. Thinking and examining, on the other hand, can be either visual or verbal, depending on what is being examined and to what end. The other activities, listening, writing and speaking, clearly seem to reflect a verbal bias. Thus, we can map, even if in approximate terms, the activities of designers into verbal and visual processing categories.

The *activity based* model of the design process is, in and of itself, inadequate to assess the relative value for design and design creativity. To accomplish this, we also need to understand what, within a designer's activities and products, signifies successful designs. We could then proceed to find out how these activities correlate with successful designs. These are the remaining tasks of this paper.

2 Experiments

Here we report on the analysis of two experiments, one centrally administered for all participants of the Delft Protocols Workshop and the other administered only by the authors of this paper.

2.1 Delft Experiments

These experiments conducted by two groups of researchers from TU Delft and Xerox PARC, respectively, included five protocols, two of which represented team design and the remainder individual design activity. In this paper we analyse an individual designer's protocol, primarily because it represents a design mode with which we are most familiar. Below, we highlight some of the data which are analysed in the next section.

Encoding the protocol data supplied into our *activity based* model, is generally straightforward. Speech is the most obvious category to define. The transcriptions provide a record of this activity. Drawing is defined as 'pen on paper' and found in the documents supplied by the experimenters. It is distinguished from writing, which is also defined as pen on paper, but this time as non-alphanumeric markings. Listening is defined as the overt attention, through the direction of gaze, and inactivity, to a source of speech, such as the experimenter in the room or at one end of a telephone line. Thinking turned out to be the most difficult one to define and codify. A set of overt behaviours were used to detect this activity: repeating the same speech pattern over and over again, solving mathematical problems, stating that thinking is needed or is being done, or total inactivity. Two experimenters parsed the data independently and then correlated their interpretations in order to eliminate ambiguities and errors.

On the average, each time it was detected, examination (E) took 25.4 seconds, drawing (D), including writing (W) as a subset, 22.2 seconds, and thinking (T) 19.8 seconds. These three categories, outside of speech, adequately describe all activities of this subject. The range for E was 2–141 seconds; for D, 1–78 seconds, and for T, 2–126 seconds. T and E occupied a greater proportion of the entire protocol (38% and 34% respectively), while D occupied only 28%. The most frequently seen multi-activity patterns for triple-modes were E–D–T with 40%, D–T–E with 25% and T–D–E with 17%. Between these three, half of all possible combinations, 82% of the protocol, are covered.

To correlate the data types and activity modes we looked for ways to identify the design decisions made by the designer throughout the protocol. While these decisions were made *vis-à-vis* the activities, thinking, drawing, listening, writing and examining, the evidence that narrates these decisions is present in the speaking activity.

Based on our analysis of the design decisions and the *activity based* modalities, we were also able to distinguish segments of the protocol which are clearly different from one another. The evidence for this existed in a variety of places including the drawings produced, the nature of these drawings, and the design decisions identified through an interpretation of utterances by the subject that a decision has been reached.

The designer produced 10 drawings in all. We show these later, in the left-hand column of Figure 2.4. These drawings can also be sorted into several categories based on the scale of the drawing and the purpose it seems to serve in the course of the design process, which will be discussed later.

2.2 Data Interpolation Experiments

When we first received the data we decided to take advantage of the opportunity presented by the fact that neither of the authors was familiar with the data or the experimental setup. Since our initial premise was to explore the relationship between the verbal and the visual data, i.e. the transcript versus the video images, we setout to test the dependencies that may exist between these two types of data.

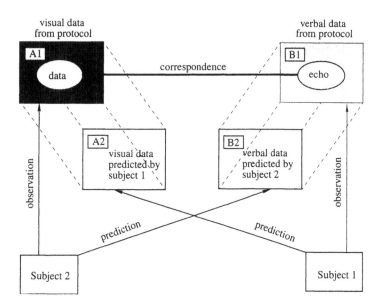

Figure 2.3
Data interpretation
experiments

We designed an experiment in which one of us took the visual data (the video without the sound track) and the other took the verbal data (the printed transcription) and set out to guess the missing half of the data (Figure 2.3). That is, Subject-1, one of us, tried to predict the drawings from the printed transcript; while the other, Subject-2, tried to predict the verbal data just by studying the video. These results are shown later in Table 2.4 and Figure 2.4, respectively.

3 Analysis

3.1 Delft Experiments:
Design Decisions

We consider design decisions to be any and all intentional de-clarations of information as valid for the design problem at hand. For example, in Table 2.1, when the designer expresses an intention to do something (1:06:17 'making the frame stiff...'), asserts a relationship between entities (1:06:33 'that seems to be a big issue'), or uses words that semantically imply the making of a decision (1:06:56 'let's assume that it's a quick release'), we consider that he is making a decision. Defined this way, vir-tually all verbalizations in the protocol appear to be design decisions.

In trying to discriminate design decisions further, we noticed that some decisions concern the design product while others concern the design process. For example, (1:07:05–07) 'I've already decided that I could just stretch that out and we would have a quick mount on the rear wheel' is clearly a design decision; while (1:06:47) 'we wanna decide where to keep them' is a process decision indicating an intention to make a design decision at some point.

Furthermore many decisions are clearly routine, like most of the examples provided above, while some are so non-routine or *novel* that they turn out to be critical for the progress of the entire design. In Table 2.1 there are two examples of this:

1:06:36   'Obviously we have these...'
1:07:05   'I've already decided that I could just stretch that out and we would have a quick mount on the rear wheel'

In the first case, the designer is remarking about the decision to include a triangulation of the supports of the rack, which is critical to the stiffness criteria (stated in the problem description and

Table 2.1 Data Analysis

Time-Stamp	Activity category			Transcription (speaking)	Parallel motor activity	Focus of attention	Non-routine decision
	E	D	T				
. . . .							
1:06:10			*				
1:06:11			*				
1:06:12			*			look at the	
1:06:13			*		intend to lift the	data sheet	
1:06:14			*		upper drawing sheet		
1:06:15			*		but put down again		
1:06:16			*				
1:06:17			*	big thing that er was		look at the	
1:06:18			*	one of their comments		bike	
1:06:19			*	is that the that making		look at the	
1:06:20			*	the frame stiff enough		drawing	
1:06:21			*	the carrier stiff enough			
1:06:22			*	for holding the bike			
1:06:23			*	the the backpack			
1:06:24			*				
1:06:25			*				
1:06:26			*				
1:06:27			*			glance at the	
1:06:28			*			bike	
1:06:29			*			glance at the	
1:06:30			*			drawing	
1:06:31			*			look at the	
1:06:32			*			bike	
1:06:33			*	and er that seems to			
1:06:34			*	be a big issue em			
1:06:35			*			look at the	NDD: trian-
1:06:36		*		odviously we have	drawing	drawing	gulating the
1:06:37		*		em these er			frame for
1:06:38		*					structural
1:06:39		*					stability
1:06:40		*					
1:06:41		*					
1:06:42		*					
1:06:43		*		em er we have some			
1:06:44		*		frame pieces em			
1:06:45		*		going out here			
1:06:46		*					
1:06:47		*		we wanna decide			
1:06:48	*			where to keep them	hold the pen	look at and	
1:06:49	*			er		walk to the	
1:06:50	*			let us see further out	pick up the tape	bike	RDD: sup-
1:06:51	*			they are the better			orts the pre-
1:06:52	*			they are em er first			vious NDD
1:06:53	*			off is	put the tape through	stand behind	
1:06:54	*				the spokes of rear	the bike and	
1:06:55	*				wheel	examine the	
1:06:56	*			let's assume that it's a		rear wheel	NDD;
1:06:57				quick release em			assumption
1:06:58	*						"what if?" of
1:06:59	*						quick release
1:07:00	*			wondering if er we			on real axle
1:07:01	*			can mount these			
1:07:02	*						

Table 2.1 Data Analysis (*continued*)

Time-Stamp	Activity category			Transcription (speaking)	Parallel motor activity	Focus of attention	Non-routine decision
	E	D	T				
1:07:03	*						
1:07:04	*						
1:07:05	*			in fact em alright I've			(repeat previ-
1:07:06	*			already decided that			ous deci-
1:07:07	*			I could em just			sions)
1:07:08	*			strech that out			
1:07:09	*			and we would have a			
1:07:10	*			quick mount on the er			
1:07:11	*			rear wheels			
1:07:12	*						
1:07:13	*					stand beside	
1:07:14	*					the bike and	
1:07:15	*					examine	
2:19:							

noticed by the designer earlier). While this is not evident in the transcription it is so in other activity categories such as drawing and examination. This is a good example of how several activity categories when examined separately are somewhat meaningless but gain special meaning when put together. In the second case, the designer is including a quick mount feature in his design, which is critical to the portability criteria (also noticed and stated earlier).

These are categorized as novel design decisions (NDD) because of the subsequent decisions that hinge upon these and the designer's own statements about the importance of these for the entire design. For example, in referring to the former example the subject says that he considers the triangulation idea to be the 'proprietary feature' of the entire design. Using this analysis we found 10 NDDs in the protocol. Table 2.2 contains a list of these. Furthermore, during the 'retrospection' stage of the protocol (Table 2.3), the experimenters ask about this issue, directly. The designer identifies a handful of key decisions, all of which are included in the set that we obtained independently. All of these NDDs fall into one or more patterns of the following type:

1. The NDD resolves a problem or bottleneck.
2. The NDD does not follow from previous assumptions.
3. The designer identifies the NDD as an important feature of the overall design.

3.2 Delft Experiments: Activity Categories

Based on the activity categories defined in Section 3.1, Table 2.1 contains a sample from our analysis of the data. The first column

Table 2.2 List of NDDs

NDD	Time stamp	Transcription	Description
0	00:38:00–00:46:00	telephone conversation	get rid of the front idea
1	01:06:36–01:06:47	(... big thing that er was one of their comments is that the making the frame stiff enough...) obviously we have em these er em er we have some frame pieces em going out here we wanna decide where to keep them er...	triangulating the frame for structural stability
2	01:06:48–01:07:51	... let's assume that it's a quick er release em ... in fact em alright I've already decided that I could em just stretch that out and we would have a quick mount on the er rear wheels ...	a quick release-mount on rear axle
3	01:11:58–01:12:39	... we're gonna have to a em er a thing that's going to attach to the frame ... we're gonna mount that frame right in this the em quick release er components on the bicycle ...	quick release components underseat
4	01:22:46–01:23:03	... this could be a pivot here pivot here for disassembly ...	rod-pivot arrangement to rear axle
5	01:26:09–01:27:25	(... there's an advantage of that and) ... there's a marketing strategy here ...	marketing idea (* not a physical design
6	01:39:07–01:39:23	(... whereas if I just have er a single one ...) I'm not taking as many moments I actually like this solution better ...	make pin-connection to avoid excessive moments
7	01:41:42–01:41:58	... it's gonna be very important that you have your wheels on tight these things on tight now there'll be tension-compression here I'm gonna have to have a detail here ...	refinement on NDD2: detail of quick connection to rear axle
8	01:52:05–01:52:27	... end view and there's gonna be a er plastic em clamp clip alright which the backpack will snap into now ...	connect backpack frame directly onto carrier frame using snap-on clips
9	01:54:26–01:55:36	... sorry OK so it's gonna come up to about there and then we're gonna have em a tube here that's gonna come up er there and then another ...	use a tube at the end of triangle frame for a pin connection to bars that go to axle for folding (* NDD4 was about folding of bars at the middle)

Table 2.3 Segmentation of Protocol

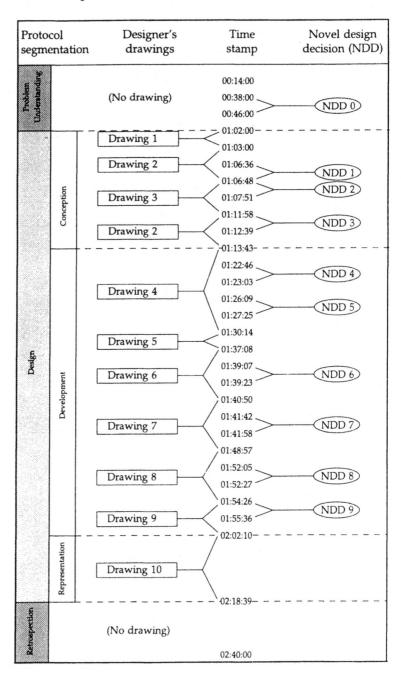

of the table shows the time stamp that corresponds to the events recorded in the other column. In referring to the data in this paper, from here on we will use the time stamp as a reference to its location in the protocol.

The next three columns show the presence of an activity, including examining (E), drawing (D) and thinking (T), by the use of an (*). The next three columns contain the transcription of the verbalizations, other motor activities, and the foci of attention of the subject, respectively. The final column shows the *novel design decisions* (NDD) represented by the data.

Note that some of the activities, such as writing and listening, defined earlier are not shown in Table 2.1. In this protocol, both occurred very infrequently. The writing instances are included with drawing. Listening, which is present only while the subject conversed with the experimenters, is left out of the analysis altogether. This latter category should be an important aspect of the group design protocol which is not analysed in this chapter.

What general patterns can we detect in these *activity based* modes that might illuminate the underlying structure of design decisions? A hypothesis we developed rather quickly was based on the emergence of NDDs. We assumed that the NDD are a function of some combination of *activity based* modalities such as E, T, and D. That is, the emergence of an NDD was likely to rely primarily on a particular *multi-activity mode* (MAM); or that all modalities were involved directly in the emergence of an NDD. In order to test this, we had to interpret the data in ways other than those commonly used in protocol analysis.

For one thing, since the modes we observed (E, T and D) are almost exclusively sequential, it is difficult to judge how they could be considered as cooperating in the development of an NDD. Even though these activities are sequential, we know from cognitive psychology that sequentiality is an artifact of the STM functions[18] and does not imply discontinuity between two sequentially appearing cognitive activities. For example, on the average, it takes 2000 ms for unrehearsed information to fade from the STM. Activities taking place within the span of two seconds then can be considered concurrent, in cognitive terms. This would imply that all adjacent activities could be considered concurrent.

When the interval of an activity is more than two seconds, that is, for the portions of the activity more than two seconds away from the boundaries to the preceding and succeeding activities,

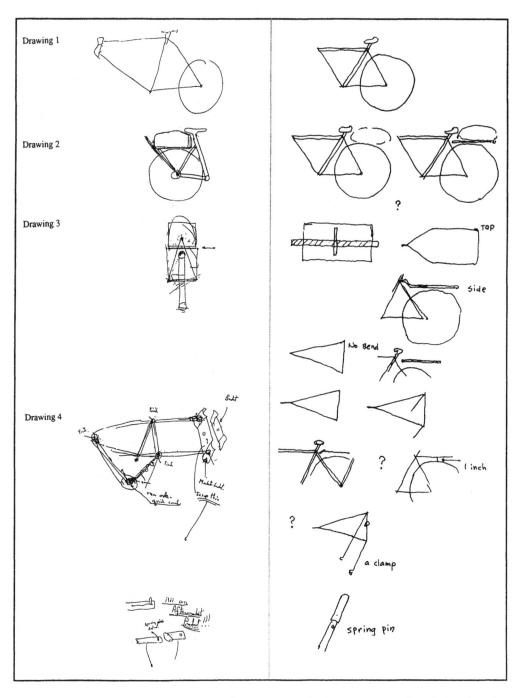

Figure 2.4 Left column: original drawings from protocol. Right column: 'predicted' drawings by Subject-1

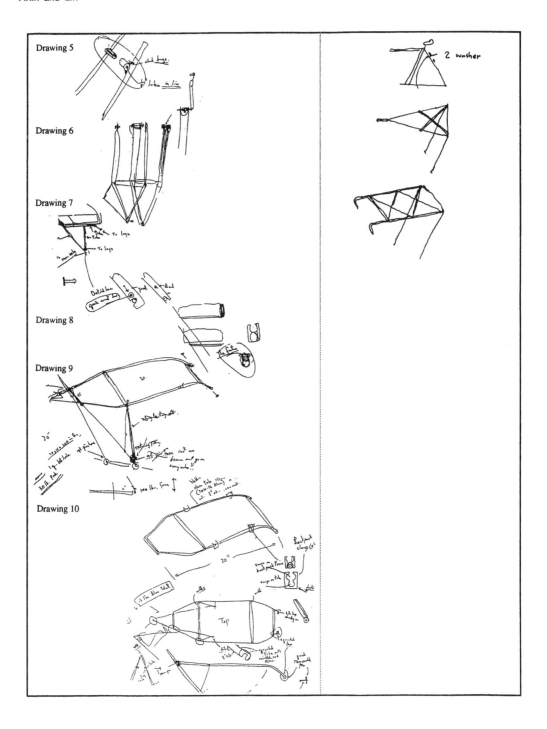

Figure 2.4 (*continued*)

the MAM should be considered a single-mode since the distance to the next activity exceeds the span. This would give a pattern of alternating single- and double-modes for a very large part of the protocol since most activity intervals (about 95%) were more than two seconds. There would hardly be any triple-mode patterns, since, given the average span of an activity (see Section 2.1), the likelihood of three different activities occurring within the two-second span is very small.

This picture would be different if we took another known cognitive threshold: the average span of time to store a chunk in LTM, which is 5000 ms. This, compared to the 2000 ms span, would yield more double-mode and some triple-mode activities. Yet, cognitively speaking, there is little justification for this assumption about the span as is the case for the previous one. This parameter of the cognitive system does not represent a 'holding pattern' for memory items in the STM. Therefore, we decided to treat the question of the appropriate span for the MAM patterns as an open question to be determined by data analysis. To illustrate our approach, let us assume that we have the following data:

1. Time stamp t1,   activity: E,   duration: 40 s.
2. Time stamp t2,   sctivity: T,a   euration: 6 s.
3. Time stamp t3,   activity: D,a   euration: 9 s .
4. Time stamp t4,   activity: T,a   euration: 21 s.
5. Time stamp t5,   activity: E,a   duration: 2 s .

Suppose for argument's sake that the span we should select at the end is 25 seconds. In these data, we have three positions for the first activity (E), at which a new modality count occurs. Shown in an approximate scale this looks like the following:

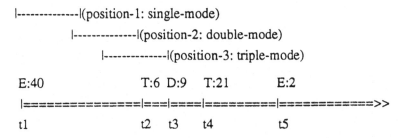

That is, first there is a single-mode (E), up to and including t1 which is equal to 40 seconds. Given that the span is 25 seconds (i.e. (40 − 25) + 1 = 16), this represents an interval of 16 seconds over which a single modality lasts. From the point t = 41 on, the modality is two. The span now bridges over into the next mode (T). Then, beyond the point t = 46, which is in activity D, the mode increases to three. The 25-second span is large enough to occupy a piece of D, T in its entirety, and some of the initial mode, E. Once the lower bound (tail) of the 25-second span leaves E, the modality count is decremented by one to two.

In this way of incrementing the modality count every time the interval's upper bound (head) crosses over into a new mode and decrementing it every time its lower bound (tail) crosses out of a former mode, we can keep track of the modality count and the intervals of the protocol in which they apply. It is important to note that for the modality count to go up, the boundary of a new mode, i.e. one that does not match the current mode types, must be crossed; and likewise to decrement. That is, when the span goes over t4 the modality still stays at three, since the new mode is T, matching one of the previous ones. At some point when the head of the activity span is still in the interval between t4 and t5, its tail will be moving out of E. At this point the modality count will be decremented by one, since this is the only E in the MAM. We developed a computer algorithm to estimate the locations and spans for all MAMs for the entire protocol, for $1 \leqslant n \leqslant 218$.

The distribution of the triple-modes throughout the protocol is an aspect of this analysis which interests us. In order to do this, we had to assume that overlapping but independent triple-modes would be considered as one, since all triple-modes consist of a combination of E, D and T. This we used in selecting a value for the span.

We show the frequency of triple-modes in the entire protocol as a function of the span in Figure 2.5. As the span increases so does the frequency of triple-modes, since the likelihood of crossing over more modal boundaries increases. This reaches a maximum value of 30 when the span is 14 seconds. After the span goes beyond 22 seconds the frequency drops sharply. This is because, as the number of triple-mode instances increases the likelihood of each instance overlapping with neighbouring instances also increases. Thus, the interval of each triple-mode consisting of many such instances increases and the frequency decreases until there is a single triple-mode instance that spans the entire protocol at about a 218-second span.

Consequently, if we want to eliminate the artificial effects of

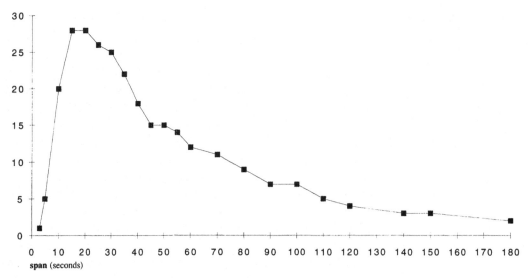

Figure 2.5   Frequency of triple-modes

triple-mode intervals running into each other due to very large span values and very few triple-mode intervals forming due to very small spans, then we should pick a value that is *optimal* between these two undesirable effects. This yields the range of 14 seconds to 22 seconds, where the frequency in Figure 2.5 is maximum. Using these as our criteria, we selected 15 seconds as our threshold span.

We also note that, given that the maximum number of triple modes is 30 and a total of 1056 seconds were spent in the triple mode, the average interval for each triple mode occurrence is 35.2 seconds.

Another aspect of the modality of activities that interests us is the frequency of different *n*-modalities that applies to each single-second interval of the protocol. This is to consider the frequency of modalities from the point of view of each atomic interval in the protocol. We could have picked a smaller interval, or a larger one for that matter, as the basis of this analysis. We used one second, for practical reasons. Any one of these atomic intervals can potentially belong to any number of modalities. For example, the interval t = 48 belongs to the single-mode interval of E and to the double-mode interval of E–T, as well as to the triple-mode interval of E–T–D. This means that the frequency of modalities for t = 48 is 3. Using this method we plotted the frequency of modalities for the entire protocol (Figure 2.6).

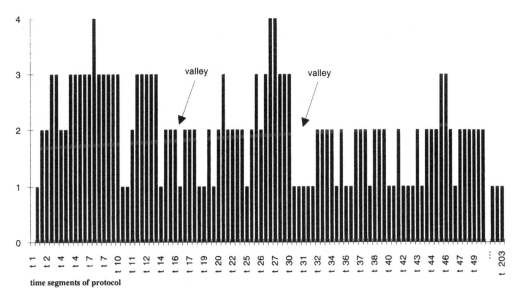

time segments of protocol

Figure 2.6 Frequency of modalities (span = 15 seconds)

3.3 Delft Experiments: Protocol Segmentation

First we segmented the protocol into three parts: problem understanding, design, and retrospection (Table 2.3). During problem understanding which takes slightly less than one hour, the designer familiarized himself with the problem, asked questions and examined documents with the apparent goal of formulating an approach to the design stage which was to follow. An insignia of this stage is the total absence of any graphic representation. This step of structuring a design problem is well documented in literature[4,8,10].

The second stage is the most complex and interesting one for our purposes. It commences with the initial efforts to formulate viable solutions to the problem (1:02:00). This includes all of the drawing activity in the protocol. It consists of three sub-segments: *conception*, *development*, and *representation*. In the conception stage, the overall approach of the solution is sketched out. It deals with the problem at a relatively large scale, and avoids the elaboration of mechanical details. The general lines of the product, a 'backpack rack', are determined with a few options on detailing that are to follow. On or about time stamp 1:13:43, the designer begins a drawing that starts to detail the design of the rack. This commences the next sub-segment. Here as well as in the previous sub-segment several NDDs are seen. In fact all of the NDDs we

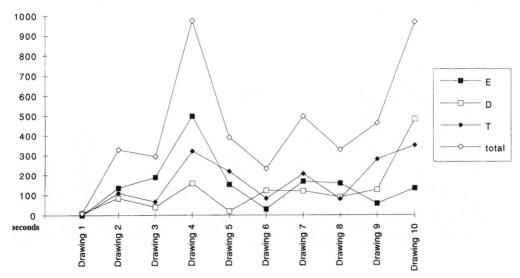

Figure 2.7   Time spent on each activity made per drawing

identified occur in these two. By the end of this sub-segment (2:02:10) the designer goes into a mode of representing the design as it is currently developed. The onset of this sub-segment appears to be more a function of the expiration of time than anything else.

The third stage is in fact a formal part of the experiment. Once the designer stops, the experimenter begins a debriefing session. This we identify as the retrospection segment. It is during this segment that the designer identifies several of the NDDs we found independently.

3.4 Delft Experiments: Drawings

First we observed that drawings 1 through 3 are large-scale drawings that characterize the conception of a sub-segment, while the others are small-scale drawings showing greater detail. We also observed that some drawings were mere presentations of already known features, thus serving a summarizing function in the protocol (such as drawings 1, 5 and 10 in Figure 2.4). These tended to appear just before new design explorations were launched by virtue of the fact that they made apparent those features that were not yet worked out or were problematic. Thus, they usually led to the alternative category of drawings that were made for the purpose of exploration or enquiry (such as drawings 3, 4 and 6 in Figure 2.4).

The times spent in three principal activities of our analysis, E, D and T, show different patterns for different segments of the protocol in which drawings are produced. Time spent in examination, for example, peaks for the early drawing segments but tapers off in the later ones (Figure 2.7). For Drawing, the peaks and valleys alternate, with the overall time-spent value building up to a maximum in the final drawings. In the case of Thinking, a combination of the two graphs is visible.

3.5 Data Interpolation Experiment

In guessing the drawings made by the designer, Subject-1 produced 20 drawings, all of which are shown in Figure 2.4. Subject-1 was instructed to correlate his drawings with the time stamps of the protocol transcription. After the drawings were matched with those of the designer, a subjective comparison was made by both authors to observe the similarities and differences between the predicted and original drawings of the protocol. Five out of 20 (representing 25% accuracy) were considered to be similar.

A similar procedure was followed in the case of the transcription statements. Subject-2 predicted 112 statements which were correlated with the original protocol transcription *vis à vis* the time stamps (Table 2.4). Based on the semantics of these statements, as subjectively interpreted by the authors, 27 statements were considered to be similar to the original data, shown as *italic* text, which represents an accuracy of about 24%.

4 Conclusions

Based on these analyses we drew several conclusions which can be organized under five categories: frequency of design decisions, designer's span of attention, activity modes, design drawings and predictability of data.

4.1 Design Decisions

In the span of more than two hours the designer makes literally hundreds of decisions. Only a very small number of these (10) are what we consider as novel or giving rise to something exceptional in the design. This indicates several interesting conclusions. One is that most of the designer's time is occupied making routine decisions. Another is that novel decisions are indeed at a premium and indicate a very large overhead (about 15 minutes of intensive work per decision). Therefore it is of value to describe as accurately as possible the precise conditions that surround these decision points.

Table 2.4   Subject-2's predictions of verbalization (Italicized text indicates matches)

Designer's verbalization data:

01:04:00

... em but er let's just em look at this here
we got a a a the wheel
we got the frame
we got the seat where is this going to be the height of
the seat
location
we got the back stay here
we have this er OK and em
what I'm going to do is I'm going to em *put in the
backpack by the way*
let's just just check see what kinda scale this thing says
here. . . .

01:08:00

... OK so em I'm drawing a picture of a wheel here em
I've drawn a picture of a er a rear axle I've drawn a
picture of a er of a er of of a pack of a backpack here
on *top end view* alright if I were to make a frame that
looked like this that would be a very poor design
because basically what I've got is that I've got a
parallelogram. . . .

1:13:00

... unfortunately I have a feeling that's going to have
to come out higher than we'd like
but that is going to be *a key feature* em
we're gonna come up to here. . . .

1:20:00

... where I'm going to try and mount I'm going to
mount something right in here
we'll have to have a clamp arrangement
... actually I'm not a great artist and then another
another half here OK
... that's going to pick up that *vertical tube down* em er
OK em I would then *bring out em two of these er tubes
two rods up here* em to provide
er torsional stiffness ... this could be *a pivot* here pivot
here for disassembly ...

1:24:00

...
I got is is a clamp there em
if I built in a clamp already OK
... these these clamps would be on the bicycle already
and ... what I'll have hee is I'm gonna have a detail
like this on this clamp ... this is gonna be a detail here
er...

1:26:00

... there's a marketing strategy here
market bracket huh market bracket alright...

.

.

02:05:00 ... I'm proposing *an aluminum tube structure*
for now ... T6 aluminum alright it's em and ... after
you after you weld it it's a welded aluminum welded
alright welded aluminum tube structure
... 02:06:00 ... (calculate)...

Subject 2's verbalization data:

(begin) 1:02:05

OK we have the bike
Rear wheel
The frame
Seat
Do you have a scaled drawing?
OK let me have it

1:04:20

This is *where it will sit*
This is *where it will fit*
So it will be supported
OK tie it to the frame

1:08:08

How to support a backpack...
This is the *bike from rear view*
There's the support for backpack
Backpack here
The support at hub
And tie it down
To secure backpack so it does not fall over to a side
This is the key to the idea

1:09:36

OK I can do it this way and secure it

1:13:10

This is where the end point of the horizontal support
is secured
This is the key element to support the back of the
backpack rack

1:20:30

OK this is the back of the seat
Fasten it here
And there are the two support bars that come out like
this
Now tie it across
Then bring it out and around
Round the end to protect injury
Bring two *support rods down to the hub*
How to avoid rocking this way?
Need some lateral bracing
We could *make these joints* sturdy
Bring another element *from the back to the hub*
Tie it to the axle
Here it will be OK because of the connection to the
back of the seat

.

.

.

2:01:45

OK *the rack like this*
A metal *(alum) bar* formed in this shape
Ends one hooked to attach to back of seat
A cross bar at one end and cross bar here where
uprights join it
A hollow tube here at the end to accommodate a pin
for the collapsible upright connections

Table 2.4 *(continued)*

02:07:00 ... *(let's say its twenty yeah say twenty twentyone inches long* so here we're gonna start from there *say twenty inches* actually ... 02:08:00 I'm gonna I'm gonna draw in just *draw a top and bottom view* ... rack ... alright ... em *tube coming up from below* ... Ok ... *em and we've got em a er er ... crosstube* em ... 02:09 ... OK it's a top view and we have a couple ... 02:10:00 (calculate and evaluate the material of backrack) ... too expensive em a half pound of aluminum costs ... OK em alright so em this is the top view and *this is the side view*	This part is also a tube for stability R = " # gauge *aluminum tube* This is *a welded connection* to cross bar How *long is this?* OK *this is 25"* It is 12" wide *Let's do a top view* Three *cross bars here* and upright supports attached are to the frame not to cross bars This is the idea This (top view) shows the construction and dimensions more accurately
02:11:00	32:11:00
er these are gonna be er squashed lugs alright and then ... is gonna come all the way back to here ... and then that's that's gonna be the end of that and then I'm gonna have em a pivot tube mounted underneath here in fact ... tube there em for these things to attach ... 02:12:00 alright and then I have this em this this these this tube here coming ... and this again the detail there em let's just see ... alright and that's gonna fit onto the bike lugs this is bike lugs	*Let's make a side view* This is the base of the rack Ends hooked down Here at the triangulation we have the uprights going beyond The base of the rack connected here with a hinged joint The hooked ends are fastened with a spring to provide lateral stability and ease of mounting and dismounting
02:13:00	2:13:30
... I'm gonna draw it like that ... OK ... here weld ... alright and then this lug here down below ... alright this is a ... 02:14:00 ... hole in lug lug weld on lug hole alright em ... em ... and em now we're going to em have a em ... a pin ... OK and so by having a lug here offset so sideview of this thing shows it's like this with a lug there *this thing here can can em can rotate up and store parallel to here* so that lug is below or it's actually gonna be there that's where the lug is on this piece so when this thing rotates ...	A quick sketch of the joinery of rack to support uprights This is # gauge aluminum This joint is pivoted so *when it folds the uprights align with rack base completely and can be part of backpack* This is 45 degree This is # gauge ... etc ... **(end)**

In the single protocol that we examined we observed the following conclusions:

1. Scarcely any NDDs (only one in this case) are made in the problem understanding segment (Table 2.3).
2. Almost all NDDs (nine in this case) are made in the conception and development sub-segments (Table 2.3).
3. NDDs occur when multi-modal activity is more subdued (Figure 2.6) but while triple-mode activities are almost exclusively present.

4.2 Designer's Span of Attention

The span of attention of the designer on any one activity (E, D, and T) is clearly limited (25.4, 22.2 and 19.8 seconds, on average, respectively). Obviously there are benefits in alternating between these multi-modal activities, as no single mode appears to be sufficient to accomplish the task of design. The correlation of the

NDD with triple-mode points in the protocol provides further support for this result.

The interesting comparison between drawing activity *versus* the other activities is that, while the *average* drawing time is comparable to that of the others, the *range* of its duration is equal to about half that of the others. This indicates that the span of the drawing activity is finer and is bunched up around a certain value rather than being spread out like the others.

4.3 Activity Modes

The *activity based* model of data is a good indicator of the entire design protocol in general and of the NDD in particular. The triple-mode for example is a predictor of the NDD. Six out of eight NDDs (excluding the one during the problem understanding segment and the one which deals with a marketing innovation, overridden by subsequent design decisions) match a triple-mode point.

The triple-mode intervals in the protocol amount to a total of 1056 seconds and occupy 1056/4470 or 23% of the data. Each triple-mode interval lasts 35.26 seconds on average (see above). The probability that each NDD, which on average has an interval of 30.3 seconds, may overlap with any part of any one of these triple-mode ranges, then, is about $(30.3 + 35.26 - 1)/4470$, or 0.0144. Since there are 30 triple-mode instances in the protocol, this probability for the entire protocol is 30×0.0144, or 0.433. Then, the probability that each of the six NDDs overlap a triple-mode is equivalent to about 0.433^6 or 6.59×10^{-3}. This shows that the coincidence between triple-mode activities and the NDD is not the result of pure chance.

4.4 Design Drawings

There are clearly different purposes for different drawings. There is a general phasing that proceeds from overall, conceptual to detailed drawings. Some drawings foster novel ideas while others are routine. Early drawings represent a different composition of the three principal activities (E, D and T) as opposed to the later ones.

4.5 Predictability of Data

Both subjects (1 and 2) were able to predict some of the data in the alternative mode (drawings and transcriptions, respectively) from one of these modes. About 25% of the predictions were accurate. This represents 25% of the data in the case of the drawings and a much smaller percentage of the total data in the case of the transcriptions as there were a lot more statements in the original transcriptions than the total number of statements predicted by

Subject-2, which was 122. This indicates that the visual and verbal data constituting a protocol have causal relationships that can be captured, probably intuitively, by experts. Consequently, we can argue that there is some predictability, if not redundancy, between these data forms. This supports our assumption that transcriptions and drawings may be echoes of each other.

5 Discussion

5.1 Modality Analysis

Our activity based model of the design protocol data provided interesting results in terms of the contribution of activity modes to the development of designs. One such result is the correlation between the triple mode (Examining–Drawing–Thinking) and novel design decisions (NDDs). Six out of a total of eight times a novel design decision was made, we found the subject alternating between these three activity modes in rapid succession (conforming to our 15-second span criteria, see Section 3.2). At this time we cannot be sure if there is a causal relationship between these; we can only conclude that a correlation exists. Therefore, we can assume neither that the triple mode is a requisite for the NDD nor the other way round. We can only say that our data suggest that designers explore their domain of ideas in a variety of activity modes (three out of three, in this case) when they go beyond routine decisions and achieve design breakthroughs.

Another result which, at face value, seems to be contrary to this one is the relatively low frequencies of alternative activity modalities (i.e. valleys in Figure 2.6) correlating with NDDs. In closer examination, all this indicates is that there were fewer alternative interpretations of the number of activity modes that apply to the data (given the 15-second threshold). That is, more decidedly than in other parts of the protocol, during the NDDs the designer was operating in the triple mode. Seen in this light this result only reinforces the first one. The designer appears to be decidedly engaged in all three major activities at the time of the NDD.

We do not know of equivalent findings in the cognitive psychology problem-solving literature. Therefore, we can only speculate about the fundamental information-processing implications of these findings. Do designers engage in multiple activities in order to verify the worth of a new idea which may have emerged from a singular or only a few activity modes? Do designers generate novel ideas as a consequence of engaging in multiple design activities? Are multiple activities useful because they cover different aspects of the design consideration at hand; or are they simply redundant with one another as a matter of ascertaining first hunches? These and related questions can be asked as a result of our findings but cannot be answered here, without further examination of their premises and indications.

5.2 Segmentation

The segmentation of the protocol into three and then the further decomposition of the middle segment into three sub-segments is consistent with previous models offered by Akin[8]. This structure is entirely consistent with the position and role of the problem structuring activity in design. What is unusual is the relatively large chunk of time taken by the designer (almost one hour of a two-hour protocol) to do the initial problem structuring, and the complete absence of graphic exploration on paper during this period. The drawing activity particularly proves to be an excellent indicator of the segments of the protocol, more than any other activity we identified.

We are not aware of other results reinforcing this point, though there is a great deal more to be said about the role of drawings, including some distinctions that can be drawn between architectural and product design activities.

5.3 Drawings

It is possible to explain and annotate each of the nuances in the drawing category individually. However, the principal implication probably is captured in Pylyshyn's characterization of the role of visual images in cognition[14]. As we are able to selectively represent different aspects of an image with symbolic encodings, the object or its model (image, drawing, etc.) is the authentic entity that embodies all of these multiple aspects and 'views' of the object. Hence the graphic entity appears to provide the road map to the various aspects of a design or design problem. Thus the drawing activity is steady throughout the protocol with the customary intensification at the end to produce the final design; while the other activities assume different levels of importance depending on the protocol segment and the design goal applicable at each moment.

Here we would like to note a particular result from the work of Goldschmidt[15] who found that architects tend to derive initial design ideas from mere doodles. The doodling activity for creative exploration appears to be non-existent in this protocol. One of the reasons for this can be the differences that exist between architectural and product design problems. Architectural design necessarily demands unique responses as the problem is more than likely rare – with a novel combination of site, client and programme. In contrast, product design exists in a domain of problem knowledge, such the manufacturing and material parameters, market conditions, etc., which is much more stable than that of architecture. Furthermore, it is more feasible to evaluate product designs for feasibility, as accurate one-to-one models can be built due to manageable scale and manufacturability and prototyping objectives. Thus, product design lends itself to more accurate functional scrutiny while architectural design has relatively less defined functional criteria of evaluation.

5.4 Data Modality
Dependencies

It may appear unusual that one would even consider predicting data with no prior knowledge of it, let alone predict about 25% of it correctly. However, neither of these results is unusual or original in cognitive psychology. In a similar experiment Charness asked chess masters to predict board positions for pieces without ever seeing these positions. His results in terms of accuracy of prediction are comparable to, if not better than, ours. The point is that there is an underlying mechanism in operation here. Experts, and our subjects would be considered experts as would be chess masters, have sufficient knowledge from previous experiences to recognize generic patterns. These patterns include what designers are likely to say when they are doing something at the drawing board, as well as what they may draw if they are uttering certain words. This is a function of the chunks that all experts possess in a particular problem domain.

What would be really unusual is if the predictions covered non-generic or idiosyncratic aspects of the protocol as well. Because there is no explanation other than pure coincidence for one individual to predict another, unfamiliar individual's personal patterns, the accurately predicted drawings and words should have no better probability than chance to match non-routine drawings and NDDs. With this hypothesis we went back to our original data and found that none of the drawings of Subject-1 matched any of the non-routine drawings made by the designer. In the case of the NDD-related statements in the protocol, we found two matches with the predictions of Subject-2. While this is unexpected, we do not consider it a significant result at this time.

In the case of the observation of dependency between the two data forms, verbalisations and drawings, we conjecture that there is probably a greater percentage of positive correlation which we were not able to demonstrate here. The *dual mode* model predicting one mode as the echo of the other probably accounts for most of this redundancy. In other words, subjects verbalizing in protocol experiments, in essence, are echoing what they are doing. Thus, in protocol analysis the primary evidence should be collected from data outside the verbalizations, and the transcriptions should only be used as a secondary source or as road maps to the former.

References

1   Eastman, C., *Explorations of the cognitive processes in design*, Department of Computer Science Report, Carnegie-Mellon University, Pittsburgh, PA (1968)
2   Krauss, R.I. and R.M. Myer, Design: a case history, in G.T. Moore (Ed.), *Emerging Methods in Environmental Design and Planning*, MIT Press, Cambridge, MA (1970), pp. 11–20
3   Foz, A., Observations on design behaviour in the *parti*, *The Design Activity International Conference*, vol. 1, University of Strathclyde, Glasgow (1973), pp. 19.1–19.4
4   Akin, Ö., How do architects design?, in J.-C. Latcombe (Ed.), *Artificial Intelligence and Pattern Recognition in Computer-Aided Design*, IFIP/North-Holland, New York (1978), pp. 65–104
5   Cuomo, D.L., *A study of human performance in computer-aided architectural design*, PhD dissertation, University of New York at Buffalo (1988)
6   Darke, J., The primary generator and the design process, *Design Studies*, **1** (1979) 36–44
7   McDermott, J., Domain knowledge and the design process, *Design Studies*, **1** (1982) 31–36
8   Akin, Ö., *Psychology of Architectural Design*, Pion, London (1986)
9   Waldron, M.B. and K.J. Waldron, A time sequence study of a complex mechanical system design, *Design Studies*, **9** (1988) 95–106
10 Akin, Ö., B. Dave, S. Pithavadian, Heuristic generation of layouts (HeGeL): based on a paradigm for problem structuring, *Environment and Planning B: Planning and Design*, **19** (1992) 33–59
11 Schön, D., *The Design Studio, An Exploration of its Traditions and Potentials* RIBA Publications, London (1983)
12 Chan, C. S., Cognitive processes in architectural design problem solving, *Design Studies*, **11** (1990) 66–80
13 Eckersley, M., The form of design process: a protocol analysis study *Design Studies*, **9** (1988) 86–94
14 Pylyshyn, Z.W., What the mind's eye tells the mind's brain: a critique of mental imagery, *Psychological Bulletin* (1973) 1–23
15 Goldschmidt, G., On visual design thinking: the vis kids of architecture, *Design Studies*, **15** (1994) 158–174
16 Kosslyn, S.M., Information representation in visual images, *Cognitive Psychology*, **7** (1975) 341–370

17 Loftus, G.R., Eye fixations and recognition memory for pictures, *Cognitive Psychology*, **3** (1972) 525–551
18 Newell, A. and H.A. Simon, *Human Problem Solving*, Prentice-Hall, Englewood Cliffs, NJ (1972)

3 The Designer as a Team of One

Gabriela Goldschmidt
Technion, Haifa, Israel

Ever since Vitruvius' first-century treatise on architecture, we accept axiomatically that a designer must know a little bit about everything because design work requires varied knowledge and an outstanding capability for mental integration and synthesis. What was always taken to be indisputable for architects is also deemed true for other kinds of designers, albeit to a lesser degree. The ancient designer was an absolute authority who could handle any professional design challenge. The habit or norm of teamwork in design is a relatively recent phenomenon, emanating from the scope and complexity of many design tasks and the need for multiple expertise and labour division.

Team design is an appealing idea and in many settings we would not consider a different work routine today, supposing that (brilliant as it might be) Vitruvius' creed is no longer appropriate. However, there is little research to support the claim that teamwork in design is superior to that of an individual designer, and if so, in what way and under what circumstances. We approach this question from the perspective of cognitive science and ask: can we access processes of design thinking so as to be able to compare the behaviour and performance of the individual to that of a team? We believe that we do have germane tools with which to study design thinking and we propose to do so in this study. We define the parameters for comparison between a team and an individual, and we present a methodology, based on protocol analysis, that we deem appropriate for the task. We show that from a cognitive perspective not much has changed since the days of Vitruvius,

whose acute and insightful observation of designing appears to
be of interest to this very day.

1 Individual and Team Problem Solving

In the engineering oriented design professions it is common
practice to initiate work on a new task with a collective 'ideation'
session, or brainstorming. The task is assigned to a team and the
team members are thus jointly involved from the very early
stages of conceptual design. This is not the case in the more art-
oriented design fields, where an individual designer is often
responsible for the conceptual design phase. He or she may
consult with colleagues and peers, but the responsibility is
personal and a team steps into the picture at a later stage, with the
chief designer's initial sketches given as a *fait accompli*.

It is therefore not surprising that there are design situations in
which it is not clear which mode of practice is desirable for the
front end of the design process: a team effort or an individual's
endeavour. Is the contribution of several minds to the conceptual
phase an asset in terms of the breadth of issues that can be
expected to surface and the number of alternative candidate
solutions that may be proposed? Or does teamwork contribute,
conversely, to a diluting of conflicting views because of pressures
towards conformity and compromise? Is the single mind less
restrained because it is free to explore unpopular directions or is it
constrained by personal biases and limited expertise?

These questions are not unique to design, and have been asked
in the context of research on problem solving and on scientific
investigation. Teamwork is generally considered more fruitful, at
least potentially, and there is evidence that in scientific research
collaborative work is constantly on the rise. In psychological
research, for example, the mean number of authors per published
article rose from 1.5 to 2.2 between 1949 and 1979[1]. This does not
mean, however, that we have evidence that collaborative under-
takings result in more satisfactory work. Kraut, Egido and
Galegher[2] have shown that psychologists opt for joint research
and authorship of articles *despite* lower satisfaction from the
quality of such work, as compared to work they carry on and
publish on their own. The reasons for seeking collaboration are
therefore independent of the results and have to do with social
and personal dimensions that we shall not venture into here.
Likewise, group problem solving processes in engineering are not
necessarily superior to individual efforts[3]. Detailed studies show

that the motivation of team members tends to decrease by as much as 30% when there is no *personal* penalty for slacking or no reward for successful performance[4]. Intergroup competition eliminates loss of productivity, but personal payoffs are still essential. Good management can take this fact into consideration by designing compensation for group performance, as is habitual in sports, for example, where important victories lead to individual bonus payments to members of a winning team.

Given the above evidence, a comparison between individuals and teams in design, and particularly at the front end of design, seems rather relevant. Industrial design protocols of a team and an individual at work on the design of a bicycle carrier are a particularly good opportunity for such a comparison for several reasons:

1. Industrial design lies somewhere between engineering and the more artistically oriented design disciplines.
2. Protocol analysis is well suited for the comparison of processes that we are interested in.
3. The equal settings of the two design sessions in question provide a considerable methodological advantage.

One rarely achieves so much control over so many parameters of a design situation that is very close to real world circumstances.

The question we must address next is that of criteria for a proposed comparison between the design team and the individual designer. We opt for *productivity* as the yardstick for commentary and analysis. But before we plunge into our study of productivity, we need to be absolutely sure that our two protocols are of the same kind and that our protocol analysis methodology can be applied to them on equal grounds.

2 Design Thinking and Verbalization

The arguments for accepting reports of think-aloud exercises as a reflection of cognitive activity are well documented and substantiated (e.g. Ericsson and Simon[5]) and require no further discussion here. But since our data consist of one think-aloud protocol and one protocol of a conversation, a question may arise regarding the compatibility of these two modes of thought verbalization. Can thinking aloud and conversing with others be seen as similar reflections of cognitive processes of the kind we want to investigate? We answer in the positive, adopting a Vygotskian view on the relation between thought and speech[6].

Vygotsky distinguishes between two planes of speech: the inner and the external. Inner speech is not a pre-linguistic form of reasoning but the semantic aspect of speech, abbreviated speech, in that it centres on predication and tends to omit the subject of a sentence and words connected to it. Inner speech is a function in itself, not an aspect of external speech, but inner and external speech together form a *unity* of speech.

Thinking aloud can be seen as being close to inner speech, whereas a conversation is certainly a sample of external speech. This is applicable to our case even though in our team protocol conversation takes place among team members who, through close association with one another and with their subject matter, achieve a fluency that is marked by numerous short-cuts. Both the inner and external planes of speech are, however, more than representations of thought. To put it in Vygotsky's words: 'Thought is not merely expressed in words; it comes into existence through them.'[6] (p. 218). We therefore accept our two protocols as equal windows into the cognitive processes involved in design thinking. We can now proceed to discuss design productivity, in terms of the design process, and will later present an analytic system with which we propose to examine the productivity of the design processes at hand, so as to be able to draw a comparison between them.

3 Design Productivity

The term productivity brings to mind issues of cost-effectiveness and profitability. But it is also related to performance, motivation, efficiency, effectiveness, production, competitiveness, quality and so on. We accept Pritchard and Watson's notion that the important issues to look at are efficiency and effectiveness[7]. We also accept their view that it is not easy to measure group productivity: interdependence among group members is necessary to achieve the group's goals. The patterns of such interdependence may be rather complex, and therefore the productivity of a group is not a simple sum of the performance of its members.

This leaves us with the need to find ways to assess effectiveness and efficiency in design thinking, as carried out by both individuals and teams. Efficiency is relevant to design thinking because it bears on creativity and expertise, among other things. Economy of thought, or the amount of mental resources that must be invested to obtain innovative ideas, is directly related to creativity[8]. Expertise has an even closer association with efficiency, because we know that the ability to take short-cuts and thereby reduce the amount of labour that is required to arrive at satisfactory solutions to problems, is a hallmark of expertise. Creativity will figure in our discussion of the protocols, but expertise is not an issue we comment on here, since we do not compare experienced and inexperienced designers in the present study. Effectiveness, which we shall more or less equate with productivity, will be our major concern.

Designing is a mode of thinking and our sense of the meaning of productive designing is akin to Wertheimer's sense of productive thinking[9]. For Wertheimer thinking is productive when it gives rise to genuine ideas, when it brings about the transition from a blind attitude to understanding, when one comes up with creative ideas, however modest the scope or the issue. It stands in contrast with rote thinking and with the following of a prescribed course of action. Productive designing, like productive thinking, is insightful and inquisitive and it is indispensable in problem solving when the problems are new and unfamiliar to the problem solver. Design problems are almost always new to some degree, and we therefore think of them as belonging to the category of ill-structured problems (e.g. Goel[10]) that require productive processes to be adequately solved. We can easily assess design products; our problem is to correlate the quality of a product with the process that brought it into being. We need an analytic system that can dissect the design process so as to understand what cognitive mechanisms lead to productivity. The system that we propose is called *Linkography* and we present it in Section 5. But before we do so, we shall briefly describe the design processes at hand, by an individual designer and by a team, so we can use these cases to instantiate our presentation of the system.

4 The Bicycle Rack Design

In the two hours allotted to the design exercise, both Dan (the individual designer) and the design team arrive at relatively complete and satisfactory proposals for the bicycle rack that they are asked to design. Our detailed analysis in this study pertains to a fraction of their work; but to understand it we first give a descriptive profile of both these processes and how they might be compared.

4.1 Solo Design: Dan

Dan is an experienced mechanical engineer, who looks at the design of the bicycle rack almost exclusively from a functional point of view (in the debriefing he says: 'It could use some more product design as far as just aesthetics go.' He also implicitly criticizes non-functional design, which he sees as too toy-like, as opposed to 'heavy-duty stuff' that he prefers). He describes himself as untidy and comments several times on his difficulties with visualization and drawing – particularly three-dimensional drawing. Yet he produces five sheets of drawings, in which an outside observer cannot detect any drawing handicap. The first actual sketch appears only 45 minutes into the exercise. The considerable portion of time before he starts sketching (close to 40% of the total time) is spent studying the problem and getting himself informed about it. This is an intentional and typical strategy, he tells us, and declares: 'there's no sense in starting from scratch if you can start at square two instead of square one or square zero.' Similar statements recur later in the process.

His exploration leads him to decide on a tubular design for his rack, and this decision is never questioned nor rated against alternative options. He takes a long time to decide on the location of the rack – front or rear of the bike, side mounted or symmetrically placed. In the debriefing after the exercise he states that the position is something he would have developed alternatives for, had this been a longer exercise.

Having decided on a tubular design and a central rear position for the rack, he spends most of the design time on joints between the rack and the bicycle and is most concerned with its stiffness. Connecting the given backpack to the rack is a secondary issue in terms of the time he spends dealing with it, and he finds a rapid solution to it which he refers to as a major *feature* of the design: plastic clamp-clips that are fitted onto the rack and into which the external frame of the backpack is snapped. Interestingly, decisions about the rack and its joinery to the bike and decisions on the backpack's joinery with the rack appear to be derived from very different categories of design criteria: the former are based on a study of existing products and the conviction that bicycle manufacturers are good with tube handling, while the latter is based on the wish to arrive at a 'proprietary product' because 'we need a feature.'

Whereas the snap-on clamp-clips are never questioned after they are first introduced, Dan does not hesitate to make amendments to various details of his design when he finds that he can achieve better results. This occurs when new information that he requests changes the picture or when, upon evaluating certain decisions, he is dissatisfied. He appears flexible enough, but only within a selected overall concept that is not challenged in any phase.

Dan's protocol was divided into 28 units by the subject matter they deal with, ranging in length from one to eight minutes, plus one unit that lasts 12 minutes. If we omit the longest unit, in which he conducted a telephone conversation, we arrive at an average of four minutes per unit.

4.2 Team Effort: Kerry, Ivan and John

The design team comprises three product design engineers who work together in practice. They have a moderate amount of experience – Kerry is somewhat more experienced than her colleagues. Although the products they typically work on are not identical, in this team effort they all act as 'general designers' with no particular areas of specialization. They appear to operate in harmony, with mutual respect for each other and an impressive ability to collaborate throughout the process. They work in a systematic manner that they seem to be practised in and which requires little discussion among them on procedure. Fifteen minutes into the session they produce a timetable in which they divide their time into well-defined design phases and they ask Ivan to be their timekeeper (Mister Schedule). They then enter a brainstorming phase in which they bring up issues and concepts and list them orderly on a whiteboard. They list the functional requirements, desired features and alternative options for the positions of the rack, concepts of joining the rack to the bike and the pack to the rack, and materials.

The process is iterative and they discuss the different items on their lists several times and make decisions to 'kill off' those which are found inferior, based on a 'design rationale' that is stated for each candidate decision. Human factors come up frequently, and a 'proprietary mounting thing' is rejected on moral grounds. The rack's position on the bike is determined easily at an early phase, and the work on joining the rack to the bike and that of connecting the pack to the rack receive about the same amount of attention, although a little more time is spent on the rack and its connections to the bicycle. Many alternative solutions are explored, especially for the way the backpack can be attached to the rack. Special dual function features such as convertibility of the rack to a lock are considered, but abandoned later on.

Questions of appearance come up, including colour: although no proprietary device is envisioned, they are all concerned with product identity and look for 'cool' ideas. None of the designers complains about difficulties in visualization, but they sketch relatively little. In addition to writing on the whiteboard, they produce four sheets of drawings, of which only two contain actual design sketches, starting approximately half an hour into the process with a sketch by Kerry. Almost all sketches are three-dimensional. In the last minutes of the session the team calculates manufacturing costs and arrives at fairly complete preliminary specifications for the product.

Although no division of labour is planned, apart from Ivan's role of timekeeper (which he performs flawlessly), a careful analysis can point to group dynamics that produces social roles in the team, and it appears that each member contributes different dimensions to the joint effort. We shall have more to say about this in Section 6.2.

The team protocol was divided into 45 units by subject matter, ranging in length from one to nine minutes, and averaging 2.66 minutes per unit.

5 Linkography: Structural Variables and Their Notation

Our analysis of the design protocol is structural; it is aimed at facilitating access to cognitive aspects of the design process, particularly those related to productivity. We parse the protocol into design *moves*, and look at the design process in terms of relationships created by the *links* among moves. The system that was developed to notate moves and the links among them, *Linkography*, is instrumental in comprehending structural patterns of design reasoning. We can give only a brief overview of the system here, and for a broader discussion the reader is referred to other papers by Goldschmidt[11,12].

Design moves   The meaning of 'move' in designing is akin to its meaning in chess: a design move is a step, an act, an operation, which transforms the design situation relative to the state in which it was prior to that move. Moves are normally small steps, and it is not always easy to delimit a move in a think-aloud protocol of a single designer. In the present experiment, we define a move in Dan's protocol by establishing beginnings and endings of coherent utterances. Wherever the recording indicates that the designer started a fresh train of thought, a new move is registered: usually such utterances start with words like 'OK' or 'alright'. The team's protocol is easier to parse, and each utterance by one of the designers is defined as one move. A small number of long utterances that could have been subdivided into separate moves, are also treated as single moves, for the sake of simplicity. Within each unit of the design process, the moves are numbered chronologically, with two kinds of exceptions: very brief or meaningless utterances such as 'yeah,' 'mm mm,' or 'em I think if if' are not numbered at all and neither notated nor included in the analysis. The second exception pertains to moves that represent a complete or partial repetition of what was just said. For example, in the team's unit 32, move 27 reads: 'so do the straps'. The following move says: 'so we would do it as a strap way OK so.' This is a secondary move that does not change the state of the design, and is numbered 27a. In the notation and analysis, secondary moves do not have an independent status and are auxiliary to the main move they follow.

Links among moves   Due to the structural nature of our investigation, no typology or classification of moves is attempted. Instead, we pose but one question: is move n linked to every one of the moves from 1 to $n-1$, in a given sequence of moves such as a design unit? We use a binary reply system of 'yes' and 'no' only, and the sole criterion used to determine linkage or its absence is common sense, in the context of the design task. To clarify how links are determined, let us turn to an example from unit 32 in the team's protocol. The following is a short sequence of moves (from 17 to 20) in which alternative ways of attaching the backpack to the rack are discussed:

17	OK bungees
18	compression straps that also snap
19	that's bun- that's bungee like
20	straps with snaps

The first move in this sequence is not linked to a previous move, by definition (in this case, move 17 is actually linked to move 2, much further down the line, which reads: 'joining concepts.'). Move 18 is not linked with move 17 – it talks about joining concepts other than bungees (move 18 is linked to a number of previous moves that mention snaps or straps). Move 19 is linked to both move 17 and move 18, because it establishes a similarity between the two different concepts brought up in them. Move 20 ignores bungees and therefore also the moves that pertain to bungees, and is linked only to move 18 in this sequence, by way of reinforcing it through repetition.

Link Index   The number of links relative to the number of moves in a given sequence is an indicator of the 'strength' of the design process, or of its productivity. We maintain that an effective design process is characterized by a high ratio of interlinking among its moves. The proportion links/moves is called L.I., or *Link Index*. In the example above, we have four moves that form three links among them; therefore in this short sequence L.I. = 0.75 (obviously, L.I. values are interesting only for longer sequences).

Linkograph   We notate links in a *Linkograph* where, contrary to habitual notation systems, the variable is the link and not the move. The Linkograph is a transformed presentation of a simple matrix, that makes it easier to read information from it. Figure 3.1 shows the Linkograph of unit 23 in Dan's protocol, and Figure 3.2 shows the Linkograph of unit 32 in the team's protocol. A Linkograph allows us to spot structural patterns at a glance: we see areas of concentrated activity and we observe linear sequences in which each move is linked to the previous move (as the example above shows, links can be established between moves that are quite far apart chronologically). Above all, we see chunks of moves that are interlinked among themselves and which have relatively few links to moves elsewhere. In our notation system chunks take the form of triangles. In Figure 3.1 we distinguish two chunks, defined by moves 1–11 and 9–19, with an overlap of three moves. In this unit the distinction between the triangles is not very sharp (if we ignored the links formed between moves 13–19 and moves 2–4, the distinction would have been more acute). In Figure 3.2 we find small detached chunks, like the one defined by moves 6–12, and semi-detached chunks, such as the

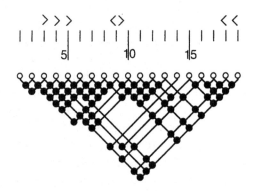

Figure 3.1
Linkograph of unit
23 in Dan's protocol

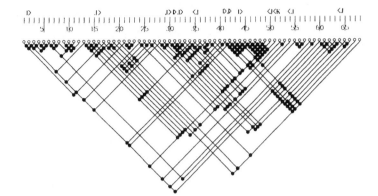

Figure 3.2
Linkograph of unit 23 in the team's protocol

one formed by moves 41–51. In some of the other design units of our protocols (not shown here) the distinction among chunks is more obvious. Chunks are evidence of systematic design thinking in which moves are not generated randomly, but according to their linkability potential, or their appropriateness in a given context.

Backlinks and forelinks When coding the protocol, we establish the links among a given move and all previous moves. We call these links *backlinks,* because they go back in time. But the graphic system of Linkography is such that at the same time, we also see all the links that subsequent moves make to a given move. These links are the move's *forelinks,* because they go forward in time. In contrast to backlinks, which can be determined at the time a move is made, forelinks can be determined only after the fact, when the entire process is completed, and as a consequence of having registered all backlinks. The two kinds of links are very different conceptually: backlinks record the path that led to a move's generation, while forelinks bear evidence to its contribution to the production of further moves. In our example above, move 17 has no backlinks and one forelink, move 18 has no backlinks and two forelinks, move 19 has two backlinks and no forelinks and move 20 has one backlink and no forelinks (the last move in a sequence has no forelinks by definition). Additional links are formed by these moves further up and down the line, as shown in Figure 3.2.

Critical Moves and critical path Moves differ greatly in the number of links they form, both backwards and forwards. If a high number of links is indicative of productivity, then we should pay special attention to moves that are particularly rich in links, in one direction or the other (and rarely, in both directions). Link-intensive moves are called CM or *Critical Moves*, and all the Critical Moves of a sequence together describe its *critical path*. We maintain that like the Link Index, a quantified critical path serves as a good, in fact the best, indicator of productivity. We therefore count the Critical Moves in each design unit and notate their percentage (out of the moves made in that unit). If the Critical Move is rich in forelinks, it is notated CM⟩ and if it is rich in backlinks, its notation is ⟨CM. A Critical Move that is rich in links in both directions is notated ⟨CM⟩ (in the present analysis there are only two such moves, one in the team's protocol and one in Dan's – move 9 in unit 23. See Figure 3.1). CM⟩ tend to occur early in a unit and ⟨CM late in the unit, but we note that there are exceptions to this rational distribution. ⟨CM⟩ occur only at overlaps between chunks. In a quantitative study, we must determine how many links qualify a move to the status of a Critical Move. The number of links we choose for this purpose is arbitrary, and depends on the grain of the analysis and on the purpose of the study. In the present study we refer to a move as critical if it generates seven links or more in one direction. The notation is CM^7 if the direction is of no interest, and $⟨CM^7$ or $CM^7⟩$ if we also relate the direction of the links. In a few cases we also inspect Critical Moves with six links, or CM^6.

We chose seven links to define the criticality of moves in this study for several reasons. First, the grain of analysis was quite dense, by which we mean that we were liberal in determining, by common sense, the existence of a link between moves. At this grain a lower number of links would have produced a very long critical path for each unit, veiling the structural information we can read off it. The structural information stems from the location of Critical Moves in the Linkograph: they tend to occur at the extremities of dense chunks. In other words, in an ideally structured process, a suggestive move is productive if it is followed by a series of moves that explore issue(s) raised by that initial move or related subjects. The exploration in turn is worthwhile if it leads to a concluding move that summarizes or evaluates points raised in the exploration. The initial move will then be a CM⟩, and the concluding move will be a ⟨CM. In highly structured processes we see this quite clearly, provided our Critical Moves are not too loosely defined, in which case they also 'frame' insignificant sub-chunks, thereby obscuring the structural pattern. We found that at the level of six or seven links, we get the clearest structural representation.

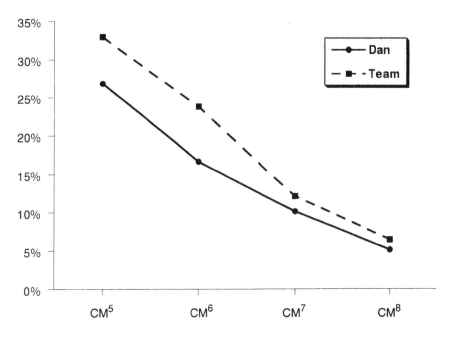

Figure 3.3
Dan and the team, %CM of total moves, by link level

The second reason for choosing the level of seven links is speculative: we hypothesize that the number of previous moves one can link to, has to do with what we remember of what had been processed shortly before we make a new move. If we accept Miller's famous theorem[13] that we can hold up to seven (plus or minus a few) items in short-termmemory, then this is the optimal number of links we can expect a link-intensive move to generate (in a few rare cases we find over ten links, but some of them can be interpreted as plain repetition).

The last and most important reason for choosing seven links as the criterion for criticality is empirically defined. We constructed critical paths for the two protocols at the levels of five, six, seven and eight links per CM. The lower the number of links, the higher the %CM. Figure 3.3 shows the relationship between CM percentage in the 'move population' and the level of links defining them, for Dan and for the team. We see that the two curves almost converge at level 7, whereas at levels 5 and 6 the difference between them is too large to allow a comparison between the two processes on the basis of Critical Moves. At level 7 the difference between the two is statistically no longer significant, as we shall see in the analysis presented in Section 6. We can therefore be sure that at level 7 there is enough of a common denominator that allows us to compare the two processes in terms of their productivity.

We are now ready to engage in a detailed analysis of the bicycle design processes, on the basis of a Linkographic analysis. We hope to show that despite methodological difficulties, quantitative protocol analysis is a powerful tool in the service of design thinking research. We agree with Lloyd and Scott[14] that qualitative analysis is useful and can lead to important insights, but we propose that it cannot replace quantitative analysis which is equally insightful when based on sound theoretical grounds.

6 Productive Designers: Dan and the Team

A comparison between the two processes, by Dan and by the team, is no easy matter. The variables that we put to the test are structural, but they are influenced by content: the actual subject matter that the designer deals with at a particular point in the process has an effect on the kinds of moves he or she generates and how extensively linked they are to one another. We are unable to process the protocols in their entirety and must therefore try to select portions from the protocols that are reasonably compatible. Compatibility is judged by content, or subject matters that are as close as possible to one another. We are interested in hard-core design activity, i.e. a stage of neither information collection or clarification, nor summaries or calculations.

We believe we found matching portions: for the team, they are units 32 through 37, where unit 34, which lasts only one minute, is not treated separately but is combined with unit 35. The entire sequence lasts 21 minutes, from 1:17 to 1:38 hours into the process. From Dan's protocol we selected a 25-minute sequence, from unit 19 to unit 23, beginning at 1:29 and ending at 1:54 hours into the process (note that the team's actual work starts at 0:05 hours and Dan starts at 0:17 hours. For ease of reference the time indications are given as marked in the protocol, but if we measure time from the moment in which work starts, we get 1:13 hours for both Dan and the team!). The two selections deal with subject matters that are as close as we could have hoped for, but the order in which these subject matters are taken up varies. Table 3.1 lists the selected units and shows the subject matter categories into which they were paired (to be discussed in Section 6.1). It also summarizes the number of moves, links, CM^7 and $CM^7\rangle$ that were found in these units in the process of linkographing them.

If we calculate the mean Link Indexes for these portions, we find that for the team L.I. = 2.75, and for Dan L.I. = 2.67. The difference is not significant ($p \geqslant 0.6$). The CM^7 percentage of the

Table 3.1 Comparable units for analysis: Dan and the team

Unit no.	Category[a]	Subject matter	Mins	Moves	Links	CM^7	$CM^7\rangle$
Dan							
20	A	Revised tubular design	5	31	63	–	–
23	B	Feature: clamp-clips	4	19	72	6	4
22	C	Snap-on plastic plate	4	22	58	3	1
19	D	Mounting points (braze-ons)	8	42	112	3	2
21	E	Bottom joint	4	24	53	2	1
		Total	25	138	358	14	8
Team							
37	A	Complete rack and joints	5	62	165	5	4
32	B	Tray and fastening devices	4	68	203	13	8
33	C	Tray features	1	16	36	1	1
34/35	D	Mounting (human factors) and mounting points	6	78	227	12	7
36	E	Features of (braze-on) joining	5	74	216	6	1
		Total	21	298	847	37	21

[a]Categories: **A** Overview revision, **B** Major features, **C** Feature alternative/assessment, **D** Essential checks (mounting points), **E** Particular details (bottom joint).

total number of moves is 12.42% for the team, and 10.14% for Dan. This difference is also not significant ($p \geqslant 0.1$) (see also Figure 3.3). The similarity between the L.I. and CM^7 values of Dan and the team are particularly interesting because of the striking difference in the absolute number of moves, links and Critical Moves that are made by Dan and by the team. Dan is slower; he takes much longer to respond to his own moves than the team members take to react to one another. Does this imply blessed thoughtfulness, that can be assumed to contribute to the linkability of moves? Our results do not suggest that this is the case. At the same time, the results do not support an assumption that the fast ping-ponging of ideas among team members necessarily generates a more cohesive body of interlinked moves than that of the slower individual designer. The similar values we get for the two variables under scrutiny lead us to the preliminary conclusion that both processes are basically equally productive.

This is an important conclusion that we wish to explore further. How are these productivity values achieved and when are they at their peak? Conversely, what correlates with a decline in productivity? In particular, we would want to find out what difference, if any, can be attributed to the singleness of Dan versus the group dynamics in the team's work? Furthermore, is Dan really talking in a single voice, and can we point to differences among the individuals that compose the team? The rest of this section takes a look at these questions.

6.1 Design Units and Productivity Correlates

The overall figures make it possible to draw a rough comparison between Dan and the team, but they teach us nothing about the reasons for design productivity. To gain some insight into what triggers fruitful designing, we must return to the units into which the processes were subdivided, and try to see what it is in them that can explain high or low values of L.I. and CM[7]. To do this, we shall first comment briefly on these units.

Category A: overview revision Dan looks at his tubular design while also thinking about joints. He is not sure of his decisions and changes them often, reassuring himself that the tubes are a wise choice: 'there's no reason why you couldn't make the tube like that...' The team makes its overview revision after having assessed partial solutions for several of the problems they have identified. They address the conflict that is raised by the need to fold the rack while also wanting to emulate the existing, non-foldable Blackburn rack.

Category B: major features. This is where both Dan and the team make major breakthroughs. Dan decides on a proprietary product 'that clearly relates to this backpack directly'. The entire unit is one long concentrated continuum. The team comes up with the idea of making a tray that can contain the backpack's straps and prevent them from 'dragging in the spokes'.

Category C: feature alternative/assessment In Dan's design, this unit precedes his 'feature' invention. He tries the idea of a vertical plastic plate over which the pack's frame would fit, thus holding it in place. He does not develop this idea. The team makes a fast evaluation of the tray idea that has been suggested in the previous unit, but they sidetrack to other issues.

Category D: essential checks: mounting points   This is Dan's earliest unit in the sequence, in which he inspects possible mounting points, requests information, tries to understand the given drawings and discovers the braze-ons. The team is interested in ease of use, which leads to an investigation of the advantages and disadvantages of mounting the rack to braze-ons.

Category E: particular details (bottom joint)   After his overview revision, Dan designs the bottom joint to the rear axle, led by the wish to provide a 'quick connect' solution. The team, having decided on braze-ons, designs the bottom joint. Their solution is basically similar to Dan's.

If we calculate L.I. values and CM^7 percentage values for each unit in the two processes and arrange them in an ascending order, we get the picture shown in Table 3.2.

The table teaches us a number of things. First, there is a partial but not full correlation between L.I. and $\%CM^7$ values in the two processes. The sequences begin and end with the same categories, but in the interim units the order differs somewhat. This means that the critical path is not an automatic consequence of high interlinkability among moves.

Second, the units of category B (see Figures 3.1 and 3.2), in which both Dan and the team made breakthroughs in terms of introducing major features of their designs, have the highest values on all counts. The boost is particularly dramatic in Dan's case. At the other end of the gamut we find two different types of unproductive design phases. In Dan's case, it is mostly a stage in which he repeatedly questions his major decisions and puts them to the test. The team, on the other hand, makes few moves in this phase before it sidetracks; the conversation goes nowhere and the

Table 3.2   Ascending sequence of L.I. and $\%CM^7$ by category (bold type and italics indicate close and relatively close values, respectively)

Dan	L.I.	A (2.03)	E (2.21)	**C (2.64)**	**D (2.67)**	B (3.79)
	$\%CM^7$	A (0)	*D (7.14)*	*E (8.33)*	C (13.64)	B (31.58)
Team	L.I.	C (2.25)	A (2.66)	**D (2.91)**	**E (2.92)**	B (2.99)
	$\% CM^7$	C (6.25)	**A (8.06)**	**E (8.11)**	D (15.38)	B (19.12)

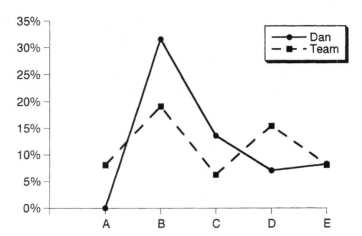

Figure 3.4
Dan and the
team, %CM^7 of total
moves, by category

team members realize that they must switch to a different subject matter in order to make progress. We learn a lot here about the cognitive nature of breakthroughs and what categorically prohibits them from happening. This point deserves thorough investigation which is beyond the scope of this discourse, but we shall add a few more remarks on this issue in the forthcoming section on creativity.

Third, when comparing Table 3.2 with Table 3.1, we notice that there is no correlation between the chronological order of units and the height of the L.I. and %CM^7 values reached in them (in Dan's case, the order is close to chronological). We must bear in mind that in both cases we selected portions from the midst of the protocols, with no clear-cut beginnings and endings. Still, we would like to propose that this finding suggests a non-hierarchical structure of the design process, at least in terms of its productivity. Likewise, there is no correlation between the values of our two variables and the lengths of the units.

Fourth, we notice that in Category E, in which Dan and the team engage in very similar searches (designing the bottom joint of the rack to the bicycle) and reach almost identical results, their %CM^7 values are almost identical. This finding, too, may give rise to assumptions regarding cognitive apparatuses, but obviously more statistical data are necessary in order to propose a general hypothesis. Figure 3.4 shows the CM^7 percentage of Dan and the team relative to the categories.

Given the above analysis, we must conclude that at least for the portions we analysed, Dan and the team are equally productive.

Not surprisingly, there are 'local' differences, which raise a question regarding the validity of the analysis sample. In other words, how can we know that the portions we dealt with are sufficiently representative of the entire process? Had we selected the units in categories A, D and E only, our results would have indicated that the team is more productive than Dan. If instead we would have selected only the units in categories E, C and B, the opposite trend would have emerged. We do not have an exhaustive answer to this question, and we find it hard to predict the results of the entire process, although we believe that differences in productivity, if any, would be minimal. Statistically, our larger sample should give superior indications, and this is particularly the case because we have reached such a good match between the selected portions. Finally, we certainly do not claim that individuals and teams are always equally productive: they cannot be, just as two different teams or two individuals are not necessarily equally productive. What we do suggest is that productive design behaviour of individuals and groups is defined by the same cognitive parameters, which can be observed and measured. The measured values of these parameters are similar for teams and individuals who reach equally productive results in the same design tasks.

6.2 The Team and Its Members

Until now we have treated the team as a single entity. But we know that it is composed of three designers, and we would like to briefly explore their design behaviour profiles and their respec-

Table 3.3 Team members' moves per unit by category

Designer	Moves	A-37	B-32	C-33	D-34/35	E-36	Total
Kerry	Moves	14	18	6	29	22	89
	CM^7	—	2	—	8	2	12
	$CM^7\rangle$	—	1	—	5	—	6
Ivan	Moves	21	21	5	23	25	95
	CM^7	1	4	—	2	2	9
	$CM^7\rangle$	1	3	—	1	1	6
John	Moves	27	29	5	26	27	114
	CM^7	4	7	1	2	2	16
	$CM^7\rangle$	3	4	1	1	—	9

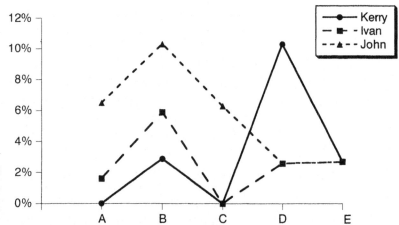

Figure 3.5
Team members,
%CM[7] of total
moves, by category

tive contribution to the team's work. A study in the framework of 'small groups' could throw light on a variety of issues such as leadership or other social roles (e.g. Hare[15,16]), but such a study is beyond the scope of the present investigation. We therefore limit our brief commentary to the cognitive parameters that we have used so far in the analysis, whose values are given in Table 3.3.

The first thing we notice is that the three designers are not equally active and make a different number of moves throughout the selection. It is therefore not surprising that the number of Critical Moves that the three generate is also not equal. However, we note that the respective contribution of each designer is not directly proportionate to his or her level of activity. Figure 3.5 shows the contribution of CM[7] by the three designers relative to the total number of moves in each unit.

If we inspect the values closely, we notice that only Ivan's contribution follows roughly the general ascendance trend of the team as a whole, as given in the sequence in Table 3.2. John's contribution is equal to or higher than that of his colleagues in all five units, and Kerry's share is equal or lower, with the exception of one unit in which she makes a very significant leap and generates a large number of critical moves. In this unit she is also much more involved than usual, generating over 37% of the moves. This is the unit in which mounting the rack to braze-ons is discussed and finally decided on. Of the three designers, Kerry is the most experienced bicycle rider and John states: 'we can assume Kerry has expert knowledge.' It is possible that the opportunity to make specific contributions when the circumstances call for it explains the sudden upward slant in Kerry's productivity curve, which is otherwise relatively low (at some point, when teased by her friends, she says jokingly: 'Help, I want out of this design exercise', which may mean that she was experiencing a problem. If so, it may have impaired her performance). We do not know, of course, if sudden productive peaks are typical of her design behaviour, but in this sample it is a predominant trait.

These results must be reviewed with great caution because, we ought to remind ourselves, we are talking about interdependency among the team members. Neither Table 3.3 nor Figure 3.5 tell us anything about preparatory moves that enabled team members to 'score' at the CM^7 level. If we check the results at the CM^6 level, for example, we get a different picture: out of 71 relevant moves, Kerry makes 20 (28.2%, as opposed to 32.4% at the CM^7 level), Ivan makes 25 (35.2%, versus 24.3%) and John makes 26 CM^6 (36.6%, versus 43.3%). We see that for the CM^6 level, results vary significantly from the ones we get for the CM^7 level.

To establish the contribution of team members to the design process, it is therefore not enough to look at partial 'bottom line' results, because interdependency within the team signifies that a particularly productive move by one designer may build on a preparatory move by another member of the team. The situation is analogous to that of a sports team, say a basketball team: the 'stars' who score most of the points cannot possibly do so without effective defence players and so on. To better understand individuals' contribution to a team effort, we must return to the protocols for a qualitative analysis of the roles they assume. The limited scope of this study forces us to do so in a cursory manner, which does not replace an in-depth study which we propose to undertake elsewhere.

John, as we already know, is the most active member of the team in this task. He is also the designer who comes up with the largest number of innovative ideas, some realistic and others less so. For example, his suggestion to redesign the backpack so that it folds in two and becomes saddlebag-like, is not realistic, but falls well within his design philosophy. In his words: 'maybe that could get us around the target price; if we can only come up with a more expensive solution but it does more stuff.' He is the one who first suggests the tray idea for the rack, which becomes the backbone of the design proposal. He convinces his friends, not without some discussion, to abandon the example of the existing

Blackburn rack, which includes several metal tubes that together guarantee stiffness and strength. He argues in favour of a single 'beefy' tube instead, so that with the help of a pivot, the rack could become foldable, as required. He wants colour bungees and colour anodizing of the aluminium tubes so as to achieve an attractive product, because to him: '...all design eventually comes down to a popularity contest.' He appears to see suggesting and cultivating new ideas (by himself and by others) as one of his major roles in the team.

Ivan is more conservative, and thinks more in terms of the precise specifications of the task: 'not constraints but we'll call them caveats or whatever I dunno just possible we have to design around.' He agrees with Kerry: 'right, there's no need in reinventing the wheel.' He finds it easy to accept other people's ideas and even when he thinks differently he is willing to consider them as options. In his role as timekeeper he actually acts as project manager in this task, and takes it on himself to make sure not only that the schedule is followed, but also that each issue is comprehensively dealt with before it is time to move on. His involvement increases when detailed design is undertaken; he makes sure that solutions are robust and that all performance criteria are strictly met. On the whole, he appears to contribute primarily in the summation of ideas and decisions, in addition to managing the process.

Kerry is mostly concerned with the 'functional spec.'. Many of her remarks pertain to functional matters and are marked by a 'no nonsense' attitude. She is in favour of an 'idiot proof [product], one way to install or one way to attach, and make it obvious too.' As noted earlier, she becomes very active when she can put particular expertise to use, in this case functionally oriented knowledge. At the same time she is also interested in 'a little bit of product identity' and is pleased when she identifies it: 'maybe there's some cool innovation here.' The profile of her design behaviour in this task is less clear than that of her colleagues.

With a gross simplification, we could say that John pushes the team's work in a creative direction, Ivan makes sure that decisions are made only after proper assessment, and Kerry steps in when she feels she can contribute expert knowledge. Needless to say, none of the designers is one-dimensional and they all participate in a multitude of ways, but they each have their own thrust nonetheless.

The question we must answer next is: how, if at all, can all of this be reflected in our quantitative productivity study? Is there anything in the statistics that correlates with the qualitative profiles we have just presented? And is there a way to compare the team with Dan in this respect? To answer this question we turn to an observation that we have not utilized in the analysis so far: the division of Critical Moves into those with forelinks, or $CM^7\rangle$, and those with backlinks, or $\langle CM^7$.

6.3 Forelinking Critical Moves: the Creativity Component of Productivity

Critical Moves with backlinks may have various meanings, from evaluation to summary to plain repetition. Critical Moves with forelinks are usually suggestive in that further moves build on them. We therefore associate $CM\rangle$ with a measure of innovation, invention and, in the extreme, with creativity, even if at a limited and local scale. We thereby substantiate and quantify Gruber's assertion (Gruber[17] pp. 177–178) that 'interesting creative processes almost never result from single steps, but rather from concentrations and articulation of a complex set of interrelated moves.' This is relevant to our discussion because innovation, invention and creativity are undoubtedly contributing factors to productivity. We therefore wish to show that our findings regarding $CM^7\rangle$ correlate with the qualitative profiles we drew from the protocols, in order to sustain our claim that productivity of processes can be observed and quantified up to this level. We do not intend this to be a study of creativity, which is one of the most complex and multi-faceted issues in cognitive science and which deserves, in the context of design research, much more attention than we can possibly devote to it here.

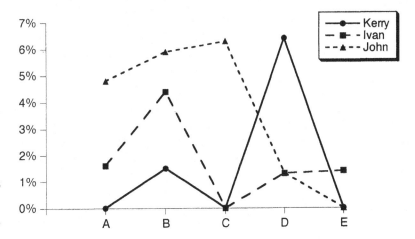

Figure 3.6
Team members, %$CM^7\rangle$ of total moves, by category

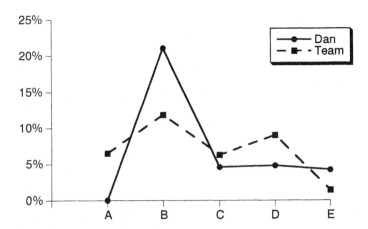

Figure 3.7
Dan and the team,
%CM^7⟩ of total
moves, by category

We want to look at CM^7⟩ at two levels. Within the team, we check the respective contribution its members make. Then we look at the team as a whole, and compare it to Dan's performance in terms of %CM^7⟩. Figure 3.6 compares team members to one another, and Figure 3.7 compares the team with Dan.

Not surprisingly, Figure 3.6 shows that, most of the time, John makes the highest percentage of CM^7⟩ (3.0% of the total number of moves), followed by Ivan and Kerry (2.0% each). Kerry's single outstandingly productive unit contains five out of the six fore-linking critical moves she makes, and during the rest of the process Ivan scores higher than she does. As in the case of the total number of CM^7, here too it is wise to look at preparatory moves as well, and we turn to CM^6⟩. We get different statistics: out of 41 relevant moves, Ivan makes 18 (43.9%, as opposed to 28.6% at the CM^7 level), John makes 14 (34.2%, versus 42.8%) and Kerry makes 9 (21.9%, versus 28.6%).

How do we interpret these results? We believe that they are congruent with the profiles we outlined above. John *is* the team member who comes up with the largest number of new ideas, but he also builds on the ideas of his team-mates. For example, John says he likes Ivan's idea of attaching a net to his tray, and proceeds to take it further. Some innovative ideas remain undeveloped, others deal with small details that require little discussion; moves that contain such ideas are reacted to by fewer than seven forthcoming moves. This does not make them unimportant to the process or less creative necessarily. Still, CM^7⟩ attract the largest number of responses, although they are certainly not all innovative or inventive, nor even necessarily the most innovative or inventive moves. They represent those ideas that played a decisive role in the design search, regardless of their inclusion in, or omission from, the final result. For example, when John says: 'so it's either a bag or maybe a little vacuum formed tray, kinda, for it [backpack] to sit in', he makes a $CM^7\rangle$ that turns out to be decisive to the process. When Ivan says: 'or you can just – OK, so lockable knobs is one option, I think, if you want to take the rack off a lot or not; and then how about just er set screws or Allen head', he expresses two alternative ideas, of which one only, the knobs, will remain with the design until its completion.

Thus, in a greatly simplified manner, we may say that John brings many fruitful notions to the foreground of the process, making forelinking moves that prove to be critical all the way to the level of $CM^7\rangle$. Ivan does the same with somewhat less grand ideas and therefore has a less impressive record at the $CM^7\rangle$ level, but is highly influential at the level of $CM^6\rangle$. Kerry, who maintains a low profile but makes a concentrated effort to resolve the all-important question of mounting points, has a low $CM\rangle$ score at levels 6 and 7 alike, with a local high peak.

Figure 3.7 compares the combined team's creativity-bound Critical Moves with those of Dan. When comparing Figure 3.7 with Figure 3.4, we see that there is a strong similarity between the two: $\%CM^7$ in both processes relate to one another as do $\%CM^7\rangle$, with small local differences. This is easy to understand when we calculate the percentage of $CM^7\rangle$ out of CM^7 for both processes in their entirety: 56.8% of the team's CM^7 are forelinking, while for Dan the figure is 57.1%, practically an identical percentage. As we see, the distribution over design units is also quite similar for both $CM^7\rangle$ and CM^7, which suggests a strong correlation between creativity and productivity.

7 A Team of One?

The stable proportion of suggestive moves in the critical path, approximately one half, was corroborated in other studies of the design process[12], and appears to be independent of a variable such as L.I. The fact that we do not reach higher values even in the most creative (or productive) portions of a design process, supports our claim that in an interactive and interdependent process even virtuoso, creative acts rely on evaluative, sometimes repetitive acts that define, constrain and clarify the problem space of the design task.

When a team acts together, implicit or explicit roles are created for the team members, along disciplinary or behavioural lines. In this respect it is immaterial to the discussion whether division of labour is or is not established in advance, along lines of expertise or other criteria. When the designer works on his or her own, with no team-mates to collaborate with, it is still necessary to produce summative and evaluative moves along with the suggestive, creative ones, for a process to be productive. The single designer must therefore assume production of all types of moves, whereas in a team situation he or she could develop a permanent or an *ad hoc* 'expertise' in the production of a certain type of moves, or in a pattern of production that takes advantage of the strongest capabilities of all participants in order to advance towards the best possible results.

We believe we have demonstrated that in Dan's process, we find precisely this behaviour. He oscillates between overviews and technical details, between functional aspects of the design product and issues related to human factors. He thinks of features, product identity and aesthetics along with stiffness, strength and ease of production. Team members do the same, but they can let a colleague answer a question they raise, or pick up someone else's line of thought and build on it. The single designer has only him or herself to rely on, and he or she must act as a team and give all the answers while also asking all the questions, often within the same move: 'Why do we want clips? Because we want to take advantage of the fact that we're using an external frame backpack. [An] internal frame can't use clips.' Dan 1 asks, Dan 2 answers, Dan 3 gives the design rationale.

One might ask whether the argument is not turned on its head: does the team not act as a single designer? We believe that Figures 3.5 and 3.6 and the related commentary suggest that this is not the case. If the team operated as an individual designer, what would the performance curves in Figures 3.5 and 3.6 correspond to in the single designer? We cannot answer this question and therefore we must assume that the team participants do not resemble different aspects of the individual designer, but rather that the individual designer is a unitary system that resembles the team.

We believe that among others, our findings may have implications for design education as well as for design management. In the least, we hope that these findings can be used to help establish a research agenda towards a possibility of such implications.

Acknowledgments The writing of this article was supported by the Technion V.P.R. fund – Edward S. Mueller Eye Research Fund, grant 020-543. The author wishes to thank Paul Hare for his helpful comments on group creativity and Maya Weil for her wise suggestions and her help with statistical matters.

References

1 Over, R. Collaborative research and publication in psychology, *American Psychologist*, **37** (1989) 996–1001
2 Kraut, R.E., C. Egido and J. Galegher, Patterns of contact and communication in scientific research collaborations, in J. Galegher, R.E. Kraut and C. Egido (Eds), *Intellectual Teamwork* Lawrence Erlbaum Associates, Hillsdale, NJ (1990), pp. 149–172
3 Lewis, A.C., T.L. Sadosky and T. Connoly, The effectiveness of group brainstorming in engineering problem solving, *IEEE Transactions on Engineering Management*, EM-22, **119** (1975) 124
4 Erev, I., G. Bornstein and R. Galili, Constructive intergroup competition as a solution to the free rider problem: a field experiment, *Journal of Experimental Social Psychology*, **29** (1993) 463–478
5 Ericsson, K.A. and H.A. Simon, *Protocol Analysis: Verbal Reports as Data*, MIT Press, Cambridge, MA (1984/1993)
6 Vygotsky, L., *Thought and Language*, MIT Press, Cambridge, MA (1986)
7 Pritchard, R.D. and M.D. Watson, Understanding and measuring group productivity, in S. Worchel, W. Wood and J.A. Simpson (Eds), *Group Processes and Productivity*, Sage Publications, Newbury Park, CA (1992), pp. 251–275
8 Perkins, D.N., *The Mind's Best Work*, Harvard University Press, Cambridge, MA (1981)
9 Wertheimer, M., *Productive Thinking*, Harper Torchbooks, New York (1945/1971)
10 Goel, V., *Sketches of Thought:* MIT Press, Cambridge, MA (1995)
11 Goldschmidt, G., Linkography: Assessing design productivity, in R. Trappl (Ed.), *Cybernetics and Systems 90*, World Scientific, Singapore (1990), pp. 291–298
12 Goldschmidt, G., Criteria for design evaluation: a process oriented paradigm, in Y.E. Kalay (Ed.), *Evaluating and Predicting Design Performance*, Wiley, Chichester (1992), pp. 67–79
13 Miller, G.A., The magical number seven, plus or minus two: some limits on our capacity for processing information, *Psychological Review*, **63** (1956) 81–97
14 Lloyd, P. and P. Scott, Discovering the design problem, *Design Studies*, **15**(2) (1994) 125–140
15 Hare, A.P., *Creativity in Small Groups*, Sage Publications, Beverly Hills, CA (1982)
16 Hare, A.P., *Groups, Teams, and Social Interaction*, Praeger, New York (1992)
17 Gruber, H.E., Afterword, in D.H. Feldman, *Beyond Universals in Cognitive Development*, Ablex Publishing, Norwood, NJ (1980), pp. 177–178

4 Ingredients of the Design Process: a Comparison between Group and Individual Work

Srinivasan Dwarakanath and **Luciënne Blessing**
University of Cambridge, UK

Engineering design research focuses more and more on detailed studies of individual and group design behaviour to increase understanding of the design process both as a social and as a cognitive process. The main reasons are the increasing need to improve the design process and hence the resulting product, and the need to develop effective tools and user interfaces. The availability of techniques developed in cognitive psychology, sociology, and computer science for analysing human behaviour, and technological developments such as video recording, support these needs by providing the methods and tools for detailed observation and analysis.

This paper discusses the results of a *descriptive, comparative study* of two design processes executed in a laboratory environment. One process involved an individually working designer who was asked to think aloud; the other involved a group of three designers. Both processes were recorded on video and transcribed into protocols. The protocol analysis focused on the differences and similarities in the ingredients of the two processes. The results provide a first step towards a better understanding of the differences between group and individual work.

A *descriptive design study* is defined as an investigation of the way in which a design process actually occurs. This investigation

Environment			Method	
Laboratory:	22 studies 13 sources 177 cases		Observation:	18 studies 9 sources 131 cases
			Combination:	4 studies 4 sources 46 cases
Industry:	32 studies 28 sources 951 cases		Observation:	11 studies 7 sources 11 cases
			Retrospective:	14 studies 14 sources 920 cases
			Combination:	7 studies 7 sources 20 cases

Figure 4.1
Context of descriptive studies in mechanical engineering (until 1993) (54 studies, 41 data sources, 1128 cases)

can be: (1) real-time, e.g. by observing designers; (2) retrospective, e.g. by means of interviews and questionnaires; (3) a combination of these and/or other techniques. The investigation can take place in industry or in a laboratory. Figure 4.1 shows a summary of descriptive design studies in the domain of mechanical engineering (for more details see Blessing[1]). Some of the 54 *studies* are based on the analysis of (part of) the same *data source*, i.e. on the same set of cases. They have been included because they address different design research issues. The actual number of data sources was 41, covering 1128 *cases* (i.e. subjects, groups or companies). Studies that only involved students were not included.

As this overview shows, the majority of cases have been analysed in *industry*. This is mainly due to the use of retrospective methods, such as questionnaires, as they allow for large numbers of cases to be investigated in one study. The disadvantages of these methods lie in the time-lapse between the event taking place and being reported, and the technique of capturing data. The consequences are fewer details of the observed process compared to real-time, introspective techniques, and possible bias of data, among others, because of the inevitable distortion that occurs in interviewing (see Ackroyd and Hughes[2] for a discussion of the latter).

The number of cases per study is far less in a *laboratory environment*. In general these studies focus on individually working designers. Of the 22 laboratory studies only four focused on design teams. Again this is mainly caused by the type of method applied. Simultaneous verbalization (where designers are asked to think aloud while designing) with video-recording provides very detailed, real-time data, but requires much effort to analyse. This is the reason that only a few cases can be addressed in a single study. Studies involving the observation of a larger number of cases in laboratory situations usually focus on one specific research aspect, and involve students only.

All the listed studies are descriptive, but of these only few are comparative. In *comparative studies* the cases are deliberately divided beforehand into two or more groups. This division might be based on the designers' experience (e.g. Schindler[3]), the amount of information in the assignment (e.g. Fricke[4]), or the tool that is being used (e.g. Blessing[1]). Comparative design studies are very difficult to execute, in particular if the aim is to generalize the results and draw consequences for practice[1]. However, they are extremely important for measuring the effects of new methods and tools, and to reveal the context in which the findings are valid. This is essential for the development of effective methods and tools for designers.

Design is a social activity. In industry, designers work a substantial part of their time with others. Hales[5] found in the project he observed in industry, that no more than 40% of the effort was carried out by people working alone. Minneman[6] observed that groups communicated not only product data but also process data and the relationship between the groups. The importance of group work is also emphasized by several other authors[7,8]. However, not all parts of a design process are equally suitable for execution in a group. One of the more interesting studies, therefore, is the comparison between individually working designers and those working in a group. Such a study may reveal which tasks can be executed more effectively or efficiently by an individual, and which ones by a group of designers. A comparison may also suggest how to support individuals and how to support groups, and may thus provide specific requirements for methods and tools for individuals and those for groups. This requires a detailed understanding of the differences and similarities between the two ways of working. This goal led the analysis of the Delft protocol data described in this paper.

1 Analysis Method

The analysis focused on the identification of the ingredients of the two design processes. Comparisons have been made between:

- the group as a whole and the individual
- the individual group members
- the individual group members and the individual.

The protocols provided by the workshop organizers were transferred to a spreadsheet and the utterances of the designers split into individual events (see Figure 4.2). Every line in the spreadsheet represents an event. An *event* is any meaningful piece in a protocol, i.e. any piece that contains information. This is related to the aim of the analysis and partially to the categories used in the classification schemes. A new event starts when a new piece of information is addressed or when something new

Figure 4.2
Abstract of the protocol

Design time	Event time	Event
0:41:26	0:00:13	OK I have spent er fortyfive minutes or so now forty minutes doing that er so em ...
0:41:34	0:00:08	(mutter) going to check every possible location
0:41:38	0:00:04	here alright so we do have one that comes up front
0:41:47	0:00:09	em really a little bit wary about backpacks on em fronts of bikes
0:41:55	0:00:08	em let's see if we can em
0:42:00	0:00:05	and there is that issue of it being off the side ...
0:42:10	0:00:10	you know from the aesthetics standpoint everybody likes things symmetric
0:42:14	0:00:04	and er this is not that big a pack
0:42:19	0:00:05	em my initial tests indicate I probably couldn't have one right like that
0:42:21	0:00:02	certainly I would not do it this way
0:42:23	0:00:02	I would do it that way

happens, e.g. when a new category is addressed in one of the classification schemes. In the group protocol a new event also starts when a different group member starts speaking. The protocol was divided into events prior to the actual analysis. During the analysis only small changes had to be made. Each event was linked to the elapsed design time in seconds, measured at the end of this event.

The duration of an event varied between one second and almost two minutes. On average an event in the individual's protocol lasted 7.2 seconds, an event in the group protocol only 2.7 seconds. The medians were two and six seconds respectively. This difference is not only caused by the fact that more statements were uttered in groups, but group members also talked at the same time, and regularly reacted with short utterances. This increased the number of events. The number of events in the individual's protocol was 1026; in the group's protocol it was 2684. The actual analysis of the design processes was based on the transcriptions, on viewing of the video tapes, and on the project data documented by the designers.

Several classification schemes are possible for an analysis. The analysis categories used in this paper are considered to represent the ingredients of a design process. They had been defined and used by the authors in earlier studies.

To analyse the processes each event was linked to the ingredient it addressed. This depended on the contents of the event, and on the context in which it was uttered. The basic ingredients were: Issues, Arguments, Alternatives, Criteria, Decisions, External information, and Others. This categorization was based on work in the area of representing design rationale, e.g. by Conklin and Yakemovic[9] and Lee and Lai[10]. It has been extended and detailed during the analysis of several other protocols by the first author (in preparation). Appendix A gives the definitions of the basic and the detailed categories as used in this paper.

Additional detail was provided by categorizing the reasons given by the designers for rejecting or accepting a solution, and by drawing a decision tree of the process. The different types of reasons used are based on the analysis of a design process in industry[1] and the work of Hubka[11]. A decision tree shows the different problems and sub-problems addressed in a design process by presenting the alternatives that were considered and the decisions made. Decision trees were used by, among others, Marples[12] and Guindon[13].

2 Findings

2.1 Basic Ingredients of the Design Processes

Figure 4.3 represents the time spent by the individual and the group of designers on the basic ingredients of the design process. Figure 4.4 represent the time spent on the basic ingredients by the individual designer and the individual group members.

Issues are topics that have to be addressed or problems that have to be solved. Arguments refer to statements that reflect upon the process or the product. Events related to Alternatives express solutions or contribute to solutions. Criteria refer to the events in which a designer mentioned or elaborated on a specific criterion without using it in an argument. The use of criteria as part of an argument is included with Arguments. Similarly, Decisions only refer to the explicitly expressed decisions. Statements labelled as External are those that were given to the group members or the individual by other people, and that convey information. Statements that do not relate to the design process are classified as Others.

The group spent more time on Issues and Arguments than did the individual. This was as expected because designers in a group discuss problems and have to explain to each other the reasons why one alternative is better than another. The individual designer spent more time on Alternatives. A possible reason is that an individual designer spends less time on arguing and thus has more time to spend on the alternatives.

The time spent on expressing explicit Decisions was similar for both the individual and the group. This seems to contradict with the expectation that in a group there is more need for explicit decisions. Analysing the number of decisions that were made in this period of time, however, showed that the group indeed made

Figure 4.3
Time spent on the basic ingredients of the design process

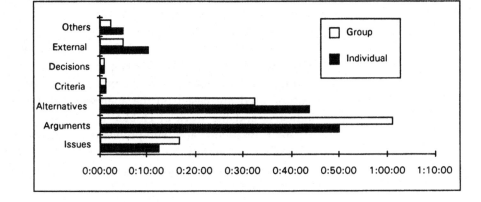

more explicit decisions (21 versus 5). This does not imply that decisions were made quicker, but only that they were stated quicker. They often concluded a discussion and could therefore be stated more briefly. Besides, group members can think while others talk, whereas the individual designer is expected to talk all the time. More details of the decision making process are given in Section 2.5.

Figure 4.3 also shows that the individual received more information from external sources compared to the group. This is mainly due to a phone call made by the individual to acquire some external information. The individual designer spent also more time on non-design activities (Others). This was mainly due to the time needed to set up the phone line.

Comparing the ingredients in the processes of the different group members, it showed that designer 'J' spent more time on all the basic ingredients than the other group members (see Figure 5.4). This confirms the impression of his role in the group obtained by looking at the video. He took the role of chief designer who set up the plan and led the process. Designer 'K' often separated from the group to work independently, which is also reflected in the time she spent expressing her thoughts. Designer 'I' acted as project manager. He kept schedule and documented the process on the whiteboard which required attention to what was happening. This is reflected in two observations. First, designer 'I' is the only group member who states Decisions, although the arguments are often delivered by the others. Second, he spent an average amount of time on the other ingredients.

Figure 4.4 also reveals that each of the group members spent less time on each ingredient than the individual designer. A plausible reason is that in a group, designers can convey their thoughts to the others only when others are not conveying *their* thoughts. Thus, in a given time span, each of the group members has less time to speak compared to the individually working designer, which implies that they have more time to think. Some observations make it likely that this individual thought process continues relatively independently of what the other group members are saying or doing. For example, the group packed more arguments in a certain time compared to the individual. This might be due to the extra time group members had to think before expressing their thoughts. Another example is the occurrence of sudden jumps in the reasoning process as captured in the

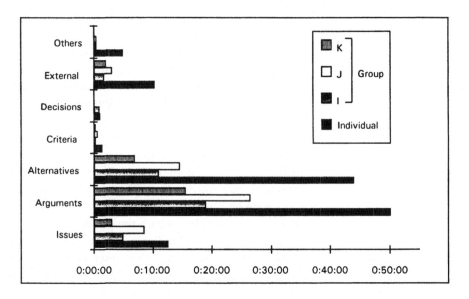

Figure 4.4
Time spent on the
basic ingredients for
each of the
designers

protocol, which might have been caused by group-independent thought processes. Sometimes these parallel processes were visible on the video when the different group members undertook different activities at the same time.

The course of each of the design processes was represented in a graph as the sequence in which the different ingredients were addressed. The graphs did not show significant differences between group and individual processes. Both the individual and the group designers continuously evaluated as and when they generated an alternative.

The following sections focus on the findings related to the basic ingredients Issues, Arguments, Alternatives, and Decisions. To enable a more detailed discussion, each basic ingredient has been divided into detail ingredients based on the way in which the basic ingredients presented themselves in the utterances of designers (see Appendix A). For example, the ingredients of type Issue could be classified as: plain Issues, Issues addressed in Previous Designs (from experience), Questions, and requests for Confirmation (see Figure 4.5).

2.2 Issues

Figure 4.5 represents the time spent on the detail ingredients of Issues by the group and the individual designer. Figure 4.6 shows the time spent by the individual designer and each group member.

The group spent more time than the individual on all the detail

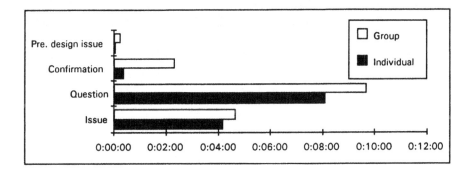

Figure 4.5
Time spent on the
detail ingredients of
Issues

ingredients of an Issue that were addressed. The largest difference is in asking questions and asking for confirmation. A Question is a minor issue which is put in the form of a question. This includes not only questions to the researcher but also questions to other group members. The latter increases the time on this ingredient for the group. The relatively large amount of time spent by the individual on asking questions is caused by the phone call he made. Answers to these questions given by the external source are captured under External Information (see Figure 4.4). The Confirmation category captures questions that include an answer. This is typical for group work. Although the group spent far more time on issues generated during previous design projects, the amount of time is too small to be relevant.

The same division of time among the designers as observed in Figure 4.3 can be observed in Figure 4.6. Designer 'J' spent more time than the rest of the group members. The individual designer spent more time than designer 'J' on asking questions and addressing issues, but not on asking for confirmation. This has been discussed in the previous paragraph.

Figure 4.6
Time spent on the
detail ingredients of
Issues for each of the
designers

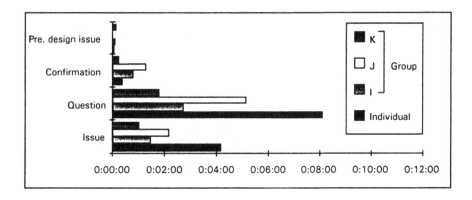

2.3 Arguments

Figure 4.7 represents the time spent on the detail ingredients of Arguments by the group and the individual designer. Figure 4.8 shows the time spent by the individual and each of the group members on these ingredients. Figure 4.9 represents the types of reasons the designers used while assessing their solutions.

From Figure 4.7 it becomes clear that the group spent more time on almost all of the detail ingredients that reflect upon the product or process, except for the Assumptions and Previous arguments. A remarkable amount of time is spent on arguing about the process in comparison to the time spent on arguing about the product, for both the individual and the group. The reason might be the time pressure in combination with the short duration of the design process. Another reason might be that the designers considered that the approaches they applied before were unsuitable for this new context.

The individual spent more time arguing about the previous prototype (Pre. argument) than the group. One reason is that he asked for this product information before generating solutions, which is also reflected in the time he spent on looking at this product (this will be discussed in connection with Figure 4.10). The group, on the other hand, did not ask for the previous prototype until after they had generated several solution concepts for each of the sub-problems. Another plausible reason why the individual designer spent more time on the previous prototype is that this was one of his few sources of solution ideas. In a group the other members provide sources for ideas.

Figure 4.7
Time spent on the detail ingredients of Arguments

The individual spent more time expressing Assumptions than the group. In a group, designers can ask questions or try to get their assumptions confirmed. This is reflected in the time spent on

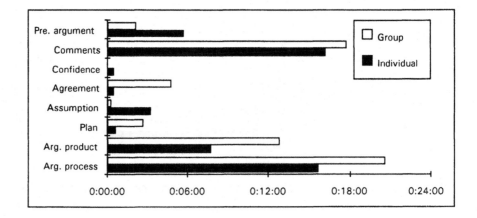

Confirmation and Questions in Figure 4.5, and in the time spent on expressing agreement with a previous statement (Agreement in Figure 4.7). The individual can only agree with the written information he receives, which explains the small amount of time spent on Agreement. Whenever the individual designer agrees with his own work or statements, this is captured as Confidence. Group members tend to express their confidence in the form of a request for confirmation (Figure 4.5). The individual did not spend much time on planning the process, e.g. arguing about the amount of time needed. The group expressed their plan more clearly and worked accordingly.

Figure 4.8 shows that the individual spent more time on all detail ingredients of Arguments than each of the individuals of the group, except on Plan and Agreement. This has been discussed in the previous paragraph. Designer 'J' spent more time arguing about the process and the product. This is typical for him in his role of chief designer. Again the contribution of 'K' is low. The role of 'I' as project manager is reflected in the comparatively large amount of time he spent on Comments and Agreements (see also the discussion related to Figure 4.4).

Figure 4.8
Time spent on the detail ingredients of Arguments by the individual designers

Figure 4.8 also shows that, although the individual designer did not spent much time on planning his process, he very often expressed arguments about the process (13% of the design time). He also commented more. Comments are defined as statements

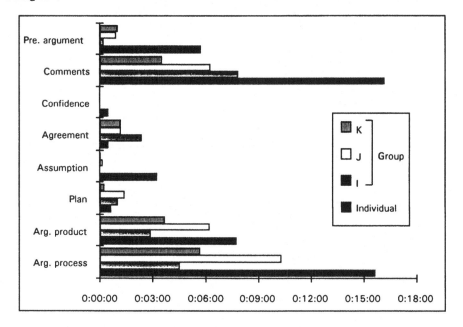

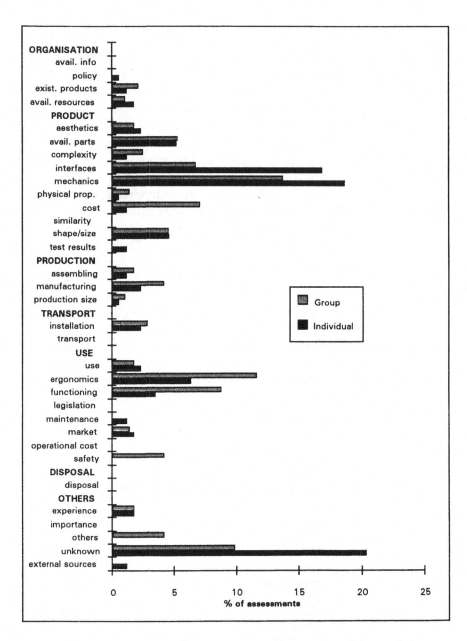

Figure 4.9 Types of reasons mentioned during the assessment of the design solution as a percentage of the total number of assessments made by the group (n = 283) and by the individual (n = 172)

which reflect the designer's understanding or mood ('I am understanding the problem now'), and include half-finished statements on the process or the product. The large amount of time the individual spent on Comments might have been caused by the fact that there is no need to express all arguments completely: he was only talking to himself. Group members used their speech in general to convey their thoughts to the other members, and Comments as defined here would therefore not be useful.

Figure 4.9 shows the different reasons the designers mentioned for accepting or rejecting a design solution. They are expressed as the percentage of the total number of assessments, rather than the absolute number. The individual designer seems to focus more on reasons within the product itself, such as the mechanics and interfaces. The group focuses more on reasons related to its interface with the user. The cost of the product is a more important reason for the group than for the individual. A possible cause is that, early on, the group planned to calculate the cost of the product. The individual designer did not calculate the cost.

About 20% of the individual's assessments did not include a clear reason (category Unknown, e.g. 'but er er not reasonable'). In the group this was only 10%. A possible reason is that, in contrast to individually working designers, group members have to communicate their reasons explicitly to contribute to the decision making process.

The group mentioned several reasons that could not be classified into a specific category. An example is 'you know theft proof its not gonner matter'. They were classified as Others. A type of reason used only by the individual designer was External, referring to the reasons he obtained from the person he contacted at the beginning of the process.

2.4 Alternatives

Figure 4.10 represents the time spent on the detail ingredients of Alternatives by the individual and the group. Apart from descriptions of solutions for stated problems (Alternatives), it includes Answers to questions, Calculations, and descriptions of Previous designs. Figure 4.11 represents the time spent by the individual and each of the group members on these detail ingredients.

Figure 4.10 shows that the individual designer spent more time than the group on almost all of the detail ingredients except for Answers. Questions asked by the individual designer were answered by external sources and therefore captured in another

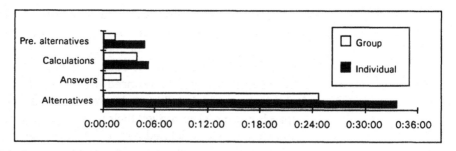

Figure 4.10
Time spent on the
detail ingredients of
Alternatives

category (see Figure 4.4). Questions asked within a group are answered within the group. However, the amount of time spent on answering is not much. The individual designer spent more time looking at the previous design alternatives and trying to understand them. This is in line with Figure 4.7 showing that the individual designer spent more time arguing about the previous design than the group. Possible reasons have been discussed in the paragraphs following that figure.

Figure 4.11 confirms the pattern in the time spent by the individual designers that was found for the other detail ingredients.

2.5 Decisions

Both the individual and the group designers evaluated their designs as and when they were generated. The individual designers made five explicit decisions, the group made 21 explicit decisions. Decision trees created for the individual and group design experiments are shown in Figures 4.12 and 4.13. The trees present only the major problems and alternatives. The sloping lines indicate the alternatives considered, the vertical lines indicate the (sub-) problems considered, and the horizontal lines

Figure 4.11
Time spent on the
detail ingredients of
Alternatives by the
individual designers

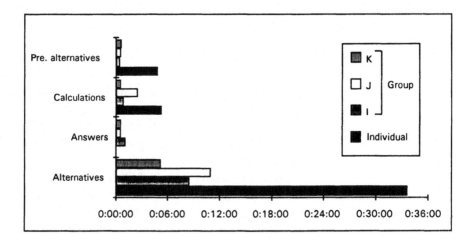

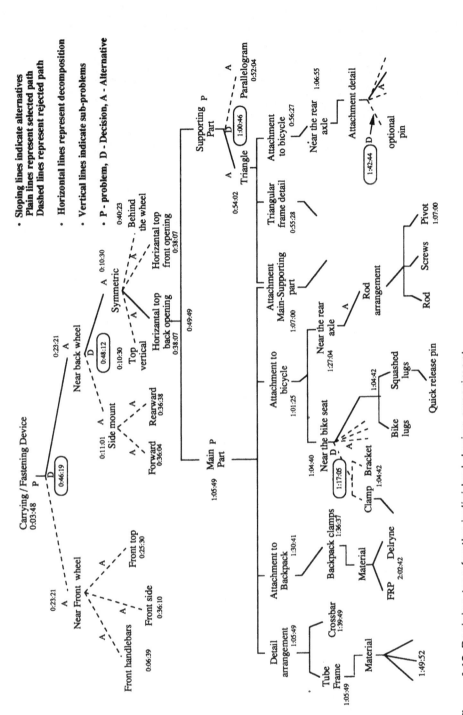

Figure 4.12 Decision tree for the individual design experiment

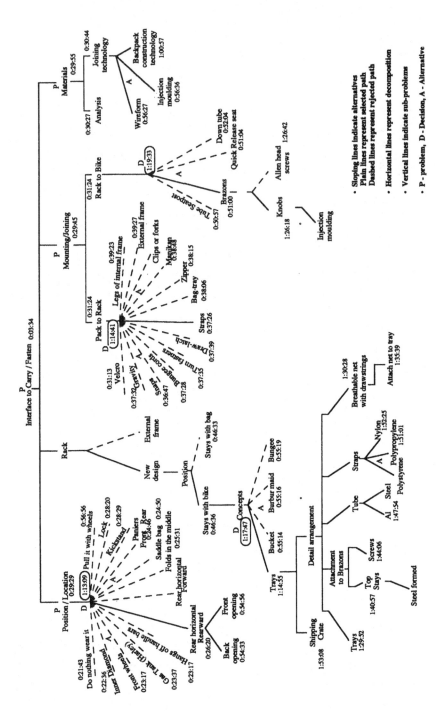

Figure 4.13 Decision tree for the group design experiment

represent the decomposition of problems into sub-problems. The broken lines indicate the rejected paths in the design process. The time (hh:mm:ss) shown in the figures indicates when during the design process a particular problem/alternative was first identified and when an explicit decision was taken. All explicit decisions in the individual's tree have been timed. In the decision tree for the group only the explicit decisions for the top level sub-problems have been timed.

Figure 4.12 shows that the individual designer started exploring solutions after understanding the problem. The timing of the decisions reveals that he made decisions at each level before going down to the next level of sub-problems. The group applied a different approach. After an initial brainstorming session in which they generated several alternatives, the group started systematically identifying the sub-problems. Then they went on to list the alternatives generated earlier and also generated new alternatives for the identified sub-problems as shown in Figure 4.13. They evaluated the generated alternatives systematically (one by one) and then made decisions for the same level sub-problems at about the same time (between design time 1:13:00 and 1:20:00). The group generated more alternatives than the individual.

The group worked more problem-oriented. The individual worked more solution-oriented. These observation have also been made during the analysis of other protocols by Dwarakanath (in preparation). A possible reason is that if a group member is deeply involved in working on a solution, the other group members are less involved and therefore in a better position to maintain an overview. Besides, each of the group members has a different perspective on the problem. An individually working designer has to come out of his involvement with the product to reflect on the problem he tries to solve.

3 Limitations

The results of the comparison described in this paper have to be viewed in the light of the limitations that are inherent to the experiment and to the analysis method.

3.1 Limitations of the Experiment

The experimental limitations are related to the request to think aloud, the number of cases, the type of assignment, the type of environment, and the subjects. Each aspect will be discussed briefly.

Thinking aloud

The individual designer was asked to think aloud to be able to record his process. Though considered to be easy to learn and not very distracting, thinking aloud might have changed the individual's way of working. For example, it might have increased his awareness of his own design process. At one time the individual designer says: 'I assume the camera is watching me'. Group work, on the contrary, requires the communication of thoughts. Therefore, there is no need to ask the designers to think aloud to record the design process, i.e. to change their way of working in this respect.

Number of protocols

The number of protocols that has been analysed is limited to two. Therefore this study can only give a first indication of differences and similarities that might be worth investigating further. The first author is currently working on a comparison of the results of this study with those of other protocol studies in which he used the same analysis method.

Assignment

In choosing an assignment for this type of experiment, a tradeoff has to be sought between time (for both subject and researcher), reality of the problem, and suitability for an individual and group of designers. The results are that the assignment is relatively small, does not require specialist knowledge, and is solvable within a few hours by an individual. More complex problems might show differences especially in group work. There might be more need to collaborate, and specialist knowledge might force group members into different roles. However, in order to compare individuals and groups, the problem should be solvable by an individual. This problem is inherent to this type of comparative analysis. The fixed, relatively short time available influenced the design process, as became clear from the statements of the designers (and from other studies). The assignment is also limited because it does not require a complete design process. Depending on the specific research question, this may have a desirable or problematic effect.

Environment

A laboratory environment restricts the number of influences on the design process and thus the variables in a study. Although this is an advantage for comparative research, it also implies that a laboratory environment is less suitable for analyses of how design actually takes place. The designers had to work in a room, not familiar to them, where every action and utterance were observed and recorded. They were not surrounded by their own 'tools' and had to work with the information provided. Besides, they could not let the problem settle overnight or while doing some other work. The individual, in addition, could not communicate with other designers. Some of the designers referred to this situation by expressing what they would normally do. This might have affected the ability to design and thus the results of the analysis. However, this is expected to be the same for every designer involved in the study. Besides, in a comparative analysis the main requirement is similarity of environment. An industrial environment cannot provide this if several experiments have to be executed.

When the results have to be used to determine how to support individuals and groups, it is also necessary to have an understanding of the differences between the two ways of working in an industrial context, as improving design processes in industry should be the ultimate aim of design research. Most of the *detailed* descriptive studies, however, focus on individual designers in laboratory situations, rather than on groups in industry. This includes two contextual differences: group versus individual, and industry versus laboratory. The latter is not addressed in this study. The results should therefore be considered as an indication of topics for further research.

Subjects

The subjects were all experienced designers, but their level and type of experience varied. The individual designer had more than 20 years of experience with an emphasis on electro-mechanical work. Two of the group members had five years of experience, the third group member eight years of experience. Their emphasis was on mechanical design. Several studies showed that the years of experience can have a considerable effect on the time needed to fulfil the assignment and on the way in which people work[1,14,15]. This could thus have affected the findings described in this paper. Another limitation is that the subjects had to work either alone or in a group throughout the design process. In practice, design consists of a combination of individual and group work. How this affects the findings is unclear. One could argue that it restricts both the individual and the group. However, the effect of having no one to work with and the effect of having to work in a group all the time might be different.

Another subject-related limitation of this type of study is what Denzin[16] calls the demand-characteristic effect. Subjects are not passive objects. They do not have to obey the role set for them in the experiment. However, 'subjects will persist in their assignment' because 'the norm of the role of subject is to be a good subject; to be a good subject demands that one interprets ambiguous meanings and assignments in a context that values participation'. This does not prevent problems, as this interpretation might cause subjects to proceed in a way that was not intended by the experimenter. In the experiment this might have resulted in designers working more systematically than usual. As this will apply to all designers, in the experiment, this is assumed not to have caused differences.

Another implication of an experimental situation is that for the subjects, the design problem is a stand-alone problem. The designers do not necessarily have experience in designing a similar kind of product. In industry it is common to have this experience at least within the company. Besides, nothing depends on the solution, for which reason the incentive to aim at an optimal solution can be low. A satisfactory solution would do. Compensating factors might have been the demand-characteristic effect (see previous paragraph) and the fact that the processes would be analysed. Again, these factors are expected to be the same for all designers in the study.

3.2 Limitations of the Analysis Method

The analysis of protocols has its specific limitations. This section focuses on the limitations related to 'parallel' events and the difference between individual and group data.

Protocol units

Video recordings and their protocols can be analysed in a variety of ways as the papers in this volume show. One of the basic activities involved in protocol analysis is the division of the protocol into units that are coded according to a classification scheme. Various schemes can be applied, depending on the aim of the study, and the sources used to develop the scheme can differ. Another clear difference is the way in which the protocol has been divided: event-based (as elementary, and very generic, events, and to a large extent prior to the actual analysis); category-based (during the analysis); and time-based (e.g. every 15 seconds). Which approach is most suitable depends on the level of detail required to achieve the aim of the study. The approach used in this paper has been mainly event-based, and to some extent category-based.

Parallel events

The use of protocol analysis for group work introduces several difficulties and, thereby, limitations compared to its use for individual work. Often the different group members are involved in different activities or talk at the same time, and they do talk more than individuals. Apart from posing problems on transcribing the video recordings, one has to be careful with the analysis. When people are speaking at the same time, and the time for each event is recorded, this will affect calculations based on the event time. In the study described in this paper, we recorded 0 seconds for each of the parallel events that consisted of short confirmations (such as 'yes', 'indeed'). Often utterances are repeated for confirmation. This might distort the results when counting specific events is important, in particular when this is done automatically.

One also has to be careful in interpreting the sequence of events in a group as the design process or approach, as is common for analyses of individual processes. As the findings in the previous sections showed, activities take place in parallel and the individual thought processes continue independent of the group process. The sequence of events in a protocol is, therefore, likely to consist of events of sometimes different processes existing in parallel and related to each group member. It seems difficult to untangle these processes.

Difference in data

A detailed comparative analysis requires the same data collection and analysis methods to be used for all cases involved. Therefore the same methods were used to study the individual and the group. However, the previous paragraphs made clear that the collected data are likely to differ, which implies that any differences or similarities that are found in an analysis of these data, might be the result of differences in collected data rather than differences or similarities between individuals and groups.

4 Conclusions

The study described in this paper reveals several differences and similarities between the ingredients of the design processes of an individually working designer and those of a group of designers. The data suggest that these might indeed relate to the number of designers. Within the group, differences were found between the members. These differences remained throughout the process. None of the findings, however, can be generalized. Several limitations of this type of experiment have been discussed in the paper, among others the number of cases and the laboratory environment. The analysis method is detailed, but limited because it looks at the process from only one point of view. Therefore, the observed differences and similarities might have been caused by factors other than the group size, or might even have been coincidental. More work has to be undertaken to understand fully the differences between group and individual design processes to be able to divide tasks and develop tools and methods in a way that takes into account the strengths and weaknesses of each type of process. A first and important contribution to this aim is the collection of papers in this book, which are all based on the same source of cases but take different points of view.

References

1 Blessing, L.T.M., *A process-based approach to computer-supported engineering design*, Dissertation, University of Twente, published in Cambridge, UK (1994)
2 Ackroyd, S. and J.A. Hughes, *Data Collection in Context*, Longman, New York (1981)
3 Schindler, M., Wissenserfassung bei Experten und Anfängern auf dem Gebiet der Konstruktion, Workshop Bamberg, Bamberg (18–19 Dec. 1991)
4 Fricke, G.A., *Konstruieren als flexibler Problemlöseprozeß – Empirische Untersuchung über erfolgreiche Strategien und methodische Vorgehensweisen beim Konstruieren*, Dissertation, Technische Universität Darmstadt, VDI Fortschrittberichte, Reihe 1, Nr. 227, VDI-Verlag, Düsseldorf (1993)
5 Hales, C., *Analysis of the engineering design process in an industrial context*, Dissertation, University of Cambridge, Gants Hill Publications, Hampshire (1987)
6 Minneman, S.L., *The social construction of a technical reality: empirical studies of group engineering design practice*, Dissertation, Stanford University, Xerox Corporation, Palo Alto Research Center, CA (1991)
7 Bessant, J.R. and B.J. McMahon, Observation of a major design decision, Design Congress, Institute of Chemical Engineering, University of Aston (Sept. 1979) 04.1–04.14
8 Evans, S., Implementation framework for integrated design teams, *Journal of Engineering Design*, 1(4) (1990) 113–120
9 Conklin, J. and B. Yakemovic, A process-oriented approach to design rationale, *Human–Computer Interaction*, 6(1) (1991)
10 Lee, J. and K-Y Lai, *A Comparative Analysis of Design Rational Representations*, CCS TR#121, Center for Co-ordination Science and MIT Artificial Intelligence Laboratory (1991)
11 Hubka, V., *Theorie Technischer Systeme*, Springer-Verlag, Berlin (1984)
12 Marples, D.L., The decisions of engineering design, *IRE Trans. Engineering Management EM-8* (June 1961) 55–71

13   Guindon, R., Designing the design process: exploiting opportunistic thoughts, *Human–Computer Interaction*, **5** (1990) 305–44

14   Waldron, M.B., Observations on management of initial design specifications in mechanical design, *Proc. Int. Conf. on Engineering Design*, Harrogate, WDK 18, Vol. 1 (22–25 Aug. 1989) 189–200

15   Dylla, N., *Denk- und Handlungsabläufe beim Konstruieren*, Dissertation, Technische Universität München, Fakultät für Maschinenwesen, München (1990)

16   Denzin, N.K., *The Research Act: a Theoretical Introduction to Sociological Methods*, McGraw-Hill, New York (1978)

Appendix A
Ingredients of the Design Process: Definitions

This appendix gives the definitions of the basic and detail ingredients used in this paper to describe the design process.

Issue

- *Issue*: Any subject /topic that has to be resolved (that requires a way of solving it).
- *Question*: A minor issue which is put in the form of a question. In these experiments the questions asked by the designers to themselves and to others are put under this category.
- *Confirmation*: A question that includes an answer, i.e. a request for confirmation.
- *Previous design issues*: Issues generated during previous design projects.

Arguments

- *Arguments*: Evaluation statements that are made with respect to any issues, alternatives, arguments, criteria and decisions. The arguments are divided as follows:

  - Argument process: arguments related to the process of design
  - Argument product: arguments related to the product
  - Previous design arguments: arguments about the design process or the product of previous projects.

- *Plan*: Statements that reflect the planning of the design process, i.e. project plan.
- *Assumption*: Statements that reflect any assumption made by the designer(s). These assumptions may be based on the designer's (s') experience or intuition.
- *Agreement*: Statements that reflect the designer's agreement with their own statements or those made by others.
- *Confidence*: Statements that reflect the designer's confidence in their arguments.

- *Comments*: Statements that reflect the designer's understanding/mood, such as 'OK so I've read the user Trials Evaluation'. This category also includes half finished statements about the design process, such as a silence followed by 'and this happens to be a ... em...' which is not continued.

Alternatives

- *Alternatives*: They are the possible courses of action which are generated to resolve an issue.
- *Answers*: Statements made in response to questions.
- *Calculation/Analysis*: Calculations and analysis done with respect to the alternatives.
- *Previous design alternatives*: Statements about the alternatives generated during previous design projects (does not include the arguments about those alternatives).

Criteria

- *Criteria*: The yardstick against which each alternative can be compared (measured).
- *Specification/Requirement*: These are identified during the problem definition and used as guidelines to generate and evaluate alternatives.
- *Constraint*: These are criteria which are specific to the alternative(s) / issue(s).
- *Previous design criteria*: Criteria generated during previous design projects.

Decisions

Statements about selecting one or more alternatives or deferring the decision to a later stage.

- *Decisions*: When the designers either explicitly said or wrote down that they selected an alternative or rejected one.
- *Previous design decisions*: Decisions made during previous design projects.

External

Information provided by external sources.
- *Reading*: Reading from the given material.
- *External*: Information from sources other than the given material, such as the project leader.

Others

Statements that do not relate to the design process. For example, while making an external phone call: 'let's see OK er it's nine to get out?'.

5 *Investigation of Individual and Team Design Processes*

Joachim Günther[1], **Eckart Frankenberger**[2] and **Peter Auer**[3]
[1]*Technical University of Munich, Germany*
[2]*Technical University of Darmstadt, Germany*
[3]*University of Bamberg, Germany*

In industry, engineering designers are collaborating more and more in teams crossing department and even company borders. Thus, the work of a single designer is part of a complex technical and social process[1,2]. In order to support the daily work in engineering design with methodical aids and to improve design education, it is necessary to understand the interrelation of different factors determining design processes. What are these factors?

Every design process starts with a *design task*, and it is obvious that the type of task is of great importance for the process. The way in which the design task is solved depends on whether a designer has expert knowledge about a field or not. This brings up the *individual prerequisites* of each single designer such as his or her skills and knowledge. Working together in a group gives another dimension to the use of a designer's abilities. The way in which a group discusses, solves conflicts, and makes decisions may increase or decrease the performance of its members. Thus the *prerequisites of the group* are of great influence on the process and its result. Last but not least, the whole process takes place under certain *external conditions* (e.g. the company's branch, size and philosophy or internal organization[3]). Figure 5.1 demonstrates the interrelation between these factors.

To show the causal interrelations which are responsible for

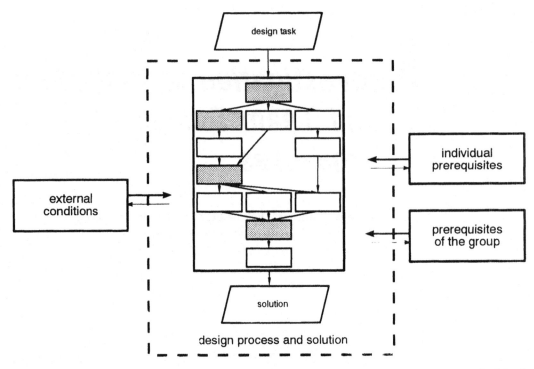

Figure 5.1
Factors determining
the design process
and its result

certain events in the design process, it is necessary to get hold of all relevant data from the above-mentioned fields. In a complex system like the engineering design process it is much too difficult to start investigating all the factors acting together. Therefore, one of the first steps we take is to understand design thinking and acting.

Experiments with designers in standardized situations in a laboratory are a suitable approach to get hold of the variety of individual procedures in engineering design. They allow us to explain observed differences in behaviour in the 'same' external situation relative to the individual prerequisites of the subjects. Some previous empirical research in this field has been conducted together with psychologists and thus provided an insight into the cognitive processes, the strategies and the tactics employed by the subjects during problem solving in design[4,5].

This study is based on the video recordings, transcripts and documents of the individual and team design process in a laboratory. Thus, we know all about the task, the process and the technical solution. We also know everything about the external conditions of the experiment. However, we do not have information concerning the individual's and the group's prerequisites, even though the video recordings allow quite a few suppositions on some of the individual prerequisites, but these are based on statements of the subjects only. Therefore, we are focusing on the design process in technical and social terms in order to demonstrate our method of record-keeping and interpretation. Further conclusions relying on a single individual and a single team process are not the aim of this paper. Accordingly, the emphasis of the following outline lies on the technique of protocol analysis and not so much on the specific processes.

1 Record-keeping Procedure

The design process can be reflected upon from different points of view. We can look at the phases, problems and subproblems of the design process. We can observe the actions of designers and their steps of analysing and making decisions in their work. But how can we comprehend all these simultaneous actions?

For the purpose of record-keeping the two design processes were classified by categories and entered into a computer record. We set the sample time of the two-hour design processes at 15 seconds and coded the categories. These were defined to describe both the *technical* development of the solution and the *social* processes. The categories used for the analysis of the 'mountain bike rack' problem were derived partly on the basis of theoretical considerations (e.g. design phases from Pahl and Beitz[6] and VDI 2221[7]), and partly on the basis of the experimental task. Existing category classes from Dylla[4] and Fricke[5] and further work in Munich[8] and Darmstadt[9] have provided the basis for the analysis which is presented here.

The category classes which were used to document the *technical process* of the individual designer and the design team are:

- *Problem solving phases*: This category class reflects the structure of the overall design process (see Section 2.1 for details).
- *Subfunctions*: This category class observes the topics in the design process with special regard to the parts of the object to be designed. The subfunctions shown in Figure 5.2 are used for the analysis.
- *Product characteristics*: This category class identifies those properties of the design object which the designers are dealing with. It consists of the categories (according to Pahl and Beitz[6]) of geometry, materials, positions, linkages, function, kinetics, forces and stress, production, assembly, safety, ergonomics, design, maintenance and costs.

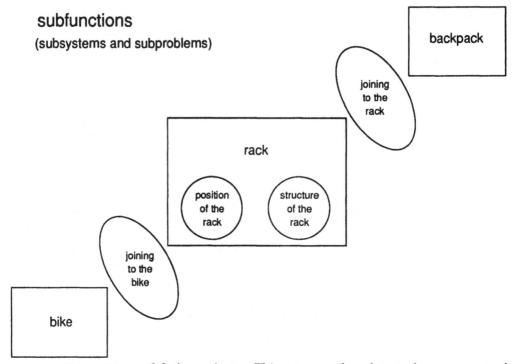

subfunctions
(subsystems and subproblems)

Figure 5.2
Subfunctions of the
design problem

- *Solution variants*:   This category class detects the occurrence of ideas and solution variants. Together with the subfunctions and the product characteristics, this class describes the development of the technical solution.

For the analysis of the team's design process it is especially important to document the part of the subject's problem-solving behaviour that is not directly linked to the technical process. Therefore, more psychologically oriented category classes were applied to record the *social* processes of the team. These are:

- *Subjects*:   This category class records which subject is responsible for a specific action, e.g. who made a statement or a suggestion or who had the idea.
- *Group processes*:   This category class observes the interaction of the group, e.g. self-organization, teamwork, individual work and polarization.
- *Deciding processes*:   This category class records the analysis of solutions, the evaluation and the decision-making of the subjects.

There is a separate column in the computer record for each of the category classes. One line in this record consists of a time

time	ph	tf	va	gp	ep	vp	text
:	:	:	:	:	:	:	:
00:18:45	1			t	f	k	real rigid attachment
00:19:00	1	jr		t	f	j	organisation: time to get off the backpack?
00:19:15	1	jr		t	f	k	thirty seconds
00:19:30	1	jr		t	f	k	sounds like a snap-in feature, wingnut
00:19:45	1	jr		t	f	k	orientation, up and down vibration
00:20:00	1			t	f	k	a lot of tension on these bolts
00:20:15	1			m	o	j	organisation, constraints: backpack, bicycle geometry
00:20:30				m	o	j	time limits, schedule
:	:	:	:	:	:	:	:

Figure 5.3
Computer-based
record

specification and a short note on the designer's spoken thoughts or actions. The coded entries of the categories are then written in the line. Figure 5.3 shows an excerpt of a computer-based record.

The record supports a largely computer-assisted interpretation of the data. Thus, it is possible to plot the procedure of the design process, to depict frequencies of categories and to determine proportions of time for a category.

1.1 Restrictions for the Analysis

Our aim is to record the design process as accurately as possible. However, the conditions of the experiment and the record-keeping process have a few limitations for the validity of any conclusions.

The requirement to think aloud interferes with the process of thought and action in the individual experiment. The single designer, Dan, is forced to formulate thoughts and ideas verbally which may differ in form and significance from what they would have been in a process without thinking aloud. The team is not thinking aloud, they are just communicating while designing. Any conclusions on the comparison of the team and individual process should consider this difference. Additionally, the external conditions were not absolutely the same: Dan used a telephone to consult with an external expert, whereas the team did not consult externally. Therefore, we have to be careful in terms of comparing the individual and the team design process.

In recording the category classes, it is usually possible to allocate observations to given categories without any doubt and without the need to interpret their meaning. The recording of subfunctions, phases and product characteristics sometimes requires a decision on the part of the record keeper as to the emphasis of the action or spoken thoughts. This, of course, leads to a lack of precision. This lack of precision is also influenced by the possible incompleteness of thoughts while 'thinking aloud'.

Another problem is change in a category during the chosen

time-scale of 15 seconds; the record keeper has to decide which is the main category to be recorded. Alternatively he can split the interval in order to record both events.

To test the reliability of the protocols, we usually compare the protocols of two or more record keepers. In spite of the above-mentioned restrictions, the presented technique of record keeping is thus a reliable basis for the analysis of the process.

2 Protocol Interpretation

The team and individual process are interpreted in order to illustrate the use of the analysis tools. This section of the paper will first focus on the overall processes in technical terms based on the problem solving phases, the subfunctions and the product characteristics. Comparing the activities of the single designer and the design team, one has to keep in mind that the examples are singular case studies of design processes. Some results of the team process, which refer to interaction between the group members, will then be presented.

2.1 The Overall Problem-solving Process

To get a rough overview of the whole technical process, it is recommended to divide the much more detailed design steps of the protocol into three main problem-solving phases. These three phases allow the description of the raw structure of the overall design processes:

- Phase 1: Clarification of the task. The aim of this phase is to understand the task and to get information about the requirements.
- Phase 2: Searching for concepts. This phase consists of the search for different principal solutions for the subfunctions of the design problem, the judgement and selection of these solutions, and their combination in order to achieve a concept for the design problem.
- Phase 3: Fixing the concept. This phase includes the further development of the concept to an optimum in accordance with the technical and economic criteria (e.g. costs, ergonomics, forces and stress). It leads to a status where embodiment design in a layout drawing can begin. The result of fixing the concept in our experiment is a hand-sketched preliminary layout.

The most obvious aspect of the individual design process in Figure 5.4 is the long time needed for the *clarification of the task* phase. Dan works for about 35 minutes with the assignment and

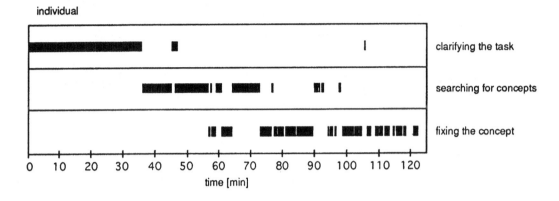

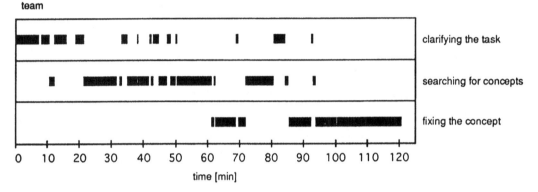

Figure 5.4
Course of the design
phases

the information system to get information on the task. He even calls other designers to get background information before he actually starts to search for concepts. The design team, however, enters the *searching for concepts* phase earlier and has more step-backs to the *clarification of the task* in their proceedings.

Both individual and team processes show iterations between *searching for concepts* and *fixing the concept*. Lack of information forces our designers to return to *clarifying the task* in the course of designing.

2.2 The Development of the Solution

The development of a solution can be described by the course and the intensity of the work on the different *subfunctions*. The subfunctions which were used to analyse the design process consist of the subproblems and subsystems as shown in Figure 5.2. The steps of working on the subfunctions by Dan and the team are illustrated in Figure 5.5.

According to this diagram, Dan's main problem in the first hour is the *position of the rack*. Most of the second hour he spends

individual

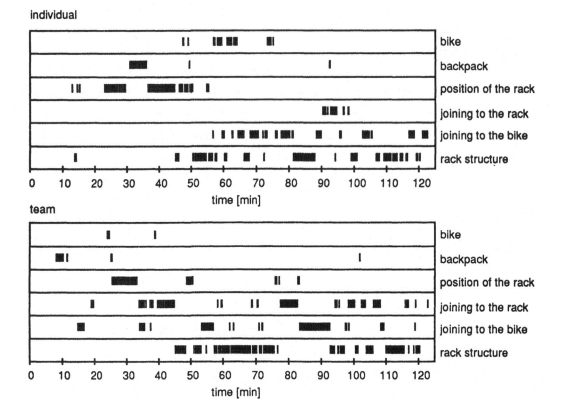

Figure 5.5
Course of the
addressed
subfunctions in the
design process

on the *rack structure* and the *rack joining to the bike*. A main difference between Dan's procedure and the team's procedure is that Dan lays no stress on the joining of the backpack to the rack. Again the statistic of the proportions of time, as shown in Figure 5.6, quantifies this observation: Dan spends only about 5% of the time on the subfunction *backpack joining to the rack*. Dan takes more time working with the bike and the backpack. For example, he very precisely calculates the weight of the backpack.

The conceptual design phase of the team starts with the search for positions of the rack (one of the main subproblems which determines the whole concept). Analysing the problem, the team sees two joining problems: *joining to the bike* and *joining to the rack*. They start to look for solution principles to join the backpack to the rack and later to join the rack to the bike. As they do not evaluate and select their solution principles for the different subfunctions at once, they have to return to them again and again. Therefore, like Dan, they often switch between the subfunctions.

During the search for a solution, the procedure can be oriented

individual

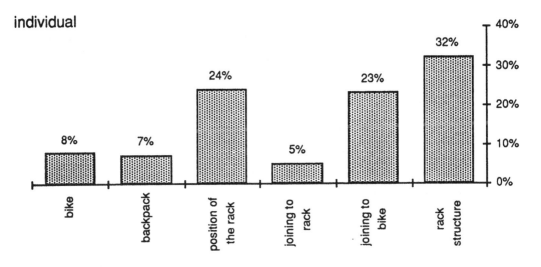

team

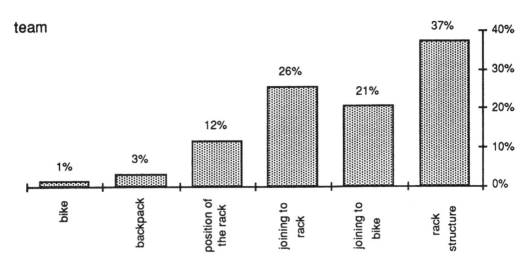

Figure 5.6
Proportions of time
spent on the different
subfunctions

to the different subproblems of the task. Another line in the design process can be the *product characteristics* such as materials, ergonomics or costs[6]. Together with the subfunctions the appearance of the product characteristics can visualize the internal guidelines of a product development. The frequencies of the different product characteristics, as shown in Figure 5.7, can help answer important questions. What kinds of product characteristics are our designers paying attention to? Is there an emphasis on certain product characteristics? Does this have an effect on the solution?

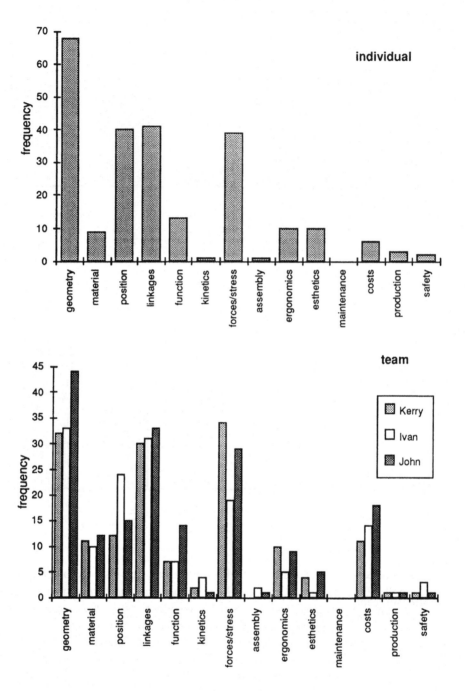

Figure 5.7 Frequency of product characteristics in the individual and team process

It is no surprise that the product characteristics *geometry, materials, positions* and *linkages* are of great importance as they are constituent elements in most design processes. The high frequency of *force and stress* in the individual and the team process can be explained by the specific requirement of the task to design a rigid and light structure. Thus both Dan's solution and the team's solution try to meet this requirement. Dan lays equal stress on *ergonomics* and *design*. *Costs* do not play an important role in Dan's process but they do in the team's process. Dan said he would have calculated the costs more exactly if he had had more time. The team spends a lot of time on an elaborate cost estimation. This reflects the more cost-oriented design of the team.

Besides the overall problem-solving process, the course of the subfunctions and the attention to product characteristics, the occurrence of the solution variants is of central importance to describing the design process and the development of the solution. The following questions arise: *when* do solution ideas come up and *who* (in a team) has the idea? Based on the example of the team's design process, Figure 5.8 shows the intensity of solution generation in five-minute intervals for each subject.

Diagrams like this allow us to visualize the frequency of specific events in the course of the process. In this example it becomes obvious how the steps back to the 'clarification of the task' (see Figure 5.4) interrupt the solution-generating activities (e.g. at minute 30). By identifying the subject who had the idea, it becomes clear that John creates most solution variants (18), followed by Kerry (12) and Ivan (6). This view of the single

Figure 5.8
Solution generating activities of the team subjects in five-minute intervals

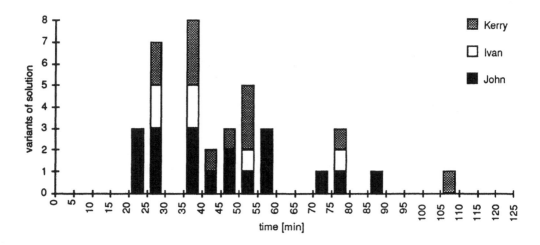

subject in the team leads to the importance of roles and social interaction in a design team.

2.3 The Social Processes in the Design Team

In order to be able to analyse social processes, it is essential to look at who in the group was responsible for a specific action. Thus, it is possible to find out how much a subject is involved in the process (Figure 5.9).

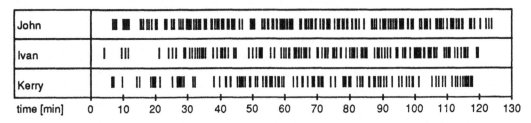

Figure 5.9
Frequencies of the team members in discussion

From Figure 5.9 one can conclude that the participation of John, Ivan and Kerry in the discussions is nearly balanced in terms of quantity. But nevertheless, there are differences in the roles they take on in the group.

In order to detect a subject's role we need to know more about the subjects' interaction. Using the protocol class *group processes* we can trace the way the subjects work together in the team. In this example, we distinguish between presentation, organization, theme-oriented working and polarization caused by opposite opinions on an item (see Figure 5.10).

As shown above, the group's discussions are mainly theme-oriented. This means that all members are concentrating on the discussed problem and the group is somehow 'guided' by the problem. These phases are interrupted in the form of frequent organization by single subjects. We can also observe a presenta-

Figure 5.10
Different ways of working in the team process

tion when John reads out the market study. Later we notice differing opinions (polarization), mainly between John and Kerry, concerning the joining technology.

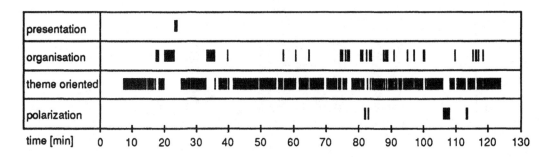

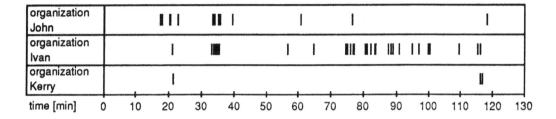

Figure 5.11
Frequencies of
organization of the
team members in the
design process

Using the class 'subjects' we can now identify who takes on certain roles in the group. For example it is of interest to know who moderates the group during the design process. Therefore, we combine the event 'organization' with the acting subject as shown in Figure 5.11.

It becomes obvious that John takes a dominant role as leader and moderator in the beginning. But when the group decides to make a time schedule (Ivan's idea, 35 minutes), Ivan takes over and moderates as 'Mister Schedule' (joke by John). Ivan also uses his role as a documentor, writing on the whiteboard to moderate the group. Kerry's part in organization activities is negligible.

In the same way we can show the subjects' roles concerning another important aspect of teamwork in design, the way a group evaluates and chooses solution variants. This procedure has a strong influence on the final solution. Therefore, the role a subject plays in the *decision processes* reflects his or her position in the team. The decision making process normally starts with an analysis of the discussed solution variant. As Figure 5.12 demonstrates, all subjects in our example take part in the analysis of solutions for about the same amounts of time.

When it comes to the question of who takes the decisions, Kerry does not play a minor role any more (as she did in organization). As Figure 5.13 shows, she and John are mainly responsible for the selection of the solution variants. Ten decisions are due to Kerry; John also took 10 decisions and Ivan 5.

Figure 5.12
Steps of analysis of
the team members

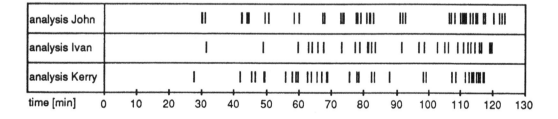

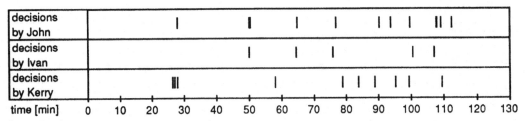

Figure 5.13 Decisions of the team members in the design process

3 Discussion

So far we have presented an overview of the interpretation of the protocols. Our general results are:

1. We are able to depict the course of the design processes with our category system.
2. Different and common aspects between the design processes can be found in comparing the data, e.g. the proportions of time of a category in different processes.
3. Aspects of the social process in the team can be described with the category system.

It is necessary to keep in mind that this approach focuses on only one half of the whole story: the design process. We are dealing with case studies without knowing the individual abilities of the designers completely and without exactly knowing how this influences the design process.

For a general understanding of engineering design, the question *why* certain events happen in the process is interesting. What are the reasons for all the observed phenomena? This refers to the factor influencing the design processes as introduced at the beginning (see Figure 5.1), especially the subjects' and the group's prerequisites. Some questions arise in combination with some of the findings:

- Why does Dan take a lot of time to clarify the task and to deal with the bike and the backpack? It may be that this behaviour is a result of his large design experience.
- Is the short time for clarifying the task in the team typical for team processes or a result of the time pressure in the experiment?
- Is Kerry's strong involvement in decisions a result of her experience in mountain biking and backpacking and her design experience with bike equipment?

- We cannot answer the question why John mainly creates the ideas and Ivan concentrates on organizing the process. Perhaps these are the jobs they are trained in or like to do most.
- Are the observed processes successful design processes under the experimental conditions? We do not have other processes under the same conditions to compare with them.

Two directions for further research can be derived from these questions:

1. Detailed information about the individual prerequisites and abilities could be a basis to explain some of the observations in the design process. Further research in cooperation with psychologists[10,11] and the development of suitable experimental equipment for the field of design could help to understand the relations between individual prerequisites and design behaviour.
2. The comparison of several case studies in teams (or with single designers) under the same external conditions and an evaluation of the resulting designs should lead to a better knowledge of successful and unsuccessful design processes.

Acknowledgments

We would like to thank the DFG (Deutsche Forschungsgemeinschaft) for funding the authors of this study (research projects: 'Denkabläufe beim Konstruieren' Ehrlenspiel/ Dörner; 'Gruppenarbeit in der Konstruktionspraxis' Birkhofer/ Dörner).

References

1   Hales, C., *Analysis of the engineering process in an industrial context*, PhD thesis, Cambridge University (1987)
2   Minneman, S. and L. Leifer, Group engineering design practice: the social construction of a technical reality, *Proc. Int. Conf. on Engineering Design ICED93* (Ed. N. Roozenburg), The Hague: Heurista, Zürich (1993) pp. 301–310
3   Blessing, L., The design process of a complex product: selected results of an analysis, *Proc. Int. Conf. on Engineering Design ICED93* (Ed. N. Roozenburg), The Hague: Heurista, Zürich (1991) pp. 342–350
4   Dylla, N., *Denk- und Handlungsabläufe beim Konstruieren*, Hanser, Vienna (1991)
5   Fricke, G.A., *Konstruieren als flexibler Problemlöseprozeß – Empirische Untersuchung über erfolgreiche Strategien und methodische Vorgehensweisen beim Konstruieren*, Dissertation, Technische Universität Darmstadt, VDI Fortschrittberichte, Reihe 1, Nr.227, VDI-Verlag, Düsseldorf (1993)

6   Pahl, G. and W. Beitz, *Konstruktionslehre*, 3rd edn, Springer-Verlag, Berlin (1993)

7   VDI-Richtlinie 2221, *Methodik zum Entwickeln und Konstruieren technischer Systeme und Produkte*, VDI-Verlag, Düsseldorf (1993)

8   Ehrlenspiel, K. and N. Dylla, Experimental investigation of designers' thinking methods and design procedures, *Journal of Engineering Design*, **4**(3) (1993) 201–212

9   Fricke, G. and G. Pahl, Zusammenhang zwischen personenbedingtem Vorgehen und Lösungsgüte. *Proc. Int. Conf. on Engineering Design ICED93* (Ed. N. Roozenburg), The Hague: Heurista, Zürich (1993) pp. 331–341

10 Auer, P. and E. Frankenberger, Vorgehensstile beim Konstruieren - Flexibilität und Invarianz beim Lösen unterschiedlicher Probleme, Bamberg: Universität, Lehrstuhl Psychologie II, unpublished memorandum (1994)

11 Badke-Schaub, P., *Gruppen und komplexe Probleme*, Strategien von Kleingruppen bei der Bearbeitung einer simulierten Aids-Ausbreitung, Peter Lang, Frankfurt am Main (1994)

6 Design Strategies

Can Baykan
Bilkent University, Ankara, Turkey

The aim of the study reported in this paper is to account for the design behaviour of the individual subject. In order to carry it out, we use an analysis of the bike rack design task and a set of hypothesized primitive information processes. These are derived from an analysis of the group design protocol. The decision to use the group protocol for identifying the primitive information processes and doing a task analysis is made before looking at the data, and is based on the expectation that the group protocol would be more complex due to interactions between the group members.

The protocols are first segmented and then encoded in terms of a set of primitive information processes based on the information that is heeded. The hypothesized information processes are task dependent. The verbalizations in the protocols indicate the inputs and outputs to the processes which are in short-term memory, rather than the processes themselves.

These processes and information structures identified and used in the analysis are a hierarchy of goals and subgoals, generate and test actions, methods which organize them, constraints and information management and time management. Goals are the most important elements in determining behaviour. They lead to the selection of appropriate problem spaces and methods and information and time management functions.These primitive information processes are sufficient to account for the segments in the individual's protocol. Other aspects of design, such as recognition processes, using intuitive constraints and accessing information from LTM via associations, are not accounted for in this protocol analysis.

1 Background

The first comprehensive work on the assumptions and principles of cognitive science and protocol analysis methods was written by Newell[1]. The advances in protocol analysis methods since then are covered in Ericsson and Simon's work[2], which contains an in-depth exposition of protocol analysis.

The initial application of protocol analysis to design was by Eastman[3]. In this work he analysed the behaviour of a designer working on a simple spatial layout problem, the configuration of a bathroom. Akin has done extensive cognitive research on different aspects of design, including identification of information processing primitives for architectural design[4], problem structuring in design[5] and modelling design[6]. Goel and Pirolli[7] discuss the characteristics of problem spaces which follow from the nature of design problems and hypothesize that these characteristics are valid across different design domains.

2 Method

In this paper, our aim is to account for the individual designer's protocol. We do a task analysis of the bike rack design problem and formulate hypotheses about a set of primitive information processes by analysing the design team protocol. The primitive information processes are formulated on the basis of an analysis of the heeded information revealed in the protocol in the context of task analysis. We then use these hypotheses to analyse the individual designer's protocol.

Segments are verbalization units that correspond to units of heeded information, pauses and syntactic information. The segments may be sentences, clauses, phrases or words[2]. We segment the whole protocol first and try to encode each segment using the hypothesized processes. Encoding a segment sometimes requires taking its context into account when the information contained in it has references or is fragmentary, but we try to keep the context as narrow as possible.

The other, larger unit of behaviour that we use in the analysis of the protocols is the *episode*, which is a collection of segments. An episode is a succinctly describable portion of behaviour[1]. The episodes are identified after encoding the protocol. Looking at the episodes, it may be possible to see the overall structure of behaviour in an aggregate form and to account for individual differences.

The decision to use the design team protocol for doing a task analysis of the design problem used in the experiments and for formulating primitive information processes was made prior to looking at the data. This was done because we expected that analysing the problem-solving behaviour of the team would also involve looking at other issues, such as group interactions. We use the team protocol for formulating the hypotheses and then use these for analysing the individual designer's data. We segmented, encoded and identified episodes in the team protocol. While doing this we did not try to account for everything in the protocol – but we tried to account for most of it. The process of segmenting, encoding and aggregating into episodes was repeated for the individual designer's protocol.

3 Task Analysis

Task analysis produces a description of the constraints on behaviour that must be satisfied to solve the problem at a given level of intelligence, where intelligence is defined as the ability to use knowledge to solve problems. Certain paths in the task environment are not available as paths to the goal for a given level of intelligence because they are too difficult[1]. The definition of intelligence given above supposes that the problem solver has knowledge and can bring it to bear on the problem. Expertise in some domain means having extensive and structured knowledge about the domain that can be applied easily.

One of the difficulties of research on design thinking is that task analysis is not definitive, for example as in chess, cryptarithmetic or other puzzles. We don't really know what it takes to solve the task. The same is true of evaluating designs. The solution is to rely on the opinions of experts in that field of design for resolving such issues. Since we are not experts in bike rack design, we use the design team protocol to do a simple task analysis.

The design problem used in the experiments involves finding a position for carrying the backpack on the bike, locations on the bike for mounting the rack, methods for attaching the rack to the bike and the backpack to the rack, and determining dimensions, materials and production methods. Cost, weight, structural strength and stability, ease of use and aesthetics are some of the important dependent variables. The dependencies between different parts of the problem and between the variables are complex. Determining the position to carry the backpack on the bike seems to be an independent decision, which affects mounting the rack to the bike. The method for attaching the backpack to the rack depends on the orientation of the rack. Ease of use and aesthetics can be evaluated on conceptual designs, but cost, structural strength, stability and weight require the design to be quite complete before they can be determined. The methods that can be applied vary from subtask to subtask. Some aspects of the problem, such as position of the rack and determining the attaching points of the rack, require trial and error. Others, such as checking structural stability or estimating cost, have well-defined procedures that may be used.

In design tasks, the solution is not implicit in task information as in puzzles or other well-defined problems. Solving the problem requires identifying the knowledge that applies unless the problem solver is an expert in that design field. Even then, they may have to access information from outside sources. There is no correct solution to a design problem. Thus, the problem solver stops working on the problem when satisfied with the result, or more often when available resources, the most important of which is time, run out. Designers manage their time and tailor their efforts by taking available time into consideration.

4 Assumptions

The first assumption is that the subjects are information processing systems (IPS)[1]. An IPS consists of a processor which carries out some set of primitive information processing operations, a short term memory (STM), an associative long term memory (LTM), and effectors and receptors for interacting with the environment. The capacity of the STM is estimated to be seven plus or minus two chunks[1] or four plus or minus two chunks[2] at the same time. The capacity of the LTM is practically unlimited, but especially fixing, and to some extent accessing information, is slow. Due to the limited capacity of the STM and the long times required for fixing information in the LTM, an IPS needs to use external memory (EM) as an aid during problem solving. EM is a visual display which consists of the notes of the subject. It is possible to think of the STM as consisting of the STM internal to the IPS plus the portion of EM in the subject's foveal view. The task-independent primitive information processes are reading and writing to EM, accessing and fixing information in LTM and making inferences. The STM is always under the direct attention of the problem solver.

An IPS solves problems in problem space(s) internal to the IPS. The problem space encodes significant information about the objective task environment and organizes problem-solving effort.

The task environment and the characteristics of the human IPS determine the possible structures of problem spaces that will be used in problem solving[1]. A problem space consists of symbol structures, operators, an initial state, the problem and knowledge. The selection of problem spaces may imply a very large selection but not much processing, since the operators are built of familiar components[1].

The assumptions about the verbalizations in protocols are that a thought is verbalized as it is heeded. The inputs and outputs to the processes are in STM and are verbalized rather than the process itself. Types of verbalization depend on the time of verbalization and mapping from the heeded to the verbalized information[2].

5 Hypotheses

The primitive information processes and functionalities given below are formulated as a result of the analysis of the design team protocol, and will be used to analyse and explain the individual's protocol. They are goals, methods, generate and test actions, constraint and information management, time management, and focus of attention functions.

5.1 Goals

Goals come from the task or problem. They are not strictly determined by the problem definition, but have a rational relation to it. Problem decomposition takes place by setting up subgoals. As a result, there may be a hierarchy of goals and subgoals. A goal is capable of controlling behaviour by evoking a pattern of behaviour that has a rational relation to it. When a goal cannot be achieved immediately, i.e. by recognition or by a primitive information process, it requires a problem space to work in. Thus, goals define problem spaces.

The goal statement may be in the form of a question or a statement as seen in the protocol fragment in Figure 6.1. Every line in this figure denotes a separate segment, and every segment

---

00:06:00
K   what do we need?
*   I guess we should look at their existing prototype, huh?*
J   . . .
*   we could also just sort of like try to quantify the problem because*
*   what's your understanding of the problem first of all?*

Figure 6.1
Goals and planning
example

happens to be a goal. Some of the goals are attended to immediately. The first segment is the reason for the generation of the
rest. The last segment leads to behaviour aimed at defining the
problem. Based on whether goals are attended to immediately or
not, we can make a distinction between goals and plans. The first
and last segments are goals and the second and third are plans.

5.2 Generate Operation

A generate operation is another primitive information process. It
proposes an idea or a symbol structure that is a (partial) candidate
solution for achieving the goal. The second, third and fourth
segments in Figure 6.1 are verbalized as the result of generate
operations.

00:41:00
I   OK pack to rack Velcro um
       pack to rack gravity
J   gravity er
K   snaps
       straps
       you can have um bungee cords

Figure 6.2
Generating candidate
solutions – type g1

Each segment in Figure 6.2 contains a possible method for
attaching the backpack to the rack. A generated idea or symbol
structure can be expressed by a single word, or it may also
contain the reason for the structure being generated, as in Figure
6.3.

Figure 6.3
Generating candidate
solutions – type g2

01:52:00
K   I go at steel here for the for the stiffness

The two different types of generate operations, seen in Figures
6.2 and 6.3, will be termed type g1 and type g2 respectively. Type
g1 can be formalized by indicating the proposed idea and what it
is a solution for, i.e., *g1{attach backpack to rack = bungee}*, and type
g2 by indicating a reason, i.e., *g2{material of tubing = steel ; stiff-
ness}*. The reason is usually a constraint, which will be defined
below. The formal descriptions are used for detailed encoding of
the individual's protocol.

5.3 Test Operation

A test evaluates a proposed partial solution. The test may contain
no reason, may contain a reference to a constraint or may

Figure 6.4
Type t1 test example

00:29:00
K   mm mm that'd be good

Figure 6.5
Type t2 test example

01:58:00
I   OK time to remove less than thirty seconds yeah we'll be able
     to permanently remove it in less than thirty seconds

compare two solutions with respect to a constraint. These are denoted as type t1, t2 and t3 tests respectively. A test segment is identified by a verbalized result, such as yes, no, good, etc. An example of a type t1 test is seen in Figure 6.4. It is formalized as *t1{candidate solution} → result*.

The candidate solution that is being evaluated is not always specified, but it is possible to infer from context what it is most of the time. Sometimes it is possible to infer using context what the constraint used in the evaluation is, but at other times the verbalized result is based on an intuitive evaluation and the constraint cannot be specified exactly.

An example of a type 2 test is seen in Figure 6.5. It contains the result of the evaluation as well as the constraint that is used in evaluating a generated solution. Its general format is *t2{candidate solution ; constraint} → result*.

Figure 6.6
Type t3 test example

00:30:00
K   that's not as bad (inaudible)

01:52:00
J   and these things could be aluminium too they don't
     have to be steel they're easier to form if they're aluminum

A type t3 test evaluates two proposed solutions with respect to a constraint to determine which one is better. Its form is *t3{candidate solution1; candidate solution2; constraint} → result*. Two examples of a type t3 test are seen in Figure 6.6. The result of the evaluations in the protocol are binary, that is, yes or no in the case of t1 and t2, and selecting one of the two structures in the case of t3.

5.4 Methods

A method is composed of primitive information processes. It organizes generate and test actions to achieve the goal. Methods are

J   *is there any sort of um cost specification for we know the*
      sales price is fiftyfive dollars but um

. . .

X   *No, I'm asking for the my assistant to say again have to*
      estimate their own ratios OK yep you have to estimate your
      own

. . .

J   *manufactured costs will be one fourth the the MSRP*

J   *OK . . . so what's a quarter of fiftyfive bucks er twelve er*
      twelve fourteen bucks

Figure 6.7
Method for
calculating
manufactured cost
from MSRP

determined by the structure of the problem space[1]. Weak
methods are general and widely applicable but may take too
long, and they do not guarantee the finding of a solution.
Examples of weak methods are trial and error, working backwards, working forwards, and means–ends analysis. Strong
methods take the structure of the problem space into account.
Specific methods for achieving a goal, such as procedures for
calculating cost or structural strength, are examples of strong
methods.

Figure 6.7 shows the application of a method for calculating the
manufactured cost of a product given its manufacturer's suggested retail price (MSRP). In this example, the method is the
application of a simple formula, which generates the MSRP from
the sales price. It is a strong method. The manufactured cost is
used as a constraint later on.

*5.5 Constraints and
Information Management*

The knowledge required for solving a design problem is not
implicit in the problem definition as it is in the case of puzzles.
The appropriate knowledge must be accessed from external
sources or from the designer's LTM using the appropriate associations.

A constraint encodes knowledge about the problem in a form
that can be easily applied in a problem space for generating and
testing candidate solutions. For example, the constraint used in
the protocol fragment in Figure 6.5 can be expressed as follows: it
should take less than 30 seconds to remove the backpack from the
rack.

Information management is finding the knowledge that is
applicable in a problem situation and formulating it in the form of
a constraint. Knowledge sources can be existing designs, solutions

Figure 6.8
Using an existing
design for finding
out about a problem

00:19:00
K the problem I see right away with that pro with that uh is that
   these are in that orientation and that's also the orientation
   of jolting and vibration so it'd be nice if
I oh up and down right
   so you have to do it much tighter in order to
K yeah you have to put a lot of tension on to these em bolts 'cos
   that's the opening is in the direction of the
I right
K loading the force it'd be nice if you could have the forks
   coming more like that so

that have been previously generated by the designer, using simulation and scenarios, and reasoning from general principles. Using designs (existing designs or newly generated ones) for identifying relevant knowledge will be termed a solution based strategy.

New constraints are recognized most often when they are failing in a design[3]. Thus generating tentative solutions for identifying problems with it is a possible strategy. Existing designs for the same problem or for analogous problems can also be used for the same purpose.

In the protocol fragment seen in Figure 6.8, the designers are looking at the prototype rack that was designed earlier. They discover shortcomings which lead them to formulate a new constraint, that the opening of the forks should not be in the direction of jolts and vibration. There does not seem to be any difference between using previous designs or solutions generated by themselves in this respect. Thus, solution based strategies for identifying constraints operate in similar ways, independent of the origin of the solution that is used.

A scenario is a mental simulation carried out in order to identify relevant knowledge about the problem. It is a means for accessing knowledge from the designer's LTM. It helps in making associations between the design situation and the knowledge in the designer's LTM. The knowledge can be used to identify constraints or to test a proposed design. Figure 6.9 contains a section from the group protocol which shows the use of a scenario for how the bike rack may be used, which results in the identification of two constraints; the backpack has to be firmly attached to the rack, and it should be removable in under 30 seconds. We can conjecture, based on this example and others, that scenarios to

00:18:00

I   do they talk about how the people wanna use it?
    they uh do these do the vacations they take long bicycle trips
    and then take short feet off uh short trips off by foot
. . .
I   em so they use the bike to get where they're going and then do
    a little hiking sounds like the bike becomes the
. . .
I   it sounds like they oughta really ride the bicycle and just
    temporarily go to work or something but you wanna be able
    to ride the bicycle
. . .
K   ride it through the country and then you get to the base of the
    hill and you wanna take your backpack and summit the
    mountain or something
. . .
K   and it's an off-road bike so you'd need a real rugged rugged
    attachment or a rigid attachment
. . .
J   so what's a reasonable time to like allow somebody to take
    this off their bike?
    should it take like under five seconds or under thirty seconds?
. . .
K   thirty seconds?
I   yeah I would say thirty seconds as well

Figure 6.9
The use of a
scenario for
identifying
information

fit every situation do not exist in the minds of designers, but are constructed from general knowledge when needed.

Other means of information management are reasoning from general principles and making simulations (for example, when the designers in the group stuff the backpack to get an idea of how much it will hold). The above strategies of information management deal with identifying the relevant knowledge and information for the purpose of identifying constraints. As design progresses, constraints may be modified or removed based on new knowledge that is gained. This is called constraint management.

5.6 Time Management

Design is open ended, thus designers are required to manage their resources. The most important resource that we see in the group protocol is time. Time management operates at the level of goals. Time management determines how much time one should spend on achieving a goal, when to stop working on a goal to move on to a new one and when to omit a goal altogether. Goals are accompanied by estimates about how long it will take to

00:20:00

. . .

J OK you you were talking about schedule stuff before
   do you wanna just set some time limits for ourselves?
I yeah I think we should uh just figure out how how much we
   wanna spend on each thing yeah so we can just move on
J so so this is kinda quantifying the problem or wherever we
   are right now (laugh)

. . .

I right now it's uh we have uh an hour and forty minutes so it's
   two it's four forty right now four twenty excuse me
J when do you wanna like stop conceptualizing
I well I think we should stop conceptualizing at uh say forty
   minutes from now

Figure 6.10
Time management

achieve them. These are used to decide whether to pursue that goal or not. Since goals define problem spaces, and the structure of the problem space determines the methods that may be used, goals have sufficient information associated with them to enable time management decisions to be made.

The protocol fragment seen in Figure 6.10 is an example of time management. Making a schedule is an explicit goal. The elements being scheduled, i.e., quantifying the problem and conceptualizing, are also goals.

5.7 Organization of Problem Solving

The order in which the goals are to be considered is not strictly determined by the task, and there is some jumping around and switching back and forth during problem solving. The focus of attention of the designer is controlled by the recognition of relevant aspects during problem solving, the knowledge that they access from LTM via associations or the problem specifications that they notice, rather than a predetermined order for attending to the goals. This is called opportunism.

In the example seen in Figure 6.11, the designers notice that weight is an important consideration as the result of comparing two solutions. Even though the current goal is not satisfied, they switch their focus of attention to a new goal, which is determining the weight of the backpack. They return to the goal that they gave up, which is mounting, later on.

Attention to control and focus of attention is indicated in the protocol by verbalizations such as 'let's see', 'where was I?', etc.

0:50:00

. . .

J   yeah so there there seems like two solution sets
*   there's one where the whatever it is stays with the pack*
*   the other solution says the stuff that stays with the bike*
K   and that's probably better for hiking
J   yeah yeah
K   then you don't have to lug around that weight
J   don't have to lug that weight so and . . .
*   there's one of the things on our spec that we didn't have is*
*   weight*

. . .

J   do we have any information that would tell us what's a
*   reasonable weight for the product?*

Figure 6.11
Opportunism

6 Analysis of Individual's Protocol

The individual designer's protocol has been segmented and coded using the primitive information-processes and other information processing functionalities hypothesized above. The aim of analysing the individual designer's protocol is to account for as much of the protocol as possible and to understand his strategies for structuring and using these primitives.

There is a difference between the two protocols with respect to the aims of the verbalizations, which is also reflected in the instructions given to the subjects for the experiment. In the design team protocol, the aim of verbalization is communication between the group members so that they can cooperate while doing the task. The members of the design team work for the same product design consultancy firm and are used to working together. They communicate naturally. The individual subject is verbalizing his thoughts as a requirement of the experiment. Possibly as a result of this, there are some differences in the verbalization patterns in the two protocols. The design team protocol contains level 1 and 2 verbalizations, whereas the individual's protocol also contains some level 3 verbalizations. In level 1 verbalizations, the subject is reporting the information in the form it is heeded. In level 2 verbalizations information that is not originally in verbal code is translated into that form, and in level 3 verbalizations there are intermediate scanning processes or

Figure 6.12
Level 3
verbalization

00:26:00

OK so em I've read the User Trials Evaluation em

0016:00
You bet. So I've been given the assignment and I'm reading the assignment
I'll read it out loud I guess in part (reads brief)

00:21:00
OK ... em OK em and the first thing I would do as a quick way of getting started is to ask you for the picture of a prototype and the test report.

00:51;00
let me just sorta get some ideas here

Figure 6.13
Top level goals
identifying design
episodes

01:06:00
let's just em design a mount for the em the bike em er

inference and generative processes[2]. In the segment seen in Figure 6.12, taken from the protocol of the individual designer, he is interpreting what he has done rather than reporting on the information he is heeding.

The top-level goals of the individual designer are to understand the problem by reading the assignment, to start designing by finding the relevant information, to decide on the location of the backpack on the bike, to determine detailed location and dimensions, design the mount, design backpack to rack connection, and specify materials and details. The segments in the protocol identifying the first four of these episodes are seen in Figure 6.13.

In identifying the episodes, the task analysis of the problem and the behaviour that occurs after the goal statement are taken into account. The goal leads to generate and test actions, to subgoals or to information management. In each episode, we see that generating and testing solutions, i.e., search, alternates with information management. In the second episode, defined by the goal to get started quickly, the designer generates a possible location just to rule it out, as seen at the top in Figure 6.14. It occurs among information management functions. This also indicates that searching for a solution can possibly start without much preparation.

Just as the design problem is divided into subtasks by setting up goals, sometimes subgoals are set up in order to achieve a goal. For example, during the design of the mounting the designer first generates a solution which fails structural con-

Figure 6.14
Generate operation for ruling out a possible location right away

00:22:00
em off bat immediately my impression is hey it's nice to have putting it on the front handlebars because you uh like low inertia on the handlebars

straints because it is a parallelogram. As a result of this, the designer makes a request for drawings showing the front, side and rear views of the bike. Then he looks at a Blackburn rack, after which search continues by generating another solution. The different information management episodes are defined by subgoals, and alternate with search. In this case, a failing design alternative leads to information management; the designer looks at the front and rear views of the Batavus bike and the Blackburn design.

We see that goals and time management always occur together in the individual designer's protocol. There is no explicit scheduling as in the group protocol, seen in Figure 6.10.

Three examples of goals and time management from the individual designer's protocol are seen in Figure 6.15. In the first segment, the designer has been working on locating the backpack on the bike. In this segment the goal is not explicitly mentioned but in order to achieve the goal in a short time he changes the method he is using to exhaustive search. This is the only positive example of time management in his protocol. In the others, examples of which are seen in the next two segments, he considers a goal and then decides not to pursue it. Thus, the last two are examples of planning.

00:56:00
OK I have spent er fortyfive minutes or so now forty minutes doing that er so em . . . (mutter) going to check every possible location here

00:58:00
I might actually at this stage of the game actually do some trials of my own go out and er see if I found people who had bike racks and do a little more field testing myself but since I don't have much time I would er not do that.

01:54:00
I have a half hour left I haven't even gotten to some of the calculations but I'm not gonna worry about them right now em

Figure 6.15
Goals and time management examples

00:22:00
OK and it probably right off the bat says the backpack's too high or something like that and that bicycle stability's an issue ...

00:23:00
em eh interestingly enough em ... let me see ... I see how the backpack interfaces interestingly enough em ... doesn't directly take advantage of the eh frame ... nor the fact that there is a frame

Figure 6.16
Identifying constraints from an existing design

Information management functions make up a large part of the design process in the protocols. All of the information sources identified in the group protocol are also used by the individual designer. In Figure 6.16, he is using the prototype design to identify constraints.

The two segments seen in Figure 6.16 are tests of type t2. In the first one, the location of the backpack which is placed vertically on the rear wheel is determined to be too high, and the constraint that the centre of gravity should be kept low for stability is identified. In the second one, the constraint identified is that mounting the frame to the rack should make use of the backpack's external frame.

In the example seen in Figure 6.17, he uses the same type of test to identify the constraint that triangular forms should be used to make the rack rigid. In this example, he is evaluating the rack design he has generated. This is also an example of reasoning from first principles.

In the two segments seen in Figure 6.18, the constraints are

Figure 6.17
Identifying constraints from basic principles

01:07:00
one of the problems with em a bicycle carrier where the frame is mounted out here and it goes to that is is that you end up with a parallelogram bad thing bad thing

Figure 6.18
Identifying constraints from simulation

00:53:00
and in fact em one of the things I noticed right off the bat is that em it doesn't really stick out much more than the em than the pedal

00:54:00
mmm so right so my thighs are gonna come at least er two or three inches at least behind there my butt OK so ...

identified by actually placing the backpack on the bike and noticing things. In the first segment, the designer removes the constraint that placing the backpack at the side of the rear wheel causes the backpack to stick out and hit things on narrow trails. This constraint has been identified at 00:26 when the side location was generated and then tested using a scenario. In the second segment, the constraint that is identified is how close the backpack can come to the seat when it is placed on the rear wheel.

As a result of analysing the individual designer's protocol, we can say that the hypothesized primitive information processes are sufficient for encoding the segments. No new primitives are needed for accounting for the protocol.

Generating the hypotheses from the group protocol helped identify some implicit, non-verbalized structures in the protocol of the individual designer. Since the group members have to cooperate on designing and follow the reasoning of each other in order to agree or disagree, they are more explicit in their goals, time management and planning functions, and even in their generate and test operations.

7 Discussion

The subjects do not seem to be familiar with bicycle rack design even though they are bikers. They do not have expertise about the specific design problem they are trying to solve. They are using general design heuristics and identifying and accessing relevant information figures very strongly in their design behaviours. We can speculate that an expert doing the same design task may require less information management.

The main parts of the problem, such as, deciding on the position of the backpack, designing a mount, and designing a backpack to rack connection, may be determined by the given task. Thus we see both subjects attend to the same issues.

The group follows a brainstorming strategy; they try to make analogies, defer judgement and try to generate many alternatives. They go through these issues twice, once during what they call the 'ideation' phase, and then again during what they call the design phase. The individual designer focuses on each issue once, but during this time he goes through what can be called 'ideation' and design.

The individual uses existing designs more extensively, and by making a telephone call finds expert opinion about the positioning of the backpack on the bike. He also reasons from first principles. His comments indicate that his preferred strategy is to use expert knowledge, failing that to use simulation and scenarios and then search. This is what he does when determining the location of the backpack. In other subtasks such as when mounting the rack to the bike, he starts with search. We can speculate that he has more knowledge about this or that it will be too time consuming to seek expert knowledge.

The constraints used for defining the design problem differ between group and individual. The design group designs a tray with straps instead of taking advantage of the external frame, and removing the backpack off the rack is expected to take less than 30 seconds. The individual designer, on the other hand, attaches the backpack to the rack with clips so that it can be taken off very easily but the clips are expected to carry a pack weighing less than 15 lb. According to available information, the pack can weigh up to 50 lb, but this information was not seen by the individual.

As a conclusion we can say that it is possible to account for some aspects of design using simple structures and diversified knowledge from very different sources. These elements are used very flexibly in different combinations. Information management alternates with search. The most important differences between group and individual are in information management. Other aspects of design, such as recognition processes, using intuitive constraints, aesthetics, and accessing information from LTM via associations, are not accounted for in the protocol data.

8 Conclusion

The aim of this study is to account for the design behaviour of the individual subject. In order to carry it out, we use an analysis of the bike rack design task and a set of hypothesized primitive information processes. These are derived from an analysis of the group design protocol. The protocols are first segmented and then encoded in terms of primitive information processes and the information that is heeded. The hypothesized primitive information processes are based on the design task and are task dependent. The verbalizations in the protocols indicate the inputs and outputs to the processes rather than the processes themselves.

These processes and information structures are a hierarchy of goals and subgoals, generate and test operations, methods that organize them, constraints and constraint management, information management and time management. Goals are the most important elements in determining the behaviour. They lead to the selection of appropriate problem spaces, methods and information and time management functions. The two types of generate operations and the three types of tests are used by both the individual and the group. These primitive information processes are sufficient to account for the segments in the individual's protocol. Other aspects of design, such as recognition processes, using intuitive constraints and accessing information from LTM via associations, are not accounted for.

The similarities between the two protocols can be attributed to the requirements of the design task and to the fact that the subjects are not experts in the task. Also, some of these are due to general design heuristics used by both the individual and the group. The differences may be due to designing in a group versus individually, and to differences in the subjects' backgrounds and personal preferences.

References

1   Newell, A. and H.A. Simon, *Human Problem Solving*, Prentice Hall, Englewood Cliffs, NJ (1972)
2   Ericsson, K.A. and H.A. Simon, *Protocol Analysis*, MIT Press, Cambridge, MA (1993)
3   Eastman, C.M., On the analysis of intuitive design processes, in G.T. Moore (Ed.), *Emerging Methods in Environmental Design and Planning*, MIT Press, Cambridge, MA (1970) pp. 21–37
4   Akin, Ö., *Models of architectural knowledge: an information processing view of architectural design*, PhD thesis, Carnegie-Mellon University, Pittsburgh, PA (1979)
5   Akin, Ö., A formalism for problem restructuring and resolution in design, *Environment and Planning B: Planning and Design*, **13** (1986) 223–232
6   Akin, Ö., Heuristic generation of layouts (HeGel): based on a paradigm for problem structuring, *Environment and Planning B: Planning and Design*, **19** (1992) 33–59
7   Goel, V. and P. Pirolli, Motivating the notion of generic design within information processing theory: The design problem space, *AI Magazine*, **10**(1) (Spring 1989) 18–36

7 Understanding Information Management in Conceptual Design

Vinod Baya and **Larry J. Leifer**
Stanford University, Palo Alto, CA, USA

Use of verbal protocol analysis as a method to better understand the design process is gaining popularity within the design community. In this paper we will be using this method to understand the information handling behaviour of an individual designer. The behaviour will be described within the framework of three informational activities, namely *generate*, *access* and *analyse*, and an information classification framework which categorizes information based on its *descriptor*, *subject-class*, *form* and *level of detail*. A measure for the amount of design information that designers handle is introduced and will be an important variable in understanding information handling behaviour. The research is based on the belief that understanding of information handling behaviour is a prerequisite to building tools and methods to support information management.

1 Information Management

Design is an information driven process. Over the course of a design process designers handle large amounts of information. Therefore, the quality of designs and the overall productivity of the design process depend heavily on the information management skills of designers. Information management is the process of capturing and organizing design information in such a manner that it can be retrieved and reused at a later time. In recent years there has been an increasing deployment of computer-aided tools to support the design process in areas such as CAD (Computer-

Aided Design), CAE (Computer-Aided Engineering) and CAM (Computer-Aided Manufacturing). However, in most cases the tools apply to very specific domains, and are usable only in the detailed design stage and provide limited or no support for information management. Thus the burden of information management lies primarily on the designers, which often results in reduced productivity (when designers lose some of their time doing information management) or loss of valuable information (when little information is captured and organized).

Existing tools for information management integrate poorly with the design process and are cumbersome to use, which results in designers spending considerable time away from design. Therefore for improved productivity there is a need for developing tools, methods and technology which integrates smoothly with the design process and supports information management without being cumbersome to use.

It is hypothesized in this study that building information management tools and methods based on a better understanding of the information handling behaviour of designers will result in a natural integration of the tools into the design process. This paper reports on the verbal protocol analysis method used by us to develop a framework for understanding and analysing the information handling behaviour of designers. The analysis procedure is described in detail along with some results derived from the analysis.

1.1 Verbal Protocol
Analysis: Pros and Cons

The approach of using protocol analysis to improve the understanding of the design process is gaining momentum in design research. Design process is very complex and it is very easy to be overwhelmed by the number of variables which affect the outcome of a design process. Protocol analysis makes it possible for researchers to reduce the complexity and to look at the design process from a perspective of their interest within a controlled environment. In other words it lets researchers simulate the design process in a laboratory and collect rich and explicit data at a very fine level of detail.

However, the techniques used in conducting experiments and analysing data collected from these experiments are very subjective and far from standard. It is difficult to exclude effects of the following from the analysis of the data: (a) background of experimental subjects, (b) nature of the design problems used in the experiment, (c) nature of the experimental setup, and (d) bias of the experimenter. This makes it very difficult to generalize any conclusions that result from such analysis. Also since the time involved in performing analysis of such data is considerable, analysis of a statistically significant number of experiments is impractical.

Despite the above shortcomings, protocol analysis remains a popular method. This is because the method can provide a deep insight into the design process from a perspective of interest which often results in an improved understanding. Also design research teams all over the world are working on developing analysis methods which would reduce or eliminate the subjectivity in analysis and make it possible to draw generalizable conclusions.

This study uses the verbal protocol analysis method to look at the design process of an individual designer from an informational perspective. Clearly since the experimental setup is a small scale simulation of real design activity, not all information handling scenarios prevalent in reality are observed in the experimental data. However, the subset of scenarios observed are real and provide a rich insight into frameworks that can be used to support information handling in real world design.

1.2 Conceptual Design Phase

Among the commonly agreed phases of the design process (conceptual, layout, detail)[1], we are focusing on the conceptual phase from an informational perspective for the following reasons:

- large amounts of information are manipulated in a short time in this phase
- this is the phase which is most informal, most complex, least understood and least supported by computational tools
- decisions and information generated in this phase have a large impact on the downstream design process and the overall cost. It has been observed that upwards of 80% of the final costs are obligated during conceptual design[2].

1.3 Objectives

This study was conducted to answer questions of the following form. In this paper we will be able to address only the first two questions.

1. What are the different activities that designers perform with information?

2. How much information is handled by designers, and can this be measured?
3. How is the information used, and for what purpose?
4. Are there any correlations between informational activities and the nature or medium or level of abstraction of information dealt with in that activity?
5. What methods, tools and frameworks will make it possible to improve information management in the design process?

1.4 Related Work

On verbal protocol and research methodology

The verbal protocol method is a popular method in single subject experiments and has been used widely[3-7]. There is debate in the research community regarding the appropriateness of this method, since it is not known how the act of verbalizing affects the natural thought process and behaviour of the subject. However, in the absence of any fundamental limitations, this remains a popular method for single subject studies.

The methodology of using experimentation as a means to develop frameworks and understandings of the design process which guide the development of design tools has been used by Tang[8] and Minneman[9]. Their focus was on understanding collaborative activity and results from their studies have been used in developing tools for collaboration between individual designers[10] as well as teams of designers[11]. Our previous study[12] in developing a framework for classifying design information has guided the development of Dedal, a tool for indexing, modelling and retrieving design information[13,14]. The need for identifying requirements for engineering information systems and developing frameworks which can be used to structure and organize information in a design process were discussed in detail in an information technology workshop[15]. This study addresses many issues raised in that workshop.

On information measures

Measuring information is an old concept. It originated in the field of communications engineering, wherein the quantity of an information message is inversely related to the probability of receiving that message from a source[16]. This measure is related to the statistical properties of a communication system and does not depend on the semantic content of the message. Suh[17] suggests a definition of information content from a cybernetic perspective. Here the quantity of an information message is related to the effect of the message on a system's ability to achieve a goal. So the measure of information content is with respect to a certain goal. To use this measure one needs to define a goal and a probability distribution on information regarding its potential to bring the system closer to the goal. Neither of these two measures is relevant to the nature of analysis done in this study. This is because the analysis aims at measuring semantic information and it is not possible to put a probability distribution on semantic information with regard to a goal. This will become clear when the analysis is described. But briefly, the measure proposed in this study is more subjective than the above two measures and is independent of any goal.

Our previous work

This study is a follow-up to our research on design information reuse[12] and information handling behaviour[19]. The previous research proposed a classification for design information based on its Descriptor, Subject Class, Level of Detail and Form. This same classification (with some modification) has been used in this study to classify information fragments. We refer the reader to Baya et al.[12] for details of the classification scheme.

2 Information Handling Behaviour Framework

To better understand the notion of informational behaviour and some of the definitions below, assume that at any time in the design process there is an information space associated with the design. At the beginning, this information space may be thought of as containing the design requirements. As designers work on the design, they make changes to the information space, either by adding information to it by generating it or accessing it, or by transforming information by analysing it.

The activities that designers perform with information during design, which are referred to as informational activities in this study, can be grouped into three categories:

1. *Generate*: An action which adds new information to the information space from an unidentified source. This includes actions such as writing, drawing, talking (in verbal protocol).
2. *Access*: An action which references information within or outside the information space. This includes actions such as reading, listening, observing.
3. *Analyse*: An action which changes the form or representation of information. A fragment of information is considered to

change if its classification changes. This includes actions such as interpret, negotiate, organize, calculate, reason.

This is a minimalist view of the informational behaviour of designers. However, it was possible to classify all activities that were observed in the experiment in these three categories. They are a suitable level of granularity for the purpose of this experiment. We note that the categories are most suitable for a single designer scenario. For a team design scenario we would add information exchange categories. Some definitions listed below will help as we go forward with the results.

Information fragment   A continuous period of time during which the informational activity and the descriptor (defined below) of the information remains the same. The information fragments are interpreted as containing a descriptor information about a subject class (defined below). This is a somewhat loose definition. Another definition for this is in section 3 (Analysis Procedure), which is based on segmentation.

Descriptor   This refers to the basic constitution of the information contained in an information fragment.

Subject-class   This is the level in the hierarchy (assembly, component, feature etc.) to which the subject of an information fragment belongs.

Medium   This is the physical form in which the information fragment exists in the information space.

Level of detail   This is associated with the phase of the design process to which the information can be associated with. It is common to view the design process as comprising three phases: conceptual, configurational (or layout), and detail design.

Level of abstraction   This is an information measure which is being qualitatively defined for this study. It is common knowledge that design information exists at various levels of abstraction all through the design process. However, there is no commonly agreed categorization or understanding of these levels. Therefore a formal definition and an insight into its relationship with informational activity and other fields of information fragments

Table 7.1 Levels of abstraction and their interpretation

Level of abstraction	Interpretation
Unlabelled	Referring to subject of information fragment without a name, as in an idea or a new concept
Labelled	Referring to the subject of information fragment by a name, as in using a name for a new concept
Associative	When fragment contains information about relations or associations with other subject
Qualitative	When attributes of a subject are qualitatively addressed, or when operation or motion is qualitatively described
Quantitative	When attributes of a subject are quantitatively described

would be valuable. Table 7.1 explains our attempt at defining level of abstraction and the corresponding qualitative interpretation. 'Subject' in the table refers to the subject of the information fragment. Note that these levels are defined from the perspective of evolution of a design concept. A concept in its initial stage exists without a label (unlabelled) and when it attains a label its level of abstraction reduces. The level reduces further when the concept is associated with another concept or described qualitatively, and the level of abstraction is the least (in its evolutionary life) when the concept is quantitatively specified.

Design information measure (dim)   This is a measure of the amount of design information contained in an information fragment. It is a measure of the semantic content and is independent of what context the information is used in. This measure uses the classification for the information fragments to calculate the numbers of dims in the fragment. A pair of descriptor and a subject from the subject list contribute 1 *dim* to this measure. Thus if a fragment is classified by a descriptor and three subjects in the subject list, its content is 3 dims. There are other guidelines to calculate the number of dims under special circumstances (multiple descriptors for a fragment, special cases with some of the descriptors, etc.). An information fragment is classified by filling out the template shown in Figure 7.1, using the definitions above.

   The fields in Figure 7.1 take values from the classification listed in Table 7.2. (Refer to Baya *et al.*[12] for more details on this classification.) Table 7.2 shows the values the various fields can take. The table only lists the fields which take symbolic values. The rest

(info-fragment
    :activity *informational-activity*
    :time (*start-time end-time*)
    :descriptor *descriptor*
    :subjclass *subject class*
    :subjects (*list-of-subjects*)
    :medium *medium*
    :levels (*level-of-detail level-of-abstraction*)
    :measures (*design-information-measure*)

Figure 7.1
Template for
classifying
information
fragment

of the fields (time and design-information-measure) take numerical values.

3 Analysis Procedure

The flow chart in Figure 7.2 lays out the different steps in the analysis procedure. The objective of the analysis was to reduce the audio and video data collected in the experiment into a form where the information handling behaviour could be quantitatively observed. We received the data after the completion of steps 1–3. This data were then segmented into information fragments and classified in the information framework described above.

The purpose of segmentation was to reduce the data into smaller chunks which would be easier to handle and could be analysed from an informational perspective. The challenge in

Table 7.2 Framework for informational activity classification

Information activity	Descriptor	Subject class	Medium	Level of abstraction	Level of detail
Generate	Alternative	Assembly	Text	Unlabelled	Conceptual
Access	Assumption	Component	Graphic	Labelled	Configurational
Analyse	Comparison	Connection	Audio	Associative	Detail
	Construction	Feature/Attribute	Video	Qualitative	
	Location	Recruitment	LIst	Quantitative	
	Operation	Design-concept	Simulation		
	Performance	Other	Gesture		
	Rationale		Picture		
	Relation				
	Requirement				
	Miscellaneous				

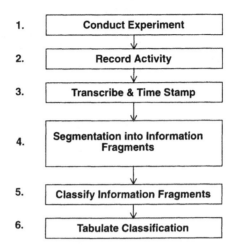

1. | Conduct Experiment
2. | Record Activity
3. | Transcribe & Time Stamp
4. | Segmentation into Information Fragments
5. | Classify Information Fragments
6. | Tabulate Classification

Figure 7.2
Flow chart of the
analysis procedure

segmenting is in objectively identifying the transitions in the protocol.Some transition points are easier to spot than others. We developed rules and guidelines for segmentation as we went along so as to maintain consistency in identifying the transition points. Typical points of transition were when there is a noticeable change in the designer's train of thought, or the designer changes his attention from one concept to another, or pauses in the protocol, or change in the informational activity, or utterance of certain phrases by the designer (such as 'and', 'OK', 'alright', 'emm'). Each segment is called an information fragment.

The next stage in the analysis is to classify each of the information fragments using the template shown in Figure 7.1. This was a very time consuming process as will be evident from the example in the next section. This also required some amount of training before we were consistent in our interpretation of the fragment. The next section shows the analysis of about one minute of the protocol data.

3.1 Analysis Example

Here is an example of segmentation and classification from about one minute of verbal protocol data. The data are shown in three different stages: after transcription, after segmentation and after classification.

After transcription:
OK I have spent er fortyfive minutes or so now forty minutes doing that er so em ... (mutter) going to check every possible location here

00:56:56:20

alright so we do have one that comes up front em really a little bit wary about backpacks on em fronts of bikes em let's see if we can em and there is that issue of it being off the side ... you know from the aesthetics standpoint everybody likes things symmetric and er this is not that big a pack em my initial tests indicated I probably couldn't have one right like that certainly I would not do it this way I would do it that way em in fact I would em maybe do it this way

00:57:51:24

em ... and em er so we have that as an option here em there would be a standard carrier

After segmentation:

00:56:56:20

alright so we do have one that comes up front

00:56:59:24

em really a little bit wary about backpacks on em fronts of bikes em

00:57:10:06

let's see if we can em and there is that issue of it being off the side...

00:57:23:26

you know from the aesthetics standpoint everybody likes things symmetric

00:57:30:05

and er this is not that big a pack

00:57:36:00

em my initial tests indicated I probably couldn't have one right like that [horizontal on the back]

00:57:42:14

certainly I would not do it this way [vertical in the back] I would do it that way [horizontal in the back] em in fact I would em maybe do it this way [horizontal in the back with opening on the outside]

00:57:51:24

After classification:

alright so we do have one that comes up front
    (info-fragment :activity ACCESS
                :time (00565620 00565924)
                :descriptor ALTERNATIVE

```
                            :subjclass DESIGN-CONCEPT
                            :subjects (front-position)
                            :medium VIDEO
                            :levels (CONCEPTUAL LABELLED)
                            :measures (1))
```
em really a little bit wary about backpacks on em fronts of bikes
em
```
        (info-fragment  :activity ANALYSE
                        :time (00565924 00571006)
                        :descriptor ALTERNATIVE
                        :subjclass DESIGN-CONCEPT
                        :subjects (front-position)
                        :medium AUDIO
                        :levels (CONCEPTUAL LABELLED)
                        :measures (1))
```
let's see if we can em and there ithat issue of it being off the
side...
```
        (info-fragment  :activity ACCESS
                        :time (00571006 00572326)
                        :descriptor ALTERNATIVE
                        :subjclass DESIGN-CONCEPT
                        :subjects (side-position)
                        :medium VIDEO
                        :levels (CONCEPTUAL LABELLED)
                        :measures (1))
```
you know from the aesthetics standpoint everybody likes
things symmetric
```
        (info-fragment  :activity GENERATE
                        :time (00572326 00573005)
                        :descriptor REQUIREMENT
                        :subjclass ASSEMBLY
                        :subjects (symmetry-for-looks)
                        :medium AUDIO
                        :levels (CONCEPTUAL LABELLED)
                        :measures (1))
```
and er this is not that big a pack
```
        (info-fragment  :activity ACCESS
                        :time (00573005 00573600)
                        :descriptor CONSTRUCTION
                        :subjclass FEATURE
                        :subjects (size-of-pack)
                        :medium AUDIO
```

```
                              :levels (CONFIGURATIONAL
                                 QUALITATIVE)
                              :measures (1))
          em my initial tests indicated I probably couldn't have one right
          like that [horizontal on the back]
              (info-fragment  :activity ACCESS
                              :time (00573600 00574214)
                              :descriptor ALTERNATIVE
                              :subjclass DESIGN-CONCEPT
                              :subjects (back-horizontal)
                              :medium VIDEO
                              :levels (CONCEPTUAL UNLABELLED)
                              :measures (1) )
          certainly I would not do it this way [vertical in the back]
              (info-fragment  :activity ANALYZE
                              :time (00574214 00574822)
                              :descriptor ALTERNATIVE
                              :subjclass DESIGN-CONCEPT
                              :subjects (back-vertical)
                              :medium VIDEO
                              :levels (CONCEPTUAL UNLABELLED)
                              :measures (1))
          I would do it that way [horizontal in the back] em in fact I
          would em maybe do it this way [horizontal in the back with
          opening on the outside]
              (info-fragment  :activity GENERATE
                              :time (00574822 00575124)
                              :descriptor COMPARISON
                              :subjclass DESIGN-CONCEPT
                              :subjects (back-horizontal back-horizontal-2)
                              :medium VIDEO
                              :levels (CONCEPTUAL UNLABELLED)
                              :measures (2))
```

4 Results

Results from the above analysis can be presented in many different ways. Since we were mainly interested in understanding the information handling behaviour of designers, we will discuss results which focus on those aspects.

4.1 On Informational Activities

Figure 7.3 shows the time varying behaviour of the designer between the three informational activities. For the sake of clarity

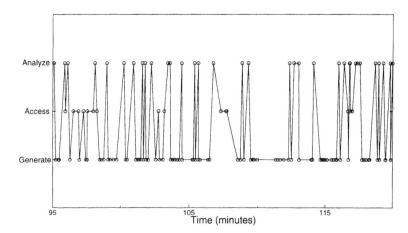

Figure 7.3
Movement of the
subject between the
different
informational
activities

the graph shows the activity between the times of 95 minutes and
120 minutes into the design experiment. A graph of the full 135
minutes would have been illegible here. However, Figure 7.3 is
representative of the activity in the other times as well. From the
graph it is clear that the designer moves fluidly between the three
stages, jumping back and forth, not spending much time on any
one type of activity. This lays down the requirement that information management tools should allow designers to move
fluidly between different informational activities.

Figure 7.4 shows the distribution of time among the three
activities. More than half the time is spent in *generating* information and more than a quarter in *accessing* information. This
confirms the belief that the conceptual design phase is mainly
generative and the designer spends a considerable amount of

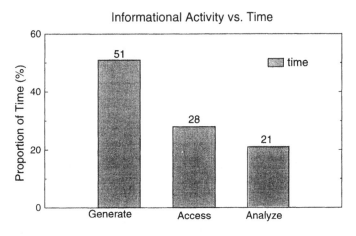

Figure 7.4
Distribution of
designer's time
between the three
activities

time accessing information. The behaviour is characteristically different between the times of 32 minutes and 45 minutes (not seen in Figure 7.3). Much of this time the designer was on the telephone and since we did not have the transcript of the person on the other end of the phone, we were unable to analyse this portion.

4.2 On Design Information Measure (dim)

The subject handled a total of 689 *dims* of information over a total time of 79.5 minutes. This time is cumulative of the duration of each information fragment (there were a total of 410 information fragments) and thus only includes the times during which the designer was actively involved in an informational activity.

The information handling rate for handling 689 *dims* over 79.5 minutes is 8.7 *dims*/min. This number compares well with the information handling rates measured by us in two other experiments (8.6 and 8.1 *dims*/min)[19]. This is an encouraging sign for the definition of information measure used in this analysis.

The similarities between the information handling rates across the experiments suggest that *dim* could be related to the information processing abilities of humans. This certainly needs further investigation; however, the knowledge of a parameter which relates to the cognitive ability of designers can be important in guiding the development of computational tools. On the other hand *dim* could simply be related to how fast the subject is talking out loud. Most humans when subjects in a verbal protocol experiment will tend to talk at similar rates so that what they are saying can be comprehended by others. However, *dim* is a semantic measure and is not measuring the words per minute being spoken by the subject. In fact the speaking rate for this experiment was about 204 words per minute, and for the two experiments we conducted earlier was 154 and 123 words per minute. Clearly the rates are quite different. This increases our confidence in the hypothesis that *dim* could be somehow related to the information processing capabilities of humans.

4.3 On Levels of Abstraction

Figure 7.5 shows the movement of the designer between the different *levels of abstraction* over time. Once again for clarity the graph only shows the behaviour over the time of 95 minutes to 120 minutes into the design experiment. It is clear from Figure 7.5 that the designer moves fluidly between the different levels of abstractions. This provides another requirement for an informa-

Time vs. Level of Abstraction

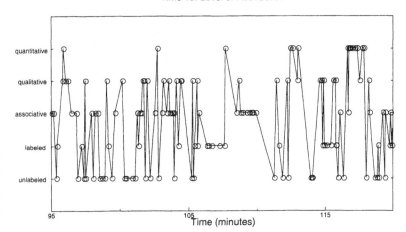

Figure 7.5
Movement of the
subject between the
various levels of
abstraction

tion management tool, which is that the tool should be able to represent and support information at all levels of abstraction.

Figure 7.6 shows the time distribution between the different levels. From the figure we can infer that designers deal with a lot of *non-quantitative* and *non-qualitative* information during the conceptual design phase. Most computational tools of today are good at representing information at the qualitative and quantitative level, but are poor at representing and dealing with information at unlabelled and associative levels. Therefore this result suggests that we need to be paying attention to representing

Level of Abstraction vs. Time

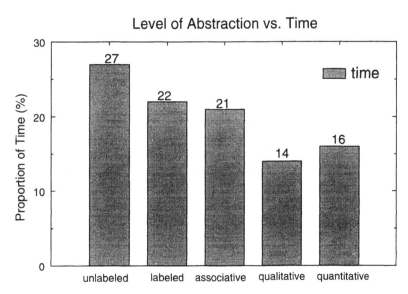

Figure 7.6
Distribution of time
among the various
levels of abstraction

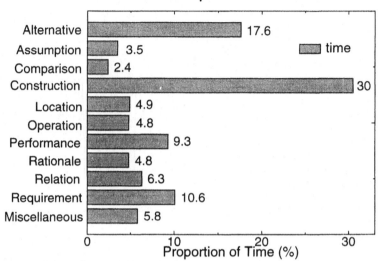

Figure 7.7
Distribution of time
among the various
descriptors

unlabelled and associative information in the development of an information management tool for conceptual design.

4.4 On Information Type

One objective in developing this analysis was to determine how designers use different types of information. Figure 7.7 gives the distribution of time among the various descriptors of information. The figure shows what type of information is handled more often than other types. Information which relates to *Alternative, Construction, Performance* and *Requirement* took the major share of the time. Knowledge of such a priority will be helpful in developing representations for the different types of information, especially when the descriptors are correlated with the level of abstraction and the medium of the information.

The results presented in this paper are from the first-order analysis of the data. Other results which we will not be able to include in this paper because of space constraint come from a second-order analysis of the data. More results come from studying the correlations between the informational activities and the various categories in the information classification framework (*Descriptor, Subject Class, Level of Abstraction*).

5 Summary

We have described a framework for analysing the information handling behaviour of designers. A definition for measuring the semantic content of design information (called *dim*, short for design information measure) has been suggested. Results pertaining to the rate at which design information is handled suggest that there can be a correlation between the rate and the information processing abilities of designers from a cognitive viewpoint. A definition for the level of abstraction of information is also introduced. Preliminary results give an encouraging insight into the information handling behaviour of designers. We have done similar analyses on two other experiments and plan to perform this analysis on further experiments, to build up a reasonable sample of experimental evidence, which can provide insights for developing information management tools needed to support conceptual design.

Acknowledgments

We would like to acknowledge many helpful discussions with Ade Mabogunje, Margot Brereton and David Cannon. We would also like to thank the AI Branch at NASA Ames Research Center for some of the equipment used to conduct this research.

References

1   Cross, N., *Engineering Design Methods*, Wiley, Chichester (1989)
2   National Materials Advisory Board, NMAB-455, *Enabling Technologies for Unified Life-Cycle Engineering of Structural Component*, National Academy Press, Washington, DC (1991)
3   Eastman, C.M., On the analysis of intuitive design processes, in G.T. Moore (Ed.), *Emerging Methods in Environmental Design and Planning*, MIT Press, Cambridge, MA (1970)
4   Akin, Ö., An exploration of the design process, *Design Methods and Theories*, **13**(3–4) (1979) 115–119. Also published in N. Cross (Ed.), *Developments in Design Methodology*, Wiley, Chichester (1984) 33–56
5   Guindon, R., Designing the design process: exploiting opportunistic thoughts, *Human–Computer Interaction*, **5** (1990) 305–344
6   Stauffer, L.A., D.G. Ullman and T.G. Dietterich, Protocol analysis of mechanical engineering design, *Proc. 1987 Int. Conf. on Engineering Design*, Boston, MA (1987)
7   Stauffer, L.A., *An empirical study on the process of mechanical design*, PhD thesis, Oregon State University (1987)
8   Tang, J.C., *Listing, drawing, and gesturing in design: a study of the use of shared workspaces by design teams* PhD thesis, Department of Mechanical Engineering, Stanford University (1989)
9   Minneman, S.L., *The social construction of a technical reality: empirical studies of group engineering design*, PhD thesis, Department of Mechanical Engineering, Stanford University (1991)
10 Minneman, S.L. and S. Bly, Experiences in the development of a multi-user drawing tool, *Proc. Third Guelph Symposium on Computer Mediated Communication*, Guelph, Ontario 154–167 (1990)
11 Tang, J. C. and S.L. Minneman, Videowhiteboard: video shadows to

support remote collaboration, *Proc. Conf. on Computer Human Interaction (CHI) '91*, New Orleans (1991) 315–322

12 Baya, V., J. Gevins, C. Baudin, A. Mabogunje, G. Toye, and L. Leifer, An experimental study of design information reuse, *Proc. 4th Int. Conf. on Design Theory and Methodology*, ASME, Scottsdale, AZ (1992) 141–147

13 Baudin, C., J.G. Underwood, and V. Baya, Using device models to facilitate the retrieval of multimedia design information, *Proc. 13th Int. Joint Conf. on Artificial Intelligence*, Chambéry, France (1993) 1237–1243

14 Baudin, C., S. Kedar, J.G. Underwood and V. Baya, Question-based acquisition of conceptual indices for multimedia design documentation, *Proc. 11th Nat. Conf. on Artificial Intelligence AAAI-93*, Washington, DC (1993) 452–458

15 Fulton, R.E. and J.I. Craig, *Report from a Workshop on 'Information Framework Technology for Integrated Design/Engineering Systems'*, Sponsored by NSF and Georgia Institute of Technology, Callaway Gardens, GA (1989)

16 Shannon, C.E. and W. Weaver, *The Mathematical Theory of Communication*, University of Illinois, Urbana (1949)

17 Kim, S.H., *Designing Intelligence: A Framework for Smart Systems*, Oxford University Press, New York (1990)

18 Suh, N.P., *The Principles of Design*, Oxford University Press, New York (1990)

19 Baya, V. and L. Leifer, A study of the information handling behavior of designers during conceptual design, *Proc. 6th Int. Conf. on Design Theory and Methodology*, ASME, Minneapolis, MN (1994) 153–160

20 Ullman, D.G., Design histories: archiving the evolution of products, *Proc. DARPA Workshop on Manufacturing*, Salt Lake City, UT (1991)

8 Analysis of Protocol Data to Identify Product Information Evolution and Decision Making Process

David G. Ullman, Derald Herling and **Alex Sinton**
Oregon State University, Corvallis, OR, USA

This paper focuses on the team design effort. In watching the videotaped session and studying the other data many types of information can seen, extracted and analysed. Experience has shown that it is necessary to analyse data of this type with an initial model of the information. The protocol information can only be used to verify, refine, refute or otherwise affect the initial model. In this reduction of the data, models of information evolution and structure, and of decision making, are tested and refined. Details on these models and their origin are given in Section 1. The procedure used in reducing the data is explained in Section 2. Section 3 presents details of the extracts of protocol studied. Section 4 contains general observations, our results. Many of these results agree with others that have been previously published so the models were verified. Additionally, there are new findings that improve the understanding of team design. Conclusions and follow-on needs are discussed in Section 5.

1 Goal

The goal of this protocol study is to verify and refine models of product information evolution and the decision making process. These are both areas of importance in the development of advanced computer aided design tools[1], an area of long term and continuing research at Oregon State University. Note that the goal focuses on both the information that describes the artifact during its evolution from a need to a final set of product specifications and the design process that supported this evolution. In earlier work based on individual protocols, the evolution of product information was studied and modelled[2]. This model has continued to evolve since the publication of the earlier paper.

Information about the product can be modelled by two basic constructs, the features of objects (object-attribute-value) and the features of relations between objects (object 1-object 2-relationship-attribute-value). Here an 'object' is defined as an assembly, a component, a feature, a human or other identifiable physical thing that is used to describe some physical aspect of the product being designed. This model differs from most others in that it emphasizes the importance of relationships between objects in the evolution of components. In fact, much of what is commonly referred to as the functionality of a component is realized through the relationships between objects[3]. Typical attributes (i.e. features) of objects include geometry, energy, information and material state, physical properties, material properties, production information, etc.

The protocol data will be used to show the evolution of a single object, the 'stay' in terms of the development of its attributes and values. For relationships between two objects, typical attributes are:

- Position Relationship: What is their relative location and orientation?
- Connection Relationship: How are they physically connected? (degrees of freedom, type of connection, rigidity of connection, part-of)
- Transmission Relationship: What is transmitted between them? (energy, materials or information transmission)

This relationship model will be used to explain part of the evolution of the stay and will be the focus of discussion about the location and orientation of the pack relative to the bicycle (i.e. its position relationship). The decisions involved in determining the location and orientation will be modelled as described below.

The second half of the goal, to study the decision making process is based on a fairly simple model of decision making. This model has been used with success by a number of researchers[4-6]. In this model, the major construct is an issue. A design issue is any question or problem identified by the designers that needs to be resolved. An issue may be the need to fulfil design requirements (i.e. criteria, constraint), the need to establish a value for some design object parameter, or the desire to determine the next part of the design problem to work on. Some important points about issues are: (1) issues are often dependent on the resolution of other issues; (2) while focused on an issue, sub-issues are developed; and (3) issues can be focused on the product (the object being designed) or on the design process (e.g. what to do next) and these two types are often interwoven with each other.

The source of the criteria has proved to be important in understanding the development of information in a design session. In earlier work[2] three sources of criteria were identified. *Given criteria* are imposed on the problem from outside the design space. They may be part of the original problem statement or imposed on the design problem by the solution of other components of the problem not under the control of the designer responsible for the problem under consideration. *Introduced criteria* are those criteria brought to the problem by the designer based on specific domain knowledge. *Derived criteria* arise as a consequence of earlier design decisions. Each change in the objects or their relationships puts new constraints on all downstream decisions. Thus, what seems like an unimportant decision early in the design process can have major effects later on in the design process.

It is interesting to note the frequency with which these different types of constraints appear during a design process. In a study that traced the evolution of a component[2], the types of criteria used during conceptual, layout and detail design activities were identified (Table 8.1). The values were based on 725 constraints used during the development of the component. As can be seen from the data in Table 8.1, given constraints rapidly give way in importance to derived constraints.

Table 8.1 Constraint types found in earlier empirical study

Design phase	Derived	Given	Introduced
Conceptual	29%	70%	1%
Layout	79%	10%	11%
Detail	84%	7%	8%
Total	75%	17%	8%

During an issue a number of *alternative solutions* are proposed. An alternative is an option generated by the design team for the resolution of a particular design issue. Any number of alternatives may be developed to resolve a design issue. Some important points about alternatives are: (1) alternatives are developed by various members of the design team; (2) the information known about an alternative is often incomplete; (3) alternatives are often dependent on the resolution of other issues or alternatives; (4) often an alternative will introduce new sub-issues; (5) alternatives to solve one issue may be inconsistent with those to solve another issue; (6) alternatives can be at different levels of abstraction (e.g. some are vague ideas and others refined, virtually completed designs) and language (e.g. text, sketches, physical objects). Additionally, each alternative is put forth with varying levels of knowledge about how well it will satisfy the criteria.

Arguments are given for and against the alternatives. An argument is the rationale for either supporting or opposing a particular alternative. Arguments perform evaluation as they are comparisons that identify the relative merits or demerits of a particular alternative. These comparisons are made either between alternatives relative to a criterion or between an alternative and a criterion (see page 169 in Reference 7 for a discussion). Some important points about arguments are: (1) arguments are made with varying levels of confidence, knowledge and abstraction; (2) arguments are posed by various members of the design team; and (3) arguments may be inconsistent with each other.

Alternatives and arguments raise new issues. These new issues increase the interdependence of the information and must be addressed in future activities. Information interdependence is most common during the design process. In a study of how a single component came into existence[2] researchers found that very early in the design process 70% of the information used in arguments for resolving issues was from the initial specifications. By the end of the project, however, 75% of this information on the shape, dimensions, tolerances, and function of the components was based on the result of design decisions. Most of the decisions that specified the final component were made on issues and their sub-issues that did not exist initially and a majority of the effort was spent on interdependent design issues.

A *decision* weighs the arguments supporting or opposing the alternatives for a particular issue. The effect of the decision is either to accept, reject, or modify an alternative, or to leave the issue pending while others are addressed. The history of the design process is the chronological sequence of decisions that relate progress on the design project[6]. The same issue revisited after gathering more information may yield a different decision. Here, the first decision on the original issue will probably be to gather more information about the alternatives and arguments, have another meeting in a couple of days and make a refined decision at that time. Thus, decision making is a dynamic activity.

Issue, alternatives and criteria can all be modelled as object-attribute-value or Object 1-Object 2-relationship-attribute-value information. This will be shown in the observations from the protocol studied here.

2 Procedure

The video tape and other material were viewed and analysed by all three authors. Based on original viewing each made a chronological list of major foci during the protocol. Based on this initial review it was decided to focus all further efforts on issues addressing the location and orientation of the pack relative to the bike and the development of the rack's rear stay.

Within the data, references to the location and orientation of the pack, or to the development of the stay were reduced to extract:

Time:
The time stamp on the video

Alternative/Criteria:
Either an alternative that is proposed as a possible design solution to the issue being discussed or a criterion to be used for evaluation. Alternatives were broken down as either positive examples (e.g. 'yeah I think what they're doing right now is most similar to the sorta child seat kind of idea') or negative examples (e.g. 'OK well (unintelligible) er some something comes to mind which threw out Velcro because it er happens to ...'). Criteria were either:

cg = criteria given
cd = criteria derived
ci = criteria introduced

Abstraction level:

For an alternative the abstraction level was labelled as:

h = highly refined example (e.g. 'they've got this em Batavus Buster')

m = medium refined example (e.g. 'it's like an old bike basket that way like the Wizard of Oz')

a = abstractly refined example (e.g. 'it'd be cool if em this rack was used for something else like you take your backpack off and then this rack you can still put stuff on it')

For a criterion the abstraction level was labelled as:

a = abstract (e.g. 'it must be quick to remove')

c = concrete (e.g. 'it must cost between $45 to $50')

Action:

Refers to what action is being taken with the information;

a = assimilating the information (i.e. bringing new information into the design space)

s = specifying (i.e. performing design work on objects or relationships)

p = planning (i.e. figuring out what to do next)

Objects:

Refers to objects directly referenced or inferred in the information being discussed (e.g. the relationship between pack and bike, the rack, the rack stay).

Information form:

Refers to the media in which the information is contained;

g = graphical, meaning drawings or sketches

m = movements by the person or physical gestures

p = physical artifact

t = textual

v = verbal

The issues for determining the location and orientation of the pack on the bike concern making decisions about two of the attributes of the relationship between these two objects. The issue of refining the rear stay concerns the evolution of this object, its attributes and values for these attributes. All information associated with these issues was drawn from the protocols at the sentence, gesture or sketch level of detail. This resulted in 91 items for the location issue, 63 for the orientation and 214 for rear stay issues.

Note that not all the information generated from the data was used in the results given in the next section. It is common with protocol reductions to generate more information than can be categorized or understood. However, a number of observations have been developed based on this limited data.

3 Reduction Results

Reduction of the protocol information led to the verification and extension of the models described in the goals section of this paper. Before giving observations based on these data, the development of the stay (an object) and location and orientation of the pack on the bicycle (attributes of a relationship) need to be detailed.

3.1 Stay Development

The development of the stay, the brace that holds the rack up was chosen for detailed study of a specific object. The best sketch of the final stay configuration is shown in Figure 8.1. This figure was extracted from an isometric layout of the entire product and shows an alternative position and estimated stay length. This is a simple part made of formed and drilled aluminium tube. To get to this final configuration 29 different issues were addressed as itemized in Table 8.2.

Table 8.2 lists the issues addressed and when they were

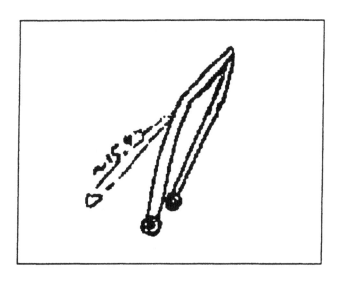

Figure 8.1
Final sketch of stay

Table 8.2 Stay development issues

Group	No.	Issue	Time (h:min)
General issues	1	Specify number of stays	:55, 1:35, 1:36
	2	Assimilate Blackburn racks	1:02, 1:32
	3	Assimilate Batavus prototype	:19, :44
	4	Assimilate Batavus bike geometry	:55, :57, 1:06, 1:27
	5	Specify general concepts for rack	1:08, 1:10
	6	Specify rack folding mechanism	:10, :50, 1:35–:36
	7	Specify stay length adjustability	:46, :57, 1:02, 1:03
Specify relationships	8	Specify ease of stay removal from bike	:37, :55
	9	Specify stays as part of pack frame	:35, :43, :52, 1:21
	10	Specify rack tray–stay interface	1:36, 1:37, 1:51
	11	Specify stay–bike frame interface	:37, :54, :55, :58, 1:08, 1:31, 1:32, 1:37
Specify stock material	12	Specify stay tube OD	1:36, 2:02
	13	Specify stay tube ID	1:36, 2:02
	14	Specify stay tube stock length	1:34, 1:43, 1:51
	15	Specify material for stay	1:00, 1:15, 1:52
Specify formed part	16	Specify stay finish	1:15, 1:53
	17	Specify manufacture of stay ends	1:52
	18	Specify manufacture of bend	1:53
	19	Specify geometry of bend	1:34
	20	Specify stay width	1:34, 1:51
	21	Specify stay length	1:50, 1:51
	22	Specify hole location	1:50
	23	Specify hole geometry	1:32
	24	Specify hole diameter	1:50
	25	Specify hole manufacture	1:53
Analyse	26	Analyse stay strength	1:05, 1:15, 1:07, 1:08
	27	Analyse stay cost	1:52, 1:53
	28	Analyse stay weight	2:02
	29	Analyse stay stiffness	1:05, 1:07

addressed during the protocol. All the issues are modelled as assimilate, specify or analyse. *Assimilate* issues are focused on bringing information into the design space. Typically assimilation issues are used to gather needed information from design specifications, examples, or library sources. *Specify* issues are the main 'design' issues where objects and relationships are created and refined. In *analyse* issues information generated by other issues is used to compare with criteria that are broader than the single attribute. Additionally, there are often planning issues focused on deciding the issues and their ordering. Planning for the development of the stay was *ad hoc*.

The issues are divided into five main groups. There is a group of general issues which focus on understanding the need for the stay (assimilating information) and specifying attributes of the stay that include considerations broader than the stay itself. Interface issues are concerned with the stay's relationship with other objects (e.g. the tray and the bike frame). Note that Issue 9 was focused on the stay as part of the pack frame. This issue was considered repeatedly, but not used in the final design. The third and fourth groups of issues were oriented to specifying the raw stock material and the final, formed part. These issues are focused on specifying the attributes or values for the features or attributes of the stay. Note that other than some work on the material and finish of the stay (issue 15) work on the actual details did not begin until time 1:34, three-quarters of the way through the session. The team had little time to generate alternatives and criteria and to refine or optimize the decisions made. The final group of issues are those focused on analysis using the values determined in other issues to generate information needed for comparison on criteria such as weight and cost.

3.2 Location and Orientation Development

The issues focusing on specification of the orientation and location attributes of the relationship between the pack and the bicycle were visited repeatedly during the :11–:59 minute period. It is interesting to note that even though the team seems to agree on the orientation during this period (the pack is horizontal with the base to the rear), it is evident in minute 1:34 that subject J does not share in this decision. At this time he seems confused about the orientation, leading one to wonder exactly what mental model he had been working within the intervening half-hour.

The issues of location and orientation attribute specification were extracted for study (Table 8.3) because (1) they are central to the product development; (2) they were easily traceable in the protocol; (3) they should indicate the evolution of new, derived criteria that affect the rest of the design activity; and (4) they demonstrate the impact of criteria development on the design process and the final product produced.

These criteria were developed from either the original problem statement (i.e. given) or the supplemental information (also given). Some were recorded on the whiteboard or the drawing paper (i.e. recorded) and others just discussed verbally (i.e. verbal).

Table 8.3 Location and orientation issue criteria listing

Issue	No.	Criterion	Time
Pack–bike location value development	1	Rack–bike position and connection relationship attributes must conform to bicycle geometry	:11, :20, :35, :40, :46, :53, :58
	2	Bike/pack/rack system centre of gravity attribute must be as low as possible	:27, :48, :49, :59
	3	Rack–bike connection relationship attribute must occur in only one way	:16
	4	Pack/rack system–bike rear wheel position relationship attribute must show clearance (no dangling parts to get caught in wheel)	:31
	5	Water picked up by his bike wheel–rack/pack system position relationship must show interference	:31
	6	User–bike type of connection relationship must yield rider comfort	:22, :27, :28
	7	Pack/rack system weight attribute must be minimal	:28
	8	User leg–bike/pack/rack system position relationship attribute must not interfere when mounting/dismounting bicycle overpack	:59
Pack–bike orientation value development	1	Rack–pack connection relationship attribute must conform to HiStar pack geometry	:11
	2	Pack–bike position relationship attribute must be constant while riding	:15
	3	Rack–bike connection relationship attribute must install/attach in only one way	:16
	4	Rack/pack system–bike connection relationship attribute must conform to bicycle geometry	:20
	5	User satisfaction attribute is high	:22
	6	Pack–contents position relationship attribute is inside pack	:42, :58

To investigate the impact of the criteria on the final product the number of references to each was recorded. For the issue of location these are divided by source and media of recording in Table 8.4. Also, in the last column is an indication of whether or not the criterion was met in the final design. For two of the criteria this is not clear and for no. 8 the researchers disagree depending on whether or not the new design is considered relative to the Batavus example or on an absolute basis. As the

Table 8.4 Location issue criteria count data

No.	Given	Introduced	Recorded	Verbal	Met?
1	1	2	2	10	yes
2	0	1	0	4	no
3	1	1	1	1	?
4	0	2	1	1	yes
5	0	0	0	1	yes
6	0	0	0	3	?
7	0	0	0	1	yes
8	0	1	0	1	yes/no

criteria were so poorly developed, the exact design target is not clear.

From Table 8.4 it can be seen that criteria 1, 2, 3, 4 and 8 are given or introduced criteria. Thus, they should have been the primary criteria used for the final design selection. But only two, criteria 1 and 4, were satisfied in the final design. This raises concerns as to the robust nature of the design selected.

The alternatives that the team proposed for the issue of location were listed on the whiteboard for consideration. They were:

Do nothing – wear it
Inner diamond
Front wheels (bike basket)
Gas tank (Harley)
Hang off handlebars
Panniers – front
Panniers – rear
Saddlebags (fold in middle)
Rear – horizontal rearward
      – horizontal forward

These alternatives were generated but no formal evaluation was made relative to the criteria. This lack raises concerns about the final design. Is it the most satisfactory design possible?

4 Observations

In this section observations made about the data are stated and then information that led to the observation is presented. Data from this protocol and protocol studies in general are too sparse to develop rules or axioms.

Observation 1. Design activities can be modelled as work on specific issues. Issues are either to assimilate information, specify objects or relationships between objects, analyse developed information, or plan.

Issues focused on the relationships between objects appear to play a major role in the design process. The location and orientation issues described above are the focus of 43% of the activity during the :11–:59 period. Even during the design of an individual component, the stay, 17/69 (25%) of the issues were focused on the relationships. In general these issues were resolved prior to extensive effort on the details of the stay itself (relationship issues occur in the :35–1:37 period, stock material and forming issues occur primarily after 1:34). This is typical of what is seen in other protocols[2]. Relationships determine how a system will interact with its given environment (the location and orientation of the pack relative to the bicycle) or how a component (or assembly) will interact with its environment.

Observation 2. Issues that focus on relationships between objects are essential parts of the design process. For the attributes of a relationship (i.e. position, connection and transmission) the model object 1–object 2–relationship–attribute–value well represents the information in the protocol.

There were 14 issues used to specify the details of the stay (nos. 12–25 in Table 8.2). There is one for each attribute of the stay. The goal of each issue is to specify a value for the attribute. For most of these, this is clear from the issue description. However, issues 17 and 18, specification of the manufacturing process, may be further refined into sub-issues before being defined sufficiently for values to be determined. In this protocol too little time was available for the team to refine these specification issues. It is to the team's credit that they at least mentioned major issues that would need to be considered if time permitted.

Observation 3. To specify design objects (systems, assemblies and components) the object–attribute–value model seems sufficient.

Assimilation issues shown in Table 8.2 (nos. 2–4) are focused on other existing hardware. These are not the only issues where information is brought into the design problem. Information was also assimilated during the execution of issues 27 and 28, where an effort was made to ascertain the cost and weight criteria.

Additionally, there was some reference to the information provided during the specification issues as well.

Observation 4. Information is assimilated into the design space during all types of issues. However, some issues are specifically focused on the assimilation.

Like many design teams, the subjects did a poor job of problem understanding. Their assimilation of information (the basis for criteria development) was *ad-hoc*, spread throughout the session and with no formal organization. It has been found that using an organized process such as Quality Function Deployment (QFD) greatly enhances the understanding of the problem and the quality of the results[6–8]. This is not to say that all requirements can be determined at the beginning of a design session, but that care in developing and refining design criteria is essential. Understanding given criteria can be far from uniform among team members. Efforts based on poorly understood criteria lead to unfocused design. For example, in the exploration of the trailer concept (1:01), although this exploration could be justified on the grounds that it assisted future design efforts, it was in clear violation of given criteria for the product and consumed valuable time.

Observation 5. Care needs to be exercised on the development and tracking of given, introduced and derived criteria.

The design team did not have a consistent picture of the criteria. For example, when they considered design solutions for the product that attaches the pack to the bike, they began exploring solutions that would require redesigning the pack configuration into a two-piece, split pack (0:29), which is clearly not what the customer wanted. Additionally, inconsistent criteria occurred when members missed data presented (0:08–0:10) and laboured under different assumptions. J thought they were going to make internal frame packs after the experimenter had stated 'the HiStar backpack which is ... to be designed for'. This lack of establishing an agreed criteria set is not uncommon. In general the use of words and design sketches can carry different meanings in the minds of the designers and criteria understanding can be far from uniform among members.

As seen on the whiteboard, criteria for the issues of location

Table 8.5 Criteria source results

	Derived	Given	Introduced
From Current Protocol, location and orientation issues	27%	60%	13%
From McGinnis and Ullman[2] Conceptual Design Phase	29%	70%	1%

and orientation were itemized as:

- Functional specifications: idiot proof (criteria location and orientation – install/attach in only one way)
- Constraints: bicycle geometry (criteria location and orientation – conform to bicycle geometry); backpack is given (criterion orientation – conform to HiStar pack)
- Features: straps in spokes (not) (criterion location – no dangling parts)
- Concept positions: kick stand (criterion location – conform to bicycle geometry)

It is worth comparing the current data to that in McGinnis and Ullman[2], the only other study to track the evolution of the design criteria. As shown in Table 8.5, there is a strong correlation between the two results. Data in the current protocol were not sufficient to compare with other data from McGinnis and Ullman.

Observation 6. Development of multiple alternatives was seen early in the design process. The number diminished as time pressure mounted.

In earlier protocol studies we noted that designers often generate only a single alternative for a specification issue[9]. In this session, however, the team did generate multiple ideas for all the general specification issues. These determined the concept that they were going to pursue. They appeared to run out of time when they specified the stock material and formed part as there was only a single alternative per issue here. Generally, better design results are found when the alternative space is explored[10].

Observation 7. Issues are dynamic in that they come into existence as needed and their relationships can change with the evolution of the product.

In most project planning the issues and their organization are considered static (e.g. project planning, CPM). In order to be able to structure a set of issues, the information needed by each and the expected results of each must be known. However, when issues are considered on a finer scale, as is done here, then the scene becomes much more dynamic. For example, issue no. 13, 'Specify stay tube internal diameter' does not even exist until 1:36 into the protocol. At this time a hollow tube is chosen for the stay material. This issue would not even have occurred if solid material or another shape had been selected. The value for the attribute 'internal diameter' is dependent on the results of issues nos. 12, 15, 17, 19 and 21. Additionally, this value is used in the analysis issues 26 and 29. The interdependency can change if one or more attributes changes.

Observation 8. The criteria evolve as the design space under consideration changes during the design effort.

The need to add a design space definition to the criteria structure is evidenced by the evolution of the criterion for a low centre of gravity. The original criterion of low centre of gravity for the design space of the bike/rack/pack system encouraged the design team to explore locating the pack in many locations on the bicycle (e.g. at the front, in the triangle area, beside the front wheel and beside the rear wheel). This exploration can be modelled as exploring the relationships between the bicycle, rack and pack to affect the value of the centre of gravity attribute of the bike/rack/pack system. As the team made the decision to locate the pack behind the seat, the criterion of low centre of gravity became one of optimizing for lowest centre of gravity for the system with a new, derived criterion (i.e. constraint). This criterion, the value of the relationship between the pack and the bicycle, fixed with the pack behind the seat and above the wheel area, now reduces the design space.

5 Conclusions and Comments on the Exercise

The analysis of this protocol clearly shows a typical example of information evolution and decision making during the design process. It demonstrates the usefulness in modelling design information based on objects, relationships, issues, alternatives, evaluations and decisions. Based on this model and extensive review of the protocol data, eight observations have been made.

Many of these have been made about other data[2] and thus this exercise has served to verify an existing model and to extend it. The validity of this model is critically important for two areas of future computer aided design tool development; recording correct and concise information for design rationale systems[11] and designing usable engineering decision software tools.

It is worth noting that at 1:22 J is recorded as stating 'I think all design eventually comes down to a popularity contest.' This is a clear challenge to the design methodology research community. Can research such as that undertaken in this workshop actually improve the design process, leading to improved products in less time?

Finally, we would like to add a few general comments on the exercise. The time given to the subjects was too short to accomplish anything realistic. They ran out of time and never really got into the detail information refinement that is the focus of most current design aids such as CAD. This means that certain types of important team behaviours and types of information development were not exemplified in this protocol. It is suggested for a further exercise that a longer, more in-depth protocol be used.

References

1 Ullman D.G., Design history, design rationale and design intent system development issues, *Proc. 6th Int. Conf.* on Design Theory and Methodology, ASME, Minneapolis, MN (1994)

2 McGinnis, B. and D.G. Ullman, The evolution of commitments in the design of a component, *Journal of Mechanical Design*, **144** (March 1992) 1–7

3 Ullman, D.G., A new view on function modeling, *Proc. Int. Conf. on Engineering Design* (ICED) The Hague, (August 1993) pp. 21–28

4 Rittel, H.W.J. and M.M. Webber, Dilemmas in a general theory of planning, *Policy Sciences*, **4** (1973) 155–169

5 Yakemovic, K.B. and J. Conklin, Report on a development project use of an issue-based information system, *MCC Technical Number STP-247-90* (June, 1990)

6 Nagy, R.L. and D.G. Ullman, A data representation for collaborative mechanical design, *Research in Engineering Design*, 3(4) (1992) 233–242

7 Ullman, D.G., *The Mechanical Design Process*, McGraw Hill, New York (1992)

8 Hunter-Zaworski, K.M. and D.G. Ullman, Mechanics of mobility aid securement and restraint on public transportation vehicles, *Transportation Research Record No. 1378*, National Academy Press, Washington, DC (1993)

9 Clausing, D., *Total Quality Development*, ASME Press, New York (1994)

10 Ullman, D.G., T.G. Dietterich and L. Stauffer, A model of the mechanical design process based on empirical data, *Artificial Intelligence in Engineering, Design, and Manufacturing*, **2**(1) (1988) 33–52, 1988

11 Dylla, N., *Denk- und Handlungsablaufe beim Konstruieren*, Doktor-Ingenieurs Dissertation, Technischen Universität München, Munchen (1990)

9 Analysis of Design Protocol by Functional Evolution Process Model

Hideaki Takeda[1], Masaharu Yoshioka[2], Tetsuo Tomiyama[2] and Yoshiki Shimomura[3]
[1]*Nara Institute of Science and Technology, Japan*
[2]*University of Tokyo, Japan*
[3]*Mita Industrial Co. Ltd., Japan*

Function is a key concept in design because ideally design is a process in which an object is realized from its functionality[1]. Although function is a well-known concept, its definition has always been vague.

There are three roles for function in design. Function is used firstly as a modelling language by which designers can compose and develop their requirements. Secondly, it also serves in object representations that can connect requirements and objects. And thirdly, after the development and deliberation of objects, function is used to evaluate objects to assess how well their purpose is satisfied. In order to support these three roles of function in design, we define function using state and behaviour (FBS: Function–Behaviour–Structure modelling)[2]. Also we operate the model using relations between functions so that function can be evolved gradually (FEP: Functional Evolution Process).

In the following sections, we will show our model of function, and then apply this model to the protocol analysis. We will explain FBS modelling in Section 2 and FEP modelling in Section 3. Then we will show results of the analysis of design process of the team design by FBS/FEP modelling.

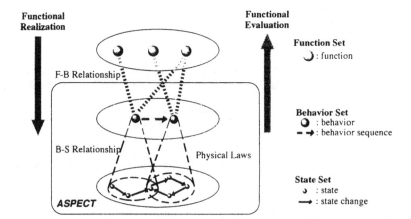

Figure 9.1
Relationship among
Function,
Behaviour, and
State

1 Function-Behaviour-State Modelling

There are many approaches to represent function, but there is an underlying problem, which is that function and behaviour are confused and mixed. Behaviour can be directly derived from the state of objects and their environment, while function is not only related to the state of objects and their environment but also related to the perception of the object by designers. For example, consider function of a car. Some people may say one of its functions is 'moving', others may say 'carrying', and others 'enclosing', even if they observe the same behaviour. Therefore we distinguish function, behaviour, and state levels in object representation (see Figure 9.1).

The state level is represented by entities, attributes of entities, and relations among entities. Entities are identifiers of objects. Attributes of the entities and relations among the entities represent structures and their states composed by the objects. A structure is part of a state that exists relatively permanently. Behaviour is then defined as the 'sequential change of states of objects'. In the physical world, changes of states of objects are governed by physical laws. We call this set of definitions of state and behaviour an *aspect*, that is a basic unit of object representation. An aspect consists of definitions of terms and entities (structure) and rules (physical laws). Design has many kinds of aspects from well-defined aspects (e.g., rigid body dynamics) to ill- or vaguely-defined aspects (e.g. manufacturability).

While behaviour is grounded on the state and within the scope of the aspect, function is indirectly related to the state and not in the scope of an aspect. We define function as 'a description of

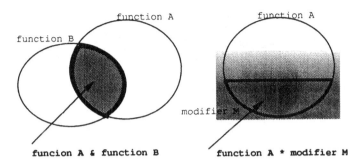

Figure 9.2
Difference between
function and
modifier

behaviour abstracted through the recognition of that behaviour for utilization'. Function is thus defined on a chunk of behaviour (or behaviour itself). There are many possible chunks of behaviour, but only some of them are meaningful to designers when they recognize and design objects.

Although function is not included in an aspect, most functions are associated with aspects, because behaviours that a function is based on often fall within a single aspect. In other words, an aspect has a set of associated functions which can serve as a description of the aspect from the viewpoint of utilization.

We define the structure of a function as follows. A *function* is represented as a combination of a *function body, objective entities*, and *functional modifiers*. A *function body* is a symbol that carries meaning of the function. A typical function body is a verb like 'move' or 'carry'. We often refer to the function body as function for short in this paper. An *objective entity* is an entity that the function occurs on or to. It should be realized as an object in a state level until the end of design. A *functional modifier* is a symbol that restricts function in order to match the functionality with the designers' intention. A typical functional modifier is an adverb like 'precisely' or 'firmly'.

The difference between a function body and a functional modifier is the degree of satisfaction (see Figure 9.2). Satisfaction of function is usually 'yes' or 'no', that is, we recognize whether a function exists in an object or not. There are no intermediate states. On the other hand, a functional modifier has a degree of satisfaction. We can judge how much 'attach firmly' is achieved in an object. We can also compare two objects by the degree of satisfaction of functional modifiers. In other words, a functional modifier is a criterion to characterize how a function is achieved. A function has as many criteria as its functional modifiers.

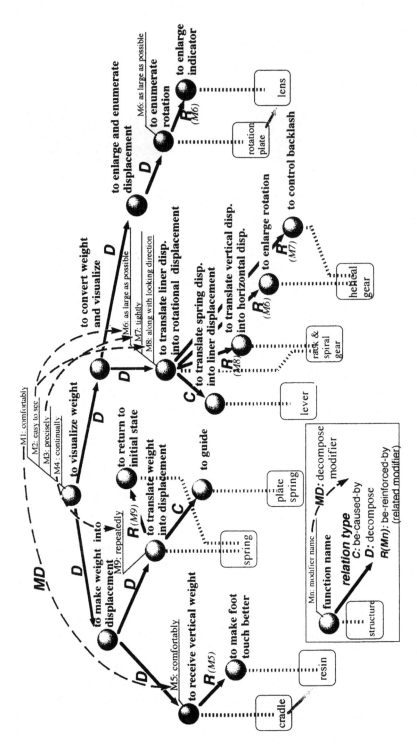

Figure 9.3 Functional evolution in design of a weighing scale

2 Functional Evolution Process

In the previous section, we clarified the role of function in object representation. The functional description of an object is its intentional representation, while behaviour and state are objective representations. In this section, we focus on how to use functions in design.

One of the three roles of function is as a language to describe requirements. Requirements are usually incomplete at the beginning of a design process. Details of requirements are realized according to detailing of object descriptions, i.e., function is also detailed during design processes. We call this detailing process of function the *functional evolution process*.

In order to represent functional evolution processes, we define relations between functions and functional modifiers, and then model how the functional evolution process is achieved in design. We provide three types of relations between functions, i.e. *decompose*, *be-caused-by*, and *be-reinforced-by* relations, and one type of relation between functional modifiers, i.e. *decompose-modifiers* relation. Functional evolution is achieved either by generating functions or functional modifiers or by forging new relations among functions and functional modifiers.

Figure 9.3 shows an example of the functional evolution graph of a weighing scale design project, obtained by protocol analysis[3].

Decompose

Designers typically divide a function into sub-functions (see Figure 9.4). For example, Function 'to visualize weight' is decomposed into Function 'to make weight into displacement' and Function 'to convert weight and visualize' (see Figure 9.3). Although this functional relation often also encourages the decomposition of the designed structures, it is not necessary to decompose structures according to the functional decomposition.

Be-caused-by

This relation means that a new function B is needed to realize a function A. In other words, B is a necessary condition for A. This relation should be supported by a causal relation on the behavioural level.

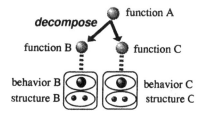

Figure 9.4
Decomposing
evolution

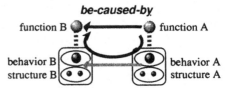

Figure 9.5
Causal evolution

This functional relation can be found as follows; first the designers would find the behaviour associated to the given function. Then they would find causal behaviours to the behaviour by using causal simulation (e.g. Qualitative Simulation[4]). Finally the designers would obtain functions associated to these behaviours (see Figure 9.5). For example, Function 'to translate weight into displacement' invokes Structure 'spring'. Then, by mental simulation, the designers find that a new structure such as 'plate spring' is needed 'to guide' spring. Function 'to guide' is then found through the behaviour and state levels (see Figure 9.3).

Be-reinforced-by

This means that a new function B is recommended to exist in order to realize function A properly. In this case, B is not a necessary condition for A, since A can exist without B. But with B, A would better accomplish its functionality (see Figure 9.6).

This relation can be generated as a result of the interpretation of functional modifiers. When a behaviour or a structure is proposed to satisfy function A, designers can evaluate this behaviour or structure using modifier M_A attached to function A. Although the behaviour or structure is a proposal to satisfy the function, its evaluation by modifier M_A as an evaluating criterion tells the designers what is missing from it. Then they can find a new function to support the missing factor, which is linked to the original function with a *be-reinforced-by* relation.

For example, Function 'to enumerate rotation' invokes the Structure 'rotation plate'. The designers examine how the rotation plate can realize the function 'to enumerate rotation' with the modifier 'as large as possible'. The modifier is a criterion to evaluate the realized function. Then the designers find that

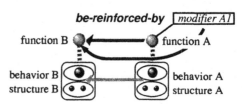

Figure 9.6
Reinforcing
evolution

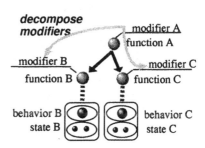

Figure 9.7
Modifier
decomposing
evolution

another function 'to enlarge indicator' is needed to accomplish the function properly. This function cannot be derived in behaviour and state levels only, but functional evaluation can generate it (see Figure 9.3).

Decompose modifiers

This relation describes that a modifier can be decomposed into some other modifiers (see Figure 9.7). This is usually caused by a decomposition of functions, but the way in which modifiers are decomposed does not correspond with the functions themselves. For example, Modifier 'comfortably (M1)' to Function 'to visualize weight' is decomposed into 'comfortably (M5)' to Function 'to receive vertical weight' which is one of the decomposed functions of 'to visualize weight' (see Figure 9.3). In this case, Modifier M1 is not distributed to all decomposed functions but to only one of them. This depends on the designers' interpretation of this modifier.

3 Analysis of the Protocol Data

3.1 Modelling Method

We used the protocol data of the team design up to time 1:49. Our reason for choosing the team design is that it is relatively easy to understand because verbalization is clearer. We believe that protocol analysis on design should be done with more than two designers[3].

We extract FBS elements in design processes, i.e. functions, functional modifiers, behaviours, and structures. In this design, the objects are usually recognized statically because it has few actions when it is used. Therefore we are concerned with structure instead of state to model the designed object. Extraction here is not a pure grammatical operation but a heuristic operation that depends on the contexts. We use grammatical information as hints to find these types of information (for example, verb words for functions).

In this paper we use Lisp-like forms *(function-body subjective objective1 ...)* to represent functions for our convenience. For example, a function description such as 'A device can carry/ fasten a backpack to a bicycle' is represented as *(carry/fasten device backpack bicycle)*. The criteria for extracting these types of information are as follows:

Function   We regard verbs used to explain objects as a functional description. In addition, some special words like *feature* also indicate functions of objects. We pick up a verb with a subjective word and objective words as a function.

Functional modifier   We interpret adverbial phrases related to verbs as functional modifiers. We also interpret additional conditions or additional explanations to function as modifiers.

Behaviour   Behaviour appears when designers invoke simulations. The problem is that such simulation is often done by non-verbal actions. For example, designers simulate some motion by operating physical objects or by a picture, or even mentally. We try to detect their simulation by observing their actions and pictures, and represent the behaviours used in this.

Structure   Since structures are what is designed, they mainly appear as noun phrases. But parts of some structures are also non-verbal, because they are expressed in figures. We gather verbal and non-verbal information to describe structures. As we restrict ourselves to symbolic representation in this analysis, we represent a structure as a set of concepts, relations among concepts, and attribute of concepts.

After the extraction of information on FBS elements, we construct descriptions of the FBS model at each step of the design process, which we call the *FBS state model*. An FBS state model should consist of the three elements of FBS, that is, function, behaviour, and structure. Since we focus our attention on transition of function and structure, we make state models in the following two ways.

First we model functions and functional modifiers at each step of the design process. Each time functions or functional modifiers are added or changed, we gather all functions and functional modifiers that are under consideration up to that time. A *function*

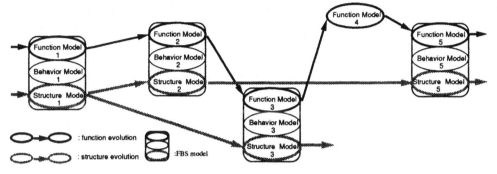

Figure 9.8
Combination of
function and
structure models

state model or *function model* consists of these functions and functional modifiers. Function models are linearly ordered according to their time of designing.

Second, we focus on structures and behaviours that the designers propose in design. Each time structures are added or changed, we gather descriptions of the structure under consideration up to that time, and construct a *structure state model* or *structure model* that consists of these structures. Since a new structure model may be based on earlier structure models, structure models can be partially ordered according to this relationship.

We can get an FBS state model at certain time by combining a function model and a structure model that is used at that time (see Figure 9.8).

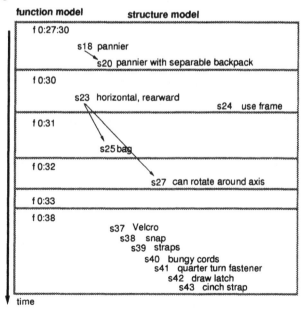

Figure 9.9
Function and
structure state
models

Figure 9.9 shows examples of relations among function and structure state models. A symbol starting with f indicates a function model; a symbol starting with s indicates a structure model. An arrowed line indicates relation between structure models. We made 42 function models and 73 structure models[5].

There were three main types of information for objects, before the design was made. The first one is the assignment with its specifications, each of which the designed product *should* satisfy. The second type is the market research and users' trials and evaluation of the prototype; these describe the criteria the product also *could* satisfy, but not necessarily. These two types of information are directly related to the product itself. The third type is details of environments of the product, i.e. details of the bicycle and the backpack that define the environment where the product is used.

Figure 9.10 shows the FBS model of the assignment. The right-hand side gives explanations of the symbols used in figures for FBS models. In this figure, we identify three independent functions, i.e.

1. *(carry/fasten device backpack bicycle)*
2. *(stack-away device)*
3. *(fold-down device)*

A thin black line indicates a relationship between a function body and an objective or subjective word. Function *(carry/fasten device backpack bicycle)* is decomposed into these two functions:

1. *(attach device backpack)*
2. *(attach device bicycle)*

A thick black arrow indicates decomposition of a function. There are four functional modifiers in the FBS model of the assignment, i.e.

1. *(easy-of-use (carry/fasten device backpack bicycle))*
2. *(a-sporty-appealing-form (carry/fasten device backpack bicycle))*
3. *(for-most-bicycles (carry/fasten device backpack bicycle))*
4. *(reasonable-price-range (carry/fasten device backpack bicycle))*

Also two modifiers *easily* are decompositions of the modifier *easy-of-use*:

5. *(easily (stack-away device))*
6. *(easily (fold-down device))*

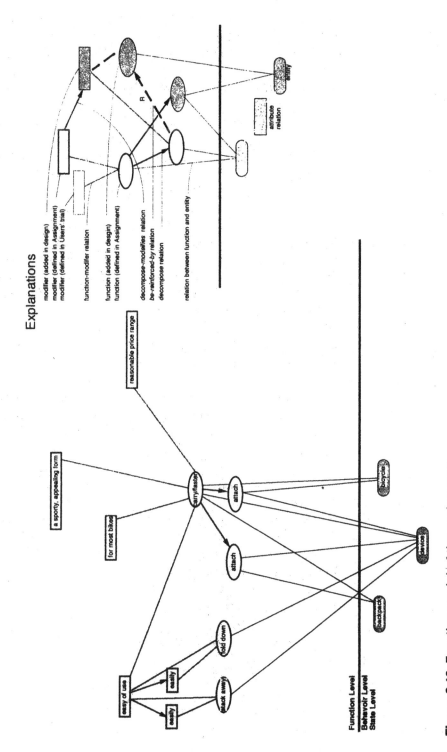

Figure 9.10 Function model of the assignment

In Figure 9.10, a box indicates a modifier, a faint line indicates relationship between a modifier and a function body, and a faint arrow indicates decomposition of a modifier.

This representation is good enough to understand what are the required functions of this device. However, it is too vague for design because this representation covers a huge number of candidates, i.e. a huge design space. Therefore the designers need to reduce the possible design space. To reduce the possible design space in this design session, they can consult the information from the users' trial and evaluation on the prototype design.

Figure 9.11 shows the FBS model of the assignment and users' trials and evaluation. Functions appearing in this figure are the same as those in Figure 9.10, but 13 modifiers are added to the FBS model of assignment. As we discussed in the previous section, adding modifiers is setting criteria to evaluate how functions are achieved. So these newly added modifiers are good information for designers to realize how the design space should be reduced. Since these modifiers are not mandatory, designers picked up some of them to consider in design session and they left others untouched. Modifiers marked with a ? did not appear in the design session.

Figure 9.12 shows the final FBS model in our analysis of the design session. A shaded oval box indicates a function added to the assignment FBS model. Sixteen new functions are added to the initial FBS model, i.e.

1. *(ride person bicycle)*
2. *(steer person bicycle)*
3. *(pedal person bicycle)*
4. *(attach bicycle attaching-device)*
5. *(attach attaching-device device)*
6. *(put-in device backpack)*
7. *(be-extensible attaching-device)*
8. *(hold device backpack)*
9. *(lock device backpack)*
10. *(lock device bicycle)*
11. *(keep-away-straps-from-wheel device)*
12. *(support device backpack bicycle)*
13. *(stand-up device bicycle)*
14. *(stand-up device)*
15. *(rack-function device)*
16. *(fender-function device)*

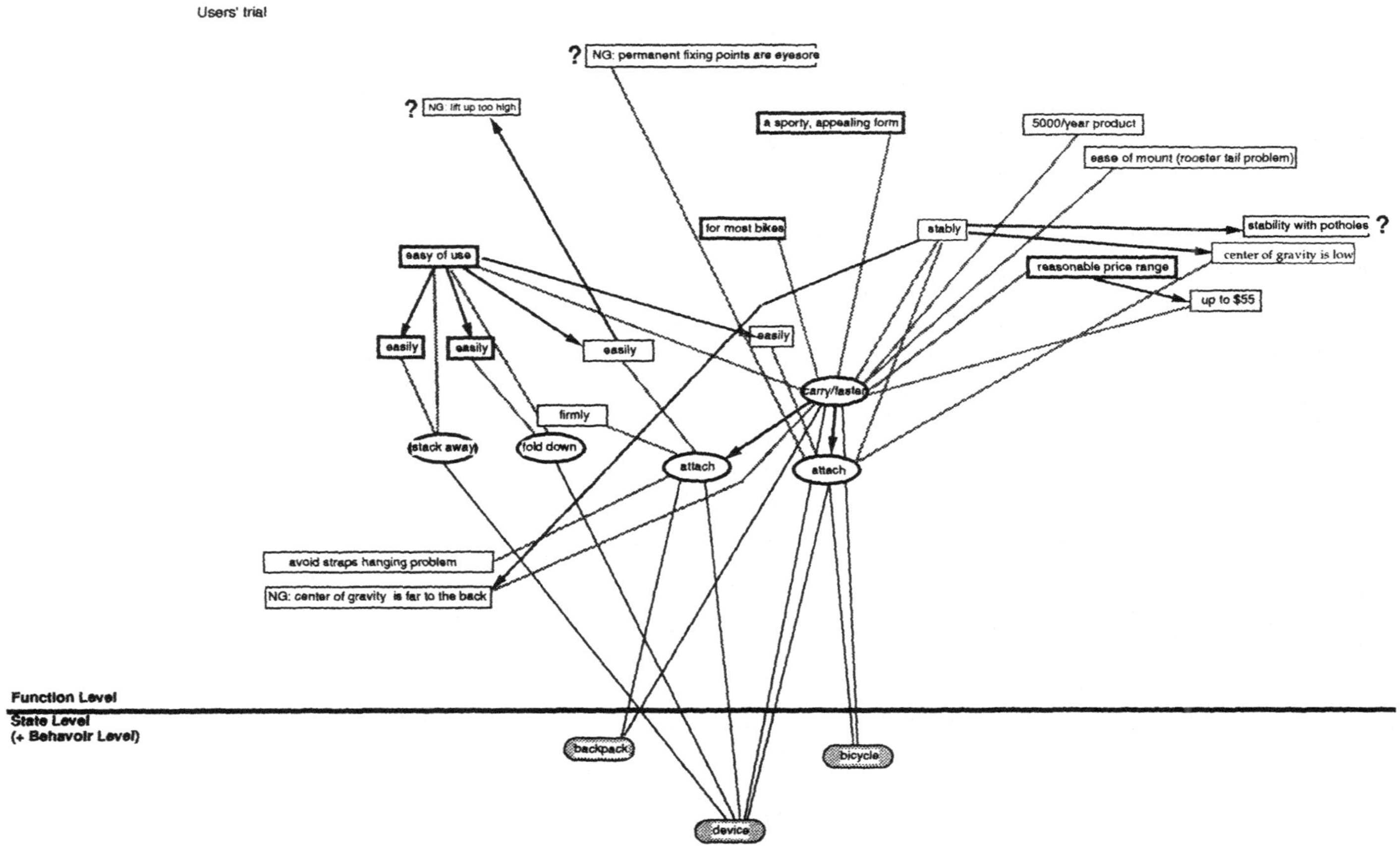

Users' trial

? NG: permanent fixing points are eyesore

5000/year product

ease of mount (rooster tail problem)

stability with potholes ?

center of gravity is low

reasonable price range

up to $55

A sporty, appealing form

stably

carry/fasten

attach

for most bikes

easily

? NG: lift up too high

easily

firmly

(fold down)

easy of use

easily

(stack away)

easily

attach

avoid straps hanging problem

NG: center of gravity is far to the back

bicycle

device

backpack

Function Level

State Level
(+ Behavoir Level)

Figure 9.11 Function model of the assignment and users' trials and evaluation

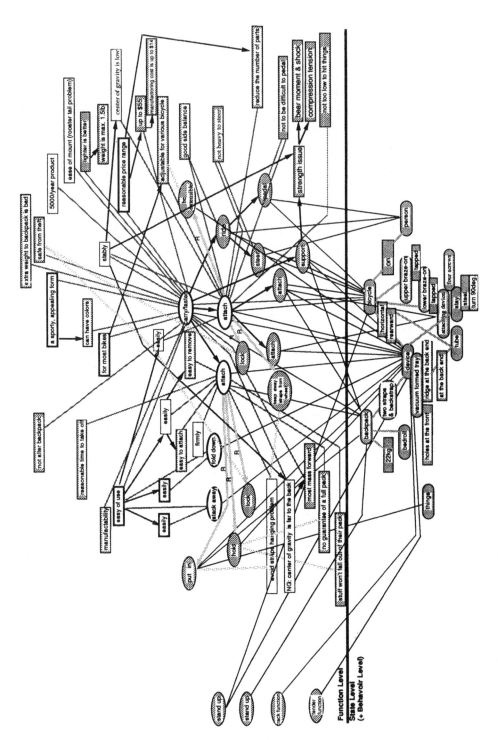

Figure 9.12 FBS model at 1:49

The first five functions are added by decomposition of functions, the next six functions are added as reinforcing functions, the next function is added by causal simulation, and the last four functions are added independently from the other functions.

This does not mean that all those functions have to be satisfied, but they are potential functions that the device may have. For example, the design solution at 1:49 will not satisfy *(put-in device backpack) (stand-up device bicycle) (stand-up device) (fender-function device)* functions. But the 'tray, net, and drawstring' design that is one of the abandoned design solutions would satisfy the function *(put-in device backpack)*.

Thirty-three modifiers are added to the assignment FBS model. Since nine modifiers appear in the Users' Trial FBS model among these modifiers, 24 modifiers are newly found in the design session. In Figure 9.12, a shaded rectangular box indicates a new modifier.

We focus our attention on function *(attach device bicycle)* here to see how functional representation is changed in the design process. Function *(attach device bicycle)* finally has two decomposed functions and three reinforcing functions. For example, a reinforcing function *(be-extensible device)* is suggested as a realization of modifier *(adjustable for various bikes)*. There are no modifiers for *(attach device bicycle)* in the Assignment FBS model. The modifiers for *(attach device bicycle)* in the Users' Trial FBS model are:

1. *easily*
2. *centre-of-gravity-is-low*
3. *stably*
4. *permanent-fixing-point-is-eyesore*

Modifiers for *(attach device bicycle)* in the final FBS model are as follows:

1. *easily*
2. *centre-of-gravity-is-low*
3. *stably*
4. *adjustable-for-most-bikes*
5. *reduce-the-number-of-parts*
6. *good-side-balance*
7. *not-heavy-to-steer*
8. *not-too-low-to-fit-things*
9. *not-too-difficult-to-pedal*

The first three modifiers are inherited from the Users' Trial FBS model, and the other six modifiers are new ones. Two of them are added by modifier decomposition: *(adjustable-for-most-bikes(attach device bicycle))* is modifier decomposition from *(for-most-bikes (carry/fasten device backpack bicycle))* by function decomposition; *(reduce-the-number-of-parts (attach device bicycle))* is modifier decomposition from *(reasonable-price-range (carry/fasten device backpack bicycle))* by function decomposition. Others are found in the construction and evaluation of new structures. Thus, the function *attach* is developed into a functional structure that consists of six functions and nine modifiers.

3.3 Analysis of Evolution Steps

In this section, we investigate how each function or functional modifier is developed in the design process. Table 9.1 shows number of appearances of each evolution type.

Adding functions

A new function can be added in the following four ways.

Adding functions by structures A new function is added by examining a newly suggested structure model. This is similar to decomposing functions, but the added functions are auxiliary or

Table 9.1 Number of appearances of evolutionary steps

Addition of function	6	
by positive examples (FAS+)		5
by negative examples (FAS–)		1
by other reasons (FDX)		0
Addition of 'decompose' relation	3	
by positive examples (FDS+)		3
by negative examples (FDS–)		0
by other reasons (FDX)		0
Addition of 'cause' relation (FC)	1	
Addition of 'reinforced-by' relation (FR)	6	
Addition of modifier		
by positive examples (MAS+)		2
by negative examples (MAS–)		8
by other reasons (MAX)		7
Addition of 'decompose modifiers' relation	12	
by positive examples (MDS+)		5
by negative examples (MDS–)		0
by other reasons (MDX)		7
Total	45	

unexpected functions; they are not relevant to existing functions. In the design session, we found six functions in this category. New functions can be found either by positive examples or by negative examples. The former means that the proposed structure has satisfied the designers' intention. In this case, a newly found function itself is added to the function model. The latter means that the proposed structure did not satisfy the intention. In this case, the negation of the newly found function is added to the function model.

For example, one of the designers suggested a propylene popup-book-like design at time 1:07. The advantage of this design, he claimed, is that it would work as a mudguard (see Figure 9.13). Although he retracted this idea immediately because of strength problems, this mudguard function is used again later as the 'fender' function when they discuss the function of the vacuumed-formed tray (at time 1:22). This tells that the *fender-function* (or *mud-guard-function*) is recognized as a function of the device.

Decomposing functions   It is claimed that decomposition of functions is the most basic procedure for handling functions in design[6]. Unless the designing domain is well investigated or well known, as in routine design, it is not easy to find a decomposition of functions. Actually in this design, decomposition of function is found only three times.

For example, at time 0:56 the designers found that attaching the carrying device to the bicycle can be realized easily by attaching an attaching device to the bicycle and attaching this attaching device to the carrying device. This process is the creation of two new functions *(attach bicycle attaching-device)* and *(attach attaching-device device)* and additions of 'decompose' relations between *(attach bicycle device)* and *(attach bicycle attaching-device)* and between *(attach bicycle device)* and *(attach attaching-device device)*.

Adding 'be-caused-by' relation   This is the procedure for making causal relations between functions. We see at time 1:06 that the designers discussed the strength issue of attaching the device to the bicycle. They simulated the dynamics of the bicycle with the device and the backpack in order to know how strength of parts would affect the behaviour. Using this simulation, they realized the function *(support device backpack)* is needed as a sub-function of function *carry/fasten* because behaviour *support* is needed for the behaviour *attach* to be realized. Since most behaviour is implicit or non-verbal, it is difficult to find such causal relations. We could find only one example.

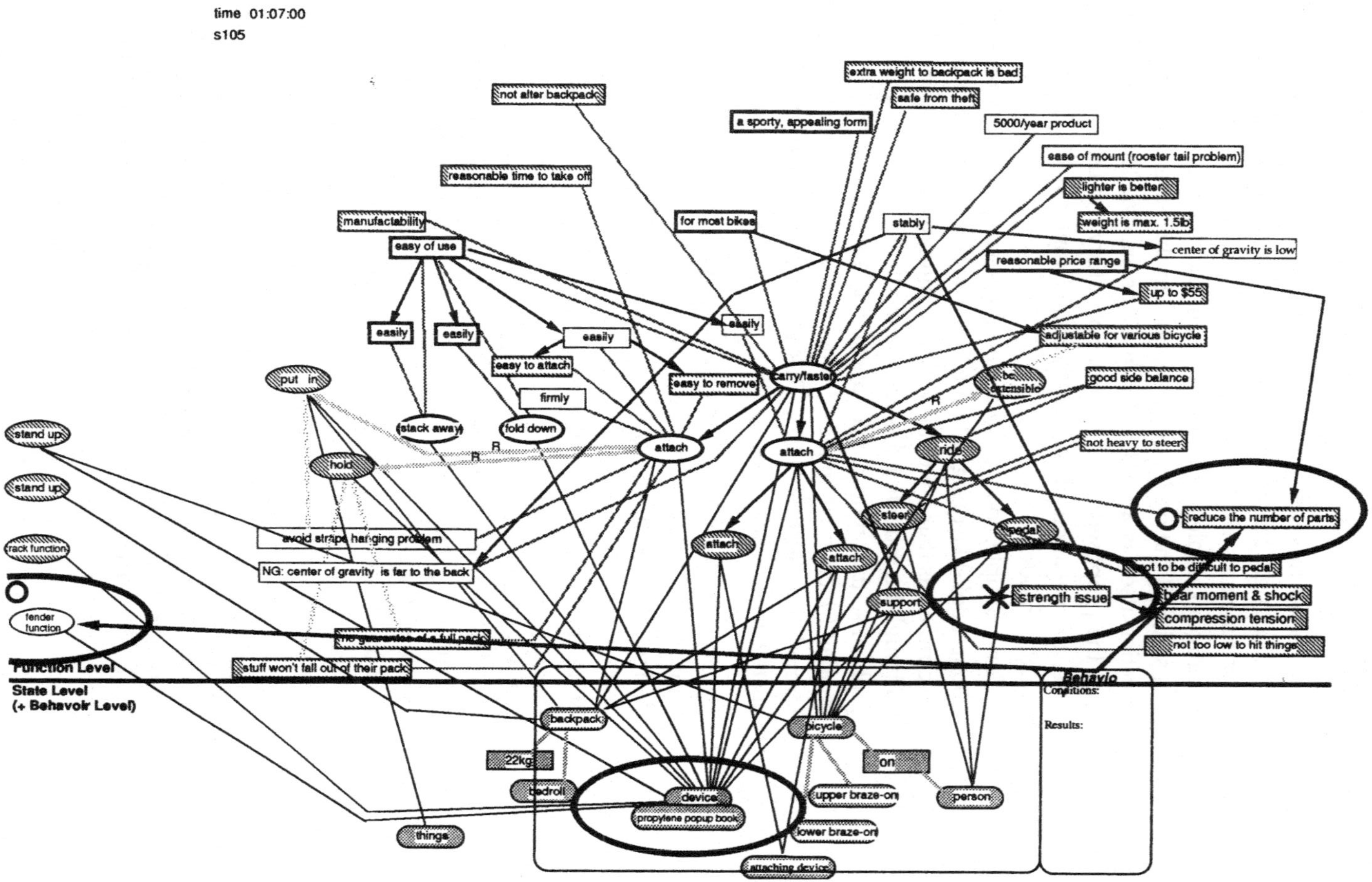

time 01:07:00

s105

Figure 9.13 An example of adding functions by structures

time 01:30:00

s136

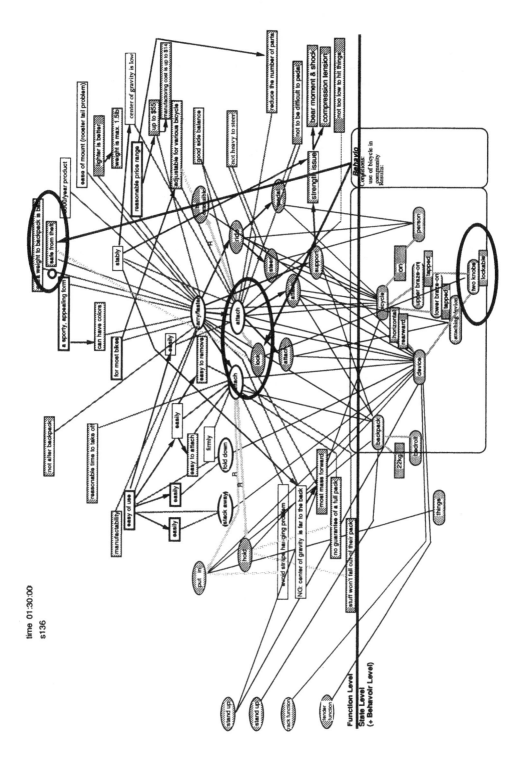

Figure 9.14 An example of adding functions by transforming modifiers

Adding 'be-reinforced-by' relation   The Assignment model and the Users' Trial model have relatively few functions and many modifiers. This implies that interpretation of those modifiers plays an important role in this design. In our model, converting modifiers into functions is achieved by adding 'be-reinforced-by' relations. A reinforcing function is found by evaluating the functionality of the proposed structure with the modifier.

For example, at time 01:30 the designers try to satisfy modifier *safe-from-theft* (this modifier is already suggested at time 00:38). In order to realize modifier *(safe-from-theft (fasten/carry device backpack))*, the designers decomposed it into *(safe-from-theft (attach device backpack))*. Then they suggested that function *lock* performed by *lockable-knob* is a realization of this modifier. This function *lock* is a reinforcing function of the function *attach* (see Figure 9.14). We could find six examples for this reinforcing evolution.

Adding modifiers

As mentioned above, there are relatively few functions and many modifiers in this design. It means that operating modifiers is an important process for evolving function models. Adding modifiers is achieved in two ways, i.e. by structures and by decomposing modifiers.

Adding modifiers by structures   A typical procedure observed in the design process is adding modifiers by examining new suggested structures. We found 17 examples.

Figure 9.15 shows how addition of modifiers is achieved. First a function model and a structure model are taken as the current model of the object (Figure 9.15(a)). Comparing the function model and the structure model, the designers suggest a new structure model (Figure 9.15(b)). Behaviours are produced by simulation on this structure model, and the function model for this structure model is created (Figure 9.15(c)). If the new function model is considered to be better than the previous one, modifiers

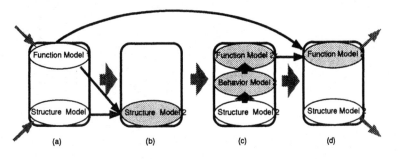

Figure 9.15
Adding modifiers by
structures

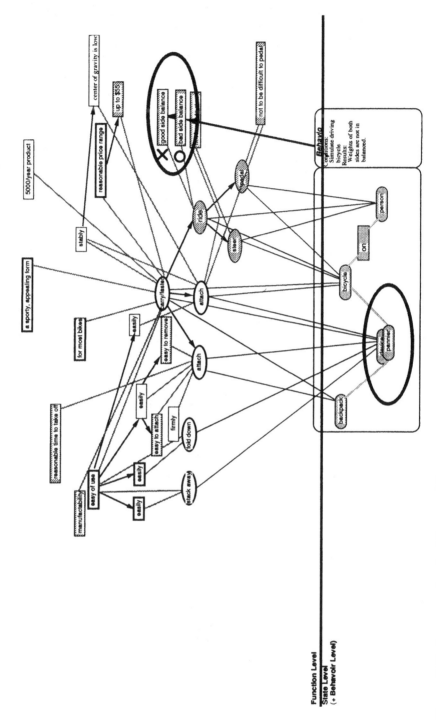

Figure 9.16 An example of adding modifiers by structures

representing the difference are to be included in the descriptions of functionality. If the new function model is considered to be worse, a negation of modifiers representing the difference has to be included (Figure 9.15(d)).

Around time 28:00, the designers suggested using panniers to attach the device to the bicycle (see Figure 9.16). Then they found that to attach the backpack on one side of the bike would be bad for the rider's balance. They decided that the pannier structure was not desirable because of this. In order to find the balancing problem, they might simulate the behaviour of the device with some aspect. The behaviour aspect of this state is the dynamics of riding a bicycle. The simulation of the dynamics of riding a bicycle with the device and the backpack tells them that weights of both sides would be not in balance. Then the designers decided that this is not good. At this moment, they discovered that riding a bike involves having a good side balance, i.e., they found a new specification to function *ride*. Thus *good-side-balance* is a new modifier to function *ride*, and is added to the functional model. The same procedure happened just after this state. They proposed a new structure, i.e. separation of the backpack, which is to satisfy *good-side-balance* modifier. Then they found that altering the backpack is not desirable. Thus *not-alter-backpack* is also added to the function model.

It is interesting that adding modifiers by a negative example appears more frequently than adding modifiers by a positive example. It suggests that contradiction or inconsistency are important for design (we discussed and modelled use of inconsistency in design elsewhere[6]).

Decomposition of modifiers   The other way to add modifiers is to decompose modifiers. We found 12 examples in this design session. Decomposition of modifiers happens along with the decomposition of functions, but to decompose modifiers needs another effort. At time 01:15, the designers suggested using aluminium for the device. The big advantage would be the colouring possibilities. Since they thought that *to-have-colours* is an idea to realize *a-sporty-appealing-form*, *to-have-colours* is a decomposition of modifier *a-sporty-appealing-form*, and can be added to the functional model as a modifier to function *carry/fasten*.

4 Summary

In this report, we showed our primary results of a protocol analysis by our functional modelling scheme. Although it is a tentative report, we can draw some remarks from these results.

Function changing and structure changing are important for tracing design processes. We have been absorbed in structure changing when we wanted to know what happened in design processes. Functional descriptions are also changing in design processes. Both changes are important, but it is impossible to model them separately. Our approach is suitable for this purpose because function and structure are represented in a single scheme.

Functional evolution is not magic, but rational in most cases. Designers often make design criteria by themselves in order to converge their design processes. Designers seem to have such criteria *a priori*. But such criteria (functional modifiers in our terms) have arisen as results of the interaction between structure and function. In our scheme we can explain why they adopt new criteria.

Function modelling is also important as a result of design. Not only the designed structure but also its intended functionality are important as the result of design, because we can evaluate how well the designed structure matches the intention of the designers. The functionality they intended is not the function given by requirements, but the function model they made during the design process.

References

1   Takeda, H., P. Veerkamp, T. Tomiyama, and H. Yoshikawa, Modeling design processes, *AI Magazine*, **11**(4) (1990) 37–48

2   Umeda, Y., H. Takeda, T. Tomiyama, and Y. Yoshikawa, Function, behaviour, and structure, in J.S. Gero (Ed.), *Applications of Artificial Intelligence in Engineering V*, vol. 1, Sringer-Verlag, Berlin (1990), pp. 177–194

3   Takeda, H., S. Hamada, T. Tomiyama, and H. Yoshikawa, A cognitive approach of the analysis of design processes, in *Design Theory and Methodology – DTM '90*, ASME (1990), pp. 153–160

4   Forbus, K.D., Qualitative process theory, *Artificial Intelligence*, **24** (1984) 85–168

5   FBS models in Delft protocol analysis, http://ai-www.aist-nara.ac.jp/~takeda/DPW/

6   Pahl, G. and W. Beitz, *Engineering Design*, The Design Council, London (1984)

7   Takeda, H., T. Tomiyama, and H. Yoshikawa, A logical and computerable framework for reasoning in design, in D.L. Taylor and L.A. Stauffer (Eds), *Design Theory and Methodology – DTM '92*, ASME (1992), pp. 167–174

10 Design Activity Structural Categories

Vesna Popovic
Queensland University of Technology, Brisbane, Australia

This analysis is centred around the cognitive processes that occurred during the design team's activity. Verbal evidence is used to demonstrate what kinds of cognitive processes took place in order for the design team members to arrive at the conceptual design solution. In order to represent the team members' interaction one has to refer to the design processes the team used. Therefore the objectives of this research are threefold:

1. To identify the design process the team used.
2. To describe the interaction of team members during the design activity.
3. To construct a schematic model of the team's design process.

The early stage of the design process is very important for the generation of good design. It is seen to be the most creative phase, its common characteristics being that it is an ill-defined task and that it encompasses visual, search and analytical tasks. The major part of the early stage of the design process is usually devoted to information translation1[1]. This means translation of information such as the brief, drawings and market research into convenient language. The aim of this translation is to determine design constraints and criteria.

The design process in this example can be viewed as a group activity whose interrelation depends on the nature of the design task and individual team members' knowledge. Design creativity depends on the quality of the knowledge that is available to the designers. Design is a prediction which concerns how things ought to be. It aims at changing an existing situation into a preferred one. The designers attempt to predict the behaviour of a proposed artifact using their knowledge and expertise.

1 Protocol Analysis

The talk-aloud (TA) protocol of the design team while working on the project was segmented into individual designers' statements. Each statement was coded and analysed and the codes are distinct to each individual team member – Ivan 'I' (1I–757I), John 'J' (1J–755J) and Kerry 'K' (1K–560K).

The TA protocol was analysed in two ways: (a) to identify the main design activity categories that constitute the early stage of the design process, and (b) to identify the design team interaction. The analysis of the design activity into its major sequences was based on the designers' model of the design process established by the team. Their model consists of seven segments that are supposed to occur sequentially (Figure 10.1).

From the protocol analysis the following activities emerged during the experiment. This includes:

- brief understanding – definition of a problem and design constraints formulation
- analysis of design constraints – translation of information, information search, knowledge representation and evaluation of information

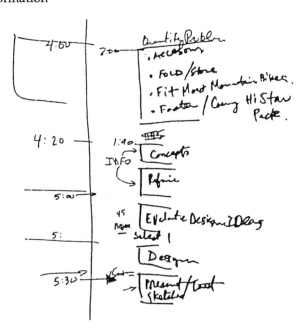

Figure 10.1
Team management planning

- generation of ideas – interpretation of design constraints, use of analogy
- evaluation
- management of the design process

I will discuss each of these categories and illustrate them with examples from the protocol.

1.1 Brief Understanding

Client brief interpretation was the first category of the design process in which the design team tried to understand task requirements. It includes statements from 1I to 36I, from 1J to 37J and from 1K to 28K. This is illustrated with statements from 6K to 8I as follows:

6K	uh huh
5I	that's what the memo says?
7K	that's what they're building
6I	it's internal internal frames they're making now
8K	they're getting busted by the internal frame folks but they think they think **an advantage would be to make this external frame also be mountable to a rack or become a rack**
7I	yeah
9K	and that would be pretty cool too
8J	they could keep selling the external frame backpacks
10K	yeah
8I	right and bicycles

It is evident from this statement that the designers tried to define and understand the problem. Statement 8K illustrates that they clarified that the client wants to have an external frame backpack to be mountable to a rack or to become a rack in itself. This formulates a design constraint that refers to an external frame.

1.2 Analysis of Design Constraints

The major part of the design activity is devoted to the identification of the design constraints to be used for the design of the bicycle rack. This consists of information translation, search from information perceptually available to the designers, and knowledge application. The designers were interacting in different ways by utilizing the appropriate knowledge domain. Evaluation processes were occurring throughout the analysis. This is illustrated with the protocol statements from 212J to 219J:

212J   I mean what what how much weight do you think somebody could realistically put in that pack **(design constraint – weight)**

136K   probably thirty fifty

00:47:00

thirty pounds **(design constraint – weight – assumption)**

213J   fill it with sand

210I   is that information we have access to um

214J   yeah what's typical weight that people carry in a backpack **(design constraint – request for information)**

211I   do we have er

X   Sorry I'm sorry you were asking me something

215J   that's OK

212I   do we have information about what er weights are that people might carry in a backpack or **(design constraint – request for information)**

216J   have they done any market surveys **(design constraint – request for information)**

213I   market surveys about

X   We do have some facts on the use of the backpack

214I   yes ... just hand over the book

217J   (laugh)

137K   (laugh)

218J   here's your book it'll be back to you in a while ... OK ... fifty-five and sixty-five litre versions (inaudible) backpack

138K   twenty-two kilograms **(design constraint – weight)**

219J   so forty-five pounds fifty pound yeah

1.3 Generation of Ideas

Idea generation was part of the analysis process and clarification of design constraints. Designers used analogy to clarify possible solutions as follows:

103J   so yeah so it's sorta like

76K   so it's more **like a pannier**

101I   a **saddlebag** yeah that's uh like a

1.4 Evaluation

Evaluation of design constraints and design ideas was occurring during the whole process. This was contrary to the initial planned design activities suggested at the beginning of the process.

The team planned the schedule to quantify actions at the beginning of the task. This is evident from the individual protocol statement lines from 19I to 23J and from 63J to 70J. This planned schedule was the representation of their planned design methodology which they follow in some parts. The methodology planned was sequential. However, during the team interaction most of the time was dedicated to the analysis and evaluation of design constraints.

19I   should we uh prepare a schedule and then just sorta stick to it or should we uh just start working?

18J   no it's probably a good idea to try to quantify our amount of time the kinda time we have left (laugh)

20I   yeah we've used fifteen minutes so we have

19J   OK so we have an hour

00:12:00

and forty-five so we need to generate some concepts and I guess refine the concepts and or whaddaya call

21I   information or

18K   yeah we wanna look at the em customer feedback or the users' testing

20J   oh yeah so maybe yeah wherever that comes in in this list and then uh … em and like evaluate evaluate design ideas and decide on a concept select one

22I   select yeah

21J   design, which everybody always forgets (laugh) oh it it's ended at that point and

23I   present

22J   present test (laugh)

00:13:00

we'll probably be able to test it on paper

24I   right

23J   paper test em

25I   so they want design sketches at the end right annotated sketches that'll be here

24J   OK so

The team also made a diagram of the design process they were planning to undertake (Figure 10.1). This process was different from what was really happening during the conceptual design stage.

The real time spent on the particular sequence of the design methodology was discrepant from their initial planning. However, during the task the designers referred several times to the time constraints and time management which demonstrates they used relevant knowledge and methods to monitor the design process.

2 Designers' Interaction

Identifying the team members' interaction during the design activity was based on the analysis of individual protocol statements in order to identify the cognitive processes which occurred.

Selected interpretation parameters were used for the analysis of the designers' interaction and were based on Ericsson and Simon[2]. The protocol interpretation revealed that the design team interacted through behavioural sequences. These sequences are identified to be the major categories for modelling of the design activity. The protocol categories that constitute behavioural sequences are:

- goal
- intention
- search – queries
- evaluation – confirmation
- inference
- self-initiated explanation
- assumption
- conflict
- action
- task
- decision

They are the main categories found to be present in the designers' behavioural sequences of interaction (Appendix A).

It has already been said that this TA protocol analysis is centred around the designers' interaction. Any protocol statement by an individual that has more than one category is regarded as a behavioural sequence. Similarly, any series of statements by different members of the team which contains more than one kind of element is regarded as a behavioural sequence. Therefore, two groups of such sequences are identified: (a) individual designers' behavioural sequences, and (b) team behavioural sequences.

Individual sequences are demonstrated with the following protocol statements:

| Statement 563J | 'J' infers (so so), searches design intent (are we straps or net over this thing what what's our what's our g- what's our idea we're gonna fit straps or) |

Behavioural sequence: *inference – search*

| Statement 568J | 'J' confirms (OK OK) infers (so er)m searches (how we we know), explains action (we're going to put product parts in different colours for presentation) |

Behavioural sequence: *inference – search – action*

| Statement 576I | 'I' confirms (yeah), infers (so), makes an intention (we'll have an open back tray), makes a self-initiated explanation (like a pick-up truck bed), makes an intention (we'll throw the em), ... infers (and) makes an intention (it'll have this little feature) |

Behavioural sequence: *confirmation – inference – intention – explanation – intention – inference – intention*

Team behavioural sequences consist of the interaction of various protocol categories. To demonstrate this I will use the following protocol statements:

Statements 573I to 575I:

573I	'I' makes an intention (let's say considering that we have about ten minutes given that)
427K	'K' makes an intention (let's design our thing)
574I	'I' makes an intention (let's do our thing yeah)
566J	'J' queries (are we designing three different things)
428K	'K' confirms (no we're designing one thing)
567J	'J' confirms (OK)
575I	'I' makes a self-initiated explanation (since this is only for the big presentation for the two people we have to do a cost analysis as well)

Behavioural sequence: *intention – intention – intention – query –*
confirmation – confirmation – explanation

Statements 578I to 432K:

578I	'I' **confirms** (OK)
431K	**'K' makes a self-initiated explanation** (about fifteen)
571J	**'J' queries** (centimetres or inches)
432K	**'K' confirms** (centimetres oh inches)

Behavioural sequence: *confirmation – explanation – query – con-*
firmation

It is a common situation that the team members' behavioural sequences are characterized by using the same protocol categories such as those shown in the statements 573I to 575I.

Interestingly, the distribution of behavioural sequences throughout the protocol shows that the individual designers' behavioural sequences are dominant during the analysis of design constraints and the generation of ideas. Team sequences are dominant during the decision making process. This is demonstrated using protocol statements from 503K to 692I:

503K	**'K' evaluates** (now we can afford our four knobs those are gonna be er about this big)
682J	**'J' design task** (just an insert)
688I	**'I' design task** (the same apart)
505K	**'K' design task** (insert moulded)
683J	**'J' confirms** (yeah insert moulded)
689I	**'I' confirms** (insert moulded)
505K	**'K' makes a self-initiated explanation** (a couple of bucks each)
684J	**'J' makes a self-initiated explanation** (you have a tool that you maybe can mould off four knobs at one shot)
506K	**'K' makes a self-initiated explanation** (boom boom family mould)
685J	**'J' assumes** (boom so maybe maybe a buck fifty total for four of them)
507K	**'K' makes a query** (a buck fifty for four?)
686J	**'J' makes a query** (does that sound too much?)

508K	'K' **evaluates** (sounds too low to me)
690I	'I' **assumes** (say two bucks)
509K	'K' **confirms** (two bucks for all four fifty cents each)
691I	'I' **confirms** (nice round number)
687J	'J' **confirms** (OK alright I give in)
692I	'I' **confirms** (OK)

Behavioural sequence: evaluation – task – task – task – confirmation – confirmation – explanation – explanation – explanation – assumption – query – query – evaluation – assumption – confirmation- confirmation – confirmation – confirmation

As already stated the design team members' interaction depends on the nature of the design task and individual team members' knowledge. This is demonstrated in the protocol statements from 327I to 230K:

327I	yes OK well maybe em OK let's look at **materials** you were talking **injection moulding**
313J	injection moulding em wire form what else comes to mind on top of these maybe like er cloth with some sewn in em diecut pieces or something for plastic reinforcement
328I	oh yeah
314J	like sorta like backpack construction technology
329I	oh
315J	emmm come on come up with some ideas (laugh)
330I	er rubber
230K	wire form plastics and injection moulded bracket things the small things to injection mould the the small bits so the tooling isn't too light

The team members were analysing possible material constraints and manufacturing methods. The members were using their knowledge in combinations, such as knowledge in a particular field (materials and manufacturing methods – 327I) and domain specific design knowledge (how material constraints may be applied – 313J, 230K). The designers responded to the situation by using the process knowledge relevant to the particular design process stage.

3 Sketch of a Design Team Model

Two distinct patterns emerged from the analysis. They are characterized as (a) individual behavioural sequences (IBS), and (b) team behavioural sequences (TBS). These are envisaged to be structural categories to model the design team process. Each group has its individual modelling space. An individual modelling space is constructed of IBS and individual designer knowledge. The team behavioural sequences modelling space is constructed of IBS and shared by all team participants. In general, design execution occurs via modelling action that corresponds with each relevant design stage.

This is the first step in attempting to model a design team activity based on team and individual behavioural sequences. Further work to follow is to develop the model and integrate behavioural sequences and design knowledge categories.

References

1   Ballay, J.M., An experimental view of the design process, in W.B. Rouse and K.B. Roff (Eds), *System Design: Behavioral Perspectives on Designers, Tools and Organisations*, North-Holland, New York (1987), pp. 65–82
2   Ericsson, K.A. and H.A. Simon, *Protocol Analysis*, MIT Press, Cambridge, MA (1993)

Appendix A Protocol Sequences of the Individual Design Team Members

The protocol sequences of the individual design team members are grouped into the main protocol categories. They demonstrate the summary of the individual designer's behavioural sequences that occurred during the design of a device for attaching a hiker's backpack to a mountain bicycle.

Goal

goal – goal interpretation
goal – goal interpretation – search

Intention

intention – inference
intention – inference – evaluation
intention – inference – intention
intention – inference – assumption
intention – inference – self-initiated explanation
intention – evaluation
intention – self-initiated explanation – inference
intention – search

Search – Queries

search – inference – conflict – search

search – inference – action – inference – action
search – inference – goal interpretation – action
search – inference – evaluation
search – inference – intention
search – intention
search – intention – self-initiated explanation
search – self-initiated explanation – inference – conflict – search – decision
search – self-initiated explanation
search – self-initiated explanation – inference – self-initiated explanation
search – evaluation
search – evaluation – intention
search – conflict
search – conflict – assumption
search – conflict – self-initiated explanation – inference – assumption
search – assumption
search – assumption – conflict
search – assumption – intention

Inference   inference – goal interpretation
inference – goal evaluation
inference – action
inference – self-initiated explanation
inference – self-initiated explanation – search
inference – self-initiated explanation – inference – conflict – self-initiated explanation
inference – self-initiated explanation – search
inference – self-initiated explanation – intention
inference – intention
inference – intention – search
inference – intention – evaluation – inference
inference – intention – inference – evaluation
inference – intention – query – assumption
inference – evaluation
inference – evaluation – intention
inference – evaluation – self-initiated explanation
inference – evaluation – self-initiated explanation – inference
inference – evaluation – assumption
inference – evaluation - search
inference – evaluation – inference – search

inference – conflict
inference – conflict – self-initiated explanation
inference – conflict – inference – intention
inference – assumption
inference – assumption – search
inference – assumption – search – error – assumption
inference – assumption – inference
inference – assumption – inference – self-initiated explanation –
inference – self-initiated explanation
inference – assumption – intention – inference
inference – assumption – evaluation
inference – search
inference – search – evaluation
inference – search – inference
inference – search – conflict – search
inference – search – self-initiated explanation – intention – inference
inference – self-initiated explanation – inference
inference – self-initiated explanation – inference – self-initiated
explanation
inference – self-initiated explanation – inference – assumption
inference – self-initiated explanation – search
inference – self-initiated explanation – assumption – self-initiated
explanation
inference – self-initiated explanation – search
inference – self-initiated explanation – evaluation – inference
inference – task – inference – task – evaluation
inference – decision

Evaluation – Confirmation   evaluation – error
evaluation – error – self-initiated explanation
evaluation – search
evaluation – search – inference
evaluation – search – inference – evaluation
evaluation – search – self-initiated explanation – inference – self-
initiated explanation
evaluation – self-initiated explanation - search – inference – self-
initiated explanation – inference – self-initiated explanation
evaluation – assumption
evaluation – assumption – inference
evaluation – assumption – inference – self-initiated explanation
evaluation – assumption – search

evaluation – action
evaluation – action – inference – evaluation
evaluation – search
evaluation – intention
evaluation – intention – inference – intention – inference – intention
evaluation – intention – evaluation – intention – assumption
evaluation – inference
evaluation – inference – search
evaluation – inference – search – inference – action
evaluation – inference – intention
evaluation – inference – intention – task – evaluation – search
evaluation – inference – intention – inference – goal
evaluation – inference – intention – inference – search – inference – intention
evaluation – inference – intention – search
evaluation – inference – action
evaluation – inference – self-initiated explanation
evaluation – inference – self-initiated explanation – conflict
evaluation – inference – self-initiated explanation – assumption
evaluation – inference – evaluation
evaluation – inference – evaluation – intention
evaluation – inference – evaluation – inference – intention – evaluation
evaluation – inference – assumption
evaluation – inference – assumption – evaluation
evaluation – inference – conflict
evaluation – inference – search
evaluation – inference – task
evaluation – inference – error
evaluation – conflict
evaluation – task

Self-initiated Explanation   self-initiated explanation – goal interpretation
self-initiated explanation – conflict
self-initiated explanation – conflict – self-initiated explanation
self-initiated explanation – conflict – self-initiated explanation – inference
self-initiated explanation – conflict – action – assumption
self-initiated explanation – inference
self-initiated explanation – inference – search
self-initiated explanation – inference – search

self-initiated explanation – inference – assumption
self-initiated explanation – inference – self-initiated explanation
self-initiated explanation – inference – evaluation
self-initiated explanation – inference – decision
self-initiated explanation – evaluation – inference – self-initiated explanation
self-initiated explanation – query – self-initiated explanation
self-initiated explanation – search – inference
self-initiated explanation – inference – conflict
self-initiated explanation – evaluation – assumption
self-initiated explanation – assumption

Assumption

assumption – inference
assumption – inference – assumption
assumption – inference – self-initiated explanation
assumption – inference – evaluation
assumption – self-initiated explanation – conflict/intention
assumption – conflict
assumption – conflict – evaluation
assumption – action
assumption – intention
assumption – search

Conflict

conflict – inference – self-initiated explanation
conflict – inference – search
conflict – evaluation – search – evaluation

Action

action – self-initiated explanation

Task

task – assumption
task – search

Decision

decision – assumption
decision – inference
decision – inference – intention
decision – evaluation

11 The Data in Design Protocols: The Issue of Data Coding, Data Analysis in the Development of Models of the Design Process

Terry Purcell, John Gero, Helen Edwards and **Tom McNeill**
University of Sydney, Australia

A video/audio protocol of a design session represents a particular form of qualitative data. More typically qualitative data takes the form of verbal data such as a record of the answers to open-ended questions, transcripts of relatively unstructured group discussions or existing text from one or more sources. However, while protocols represent a particular form of qualitative data, they share with the other forms the fact that the protocols, transcripts and so on are not the actual data on which analysis is performed. Rather the protocol or transcript has to be segmented or categorized in some way with the frequency of occurrence of each category forming the data that are interpreted and discussed. Design protocols are also different in that most of the more traditional forms of qualitative data do not have time as such an important facet of the data.

In many of the areas in which qualitative data have been gathered and analysed, the traditional way in which the original data have been segmented is by inferring the categories by a careful reading of the original record or part of the record. The frequencies of occurrence of the categories are then obtained by a reanalysis of the full record. A more recent and potentially richer approach to the analysis of qualitative data is the 'grounded theory' approach of Glaser and Strauss[1]. Here the data are not segmented with each segment then being classified into a single category. Rather the original record is repeatedly revisited with both multiple codings of the same segment possible and with differing sets of segments of finer or coarser grain being recognized.

An examination of the design protocol literature indicates that it is the traditional approach to data segmentation, analysis and interpretation which has formed the basis for much of the initial work in the area. However, more recently, there has been a shift in the analysis of design protocols towards imposing an externally derived structure onto the protocols, perhaps to a greater degree than in the traditional areas where qualitative data have been collected and interpreted. This external structure has been based on existing analyses and models of the design process[2-6]. Segmentation of the original protocol in this approach is not generated by the data but based on other criteria, such as pauses in the flow of words, or on semantic/syntactic criteria for recognizing discrete utterances. There are two consequences of this shift. One is that the number of categories used in the analysis is relatively limited. For example, in the recent paper by Lloyd and Scott[5] three categories are used – generative, deductive and evaluative utterances. While this particular experiment and others using this approach have produced interesting results, they represent a severe loss of detail from the original protocol. The second is that the results of the later experiments are difficult to compare to the earlier work and to each other. As a result, while the intention underlying the shift to imposing structure on protocols was to test existing hypotheses and models of the design process, the wide variation in the categorical structures imposed actually inhibits comparisons between experiments and the development of theory. However, a reading of this literature also reveals commonality and overlap between the categories used and tantalizing possible correspondences between the results that have been obtained.

This analysis of the nature of the data collected in a protocol study of design has formed the starting point for our analysis of the protocol of the session involving an individual designer. What we have attempted to develop is a coding scheme that brings together these various approaches and addresses the problems we have identified. First, the coding scheme embodies both data and theory generated categories. The data generated categories are derived from previous work and from this particular protocol. While these categories depend on the data, it was also apparent that many of them were similar to the more theoretical categories employed in recent work in the area. The theoretical categories are derived from the work of Gero et al.[7-9] where it is argued that design involves reasoning in three domains – function, structure and behaviour. Second, the coding scheme that we have developed is far richer than that used in most previous work. For example, Eckersley[4] developed a coding scheme that consisted of eight categories, a number which considerably exceeds that used in previous work. In contrast, our coding scheme consists of 27 categories. The danger in increasing the number of categories is that the results become too complex, potentially masking relationships and patterns in the data, and are generally too difficult to understand. While this presents problems, we have attempted to counter them by the structure that is present in the coding categories and by the methods that are used to visualize the results. Finally, while time is an important facet of the analysis of a design protocol, as noted previously, we consider that too great an emphasis on changes that occur over the period of a full design session mask significant aspects of what is occurring either simultaneously in the sense of the same segment having multiple codings or in relationships between segments; that is at a much smaller time scale than that involved in a full design session. This paper therefore reports the process of development of our coding scheme and presents a number of examples of the application of this scheme which illustrate the results of our approach.

1 The Development of the Coding Scheme

The current coding scheme was developed by Gero and McNeill[10] over a number of iterations from a version previously used to analyse conceptual electronic design sessions. Early work carried out by McNeill and Edmonds[11] concentrated on the problem domain and the designer's navigation through the problem domain. This meant that, by necessity, the analysis was problem specific and results were dependent on the particular problem being tackled by the designer. The work centred around the notion of attaching labels to events within the design episode and viewing the design episode as a sequence of inter-dependent events.

In subsequent work by Gero and McNeill[10], it was desirable to compare a number of design episodes and the need was seen for an approach that would give a result that was problem independent. The coding scheme which they developed evolved over a period of time with each iteration being examined for how well it captured significant characteristics of the design episodes. As the coding progressed new codes were added to describe events that were not appropriately covered by an existing code. The resulting scheme incorporated three broad classifications, two of which were concerned with the designer's place in the problem space while the third was concerned with the designer's activity at any time.

The coding scheme used in the analysis of the Delft individual design session was a modified version of the scheme developed by Gero and McNeill. Although the scheme was considered to be problem independent, the coders first perused the Delft individual design session in an unstructured way in order to ensure that the coding scheme would be appropriate. An important difference in experimental procedure was revealed between the Delft individual design session and the electronic engineering sessions on which the coding scheme was based. The electronic engineering sessions featured designers working on tasks that they were required to complete as part of their own everyday work. As a result, most of the work of defining and analysing the problem had already been carried out prior to the recorded design sessions. In the Delft experiment, the designer was given a design problem to which he had not previously been exposed. From the perusal of the Delft tape, it was apparent that in order to complete the task, the designer would need to analyse the problem to a greater extent than the designers in the electronic engineering sessions. The coding scheme was then revised in order to take this difference into account. The coding scheme as used in the present analysis of the Delft individual design session is presented in Table 11.1 and discussed in greater detail in the following section.

2 A Description of the Content of the Coding Scheme

The essence of the coding scheme is that each event in a design episode is categorized over three broad classifications. The first two classifications centre around the problem being solved. The third of the three classifications encoded is concerned with the designer's activity at any time and provides a problem-independent event analysis of the data. Details of the codes which constitute each of the three classifications are outlined in Table 11.1.

The first classification, *Problem Domain: Level of Abstraction*, is established as the coding progresses. This group of codes is related to the way the designer appears to be dissecting the problem as well as a reflection of the complexity of the problem itself. In the present design episode, the *System* level (0) relates to how the finished product would be used and the relation between the bicycle, backpack and backpack carrier as a whole assembly. The bicycle, backpack and backpack carrier could all be considered as subsystems in this design episode. The *Subsystems* level (1) indicates that the designer is focusing on issues which pertain to only one of these subsystems at that particular time. The *Detail* level (2) indicates that the designer is considering the characteristics of physical components.

The second classification, *Problem Domain: Function, Structure and Behaviour*, monitors the forms of reasoning used throughout the design episode. Reasoning about *Structure* (S) involves the manipulation of physical properties in order to generate a physical solution to an abstract problem. Reasoning about *Behaviour* (B) concerns the description of the object's actions or processes in given circumstances. Reasoning about *Function* (F) refers to the manner in which the designed object fulfils its purpose. The 'R' code was used to highlight when the designer was establishing or revising his or her requirements.

The third classification, *Strategy* describes the moment-to-moment activities of the designer during the design episode. The codes used can be dissected into four groups. The first group is concerned with analysis of the problem. The second group is concerned with the proposal of a solution and the third is concerned with analysis of the proposed solution. The fourth group is concerned with times when the designer made explicit reference to what he or she was doing or what he or she knows. An additional 'X' code has been included to indicate occasions when the experimenter interacted with the designer or made a comment. Some examples of how these codes were applied to the Delft individual design session are listed in Table 11.2.

Table 11.1   The coding scheme

First Classification – Problem Domain: Level of Abstraction

Describes where the designer is within the problem domain in terms of level of abstraction.

The code consists of a numeral. The numerals indicate different aspects of the problem domain.

0 – System	The designer is considering the problem from the point of view of the user.
1 – Subsystems	The designer is considering the problem in terms of the subsystems.
2 – Detail	The designer is considering the details of the subsystems.

Second Classification – Problem Domain: Function, Structure and Behaviour

The letters indicate various aspects of the problem domain.

F – Function	The designer is working with the functional aspects of the problem domain.
B – Behaviour	The designer is working with the behavioural aspects of the problem domain.
S – Structure	The designer is working with the structural aspects of the problem domain.
R – Requirements	The designer is adding to, modifying or reconsidering aspects of the initial requirements.

Third Classification – Strategy

Describes the designer's immediate actions and his short term strategy. This classification is divided into four groups of codes. The list includes examples to illustrate the meaning of the code.

Analysis Problem

Ap – Analysing the Problem	'What is the system going to need to do....'
Cp – Consulting Information about the Problem	As above but using external information
Ep – Evaluating the Problem	'That's an important feature....'
Pp – Postponing the Analysis of the Problem	'I can find that out later....'

Table 11.1 *(continued)*

Synthesizing Solution

Ps – Proposing a Solution	'The way to solve that is....'
Cl – Clarifying a Solution	'I'll do that a bit neater....'
Re – Retracting a Previous Design Decision	'That approach is no good what if I....'
Dd – Making a Design Decision	'OK. We'll go for that one....'
Co – Consulting External Information for Ideas	'What are my options....'
Pd – Postponing a Design Action	'I need to do ... later'
La – Looking Ahead	'These things will be trivial (difficult) to do'
Lb – Looking Back	'Can I improve this solution?' 'Do I need all these features?'

Evaluating Solution

Ju – Justifying a Proposed Solution	'This is the way to go because....'
An – Analysing a Proposed Solution	'That will work like this....'
Pa – Postponing an Analysis Action	'I'll need to work that out later'
Ca – Performing Calculations to Analyse a Proposed Solution	'That's seven inches times three'
Ev – Evaluating a Proposed Solution	'This is faster, cheaper etc....'

Explicit Strategies

Ka – Explicitly Referring to Application Knowledge	'In this environment it will need to be....'
Kd – Explicitly Referring to Domain Knowledge	'I know that these components are....'
Ds – Explicitly Referring to Design Strategy	'I'm doing this the hard way....'
X – All comments made by the experimenter	

Table 11.2 Examples of statements from the protocol coded according to the Strategy classification

Code	Examples from the protocol
Analysing Problem	
Ap – Analysing the Problem	'...and first of all I notice that it's pretty small but the product spec wasn't that specific...'
Cp – Consulting Information about the Problem	'is Batavus willing to add any fittings to their bicycle in their production line to accommodate this product?'
Ep – Evaluating the Problem	'that's a pretty heavy backpack'
Pp – Postponing the Analysis of the Problem	'I can calculate that and I'll do that later'
Evaluating Solution	
Ju – Justifying a Proposed Solution	'another reason I make this in aluminum is because bike works are good at welding'
An – Analysing a Proposed Solution	'any bends these rods are then going to be in tension and compression'
Pa – Postponing an Analysis Action	No equivalent example
Ca – Performing Calculations to Analyse a Proposed Solution	'forty times one inch is point nought four cubic inches per inch'
Ev – Evaluating a Proposed Solution	'certainly I would say that that sorta looks classy having a backpack in the centre....'
Synthesizing Solution	
Ps – Proposing a Solution	'immediately my impression is hey it's nice to have ... putting it on the front handlebars'

Table 11.2   (continued)

Code	Examples from the protocol
Cl – Clarifying a Solution	'OK so what we have is, we have a, let's see if I can draw a sketch here'
Re – Retracting a Previous Design Decision	'we're going to have to scrap this'
Dd – Making a Design Decision	'so that's one of the possible positions'
Co – Consulting External Information for Ideas	'these we know are designs that the chap had worked out'
Pd – Postponing a Design Action	'I can't just remember right now think about that em er'
La – Looking Ahead	'actually what I would do is run down to the local windsurfing store and buy a boom extension and I would mock it up....'
Lb – Looking Back	No equivalent example
Explicit Strategies	
Ka – Explicitly Referring to Application Knowledge	'having used a backpack on a bike in the past ... I learned very early on that....'
Kd – Explicitly Referring to Domain Knowledge	'I know that em the issue is going to be bending stiffness'
Ds – Explicitly Referring to Design Strategy	'OK it's five-ten I've gotta start getting to a more concrete design'

3 Method

All forms of qualitative research are based on subjective judgements about the data by the coder(s). As a result, there can be variation between individuals who code the same set of data. A common means of demonstrating the accuracy of the results is to utilize more than one coder and to quantify differences in their interpretation of the data by calculating inter-rater reliability. For example, where a protocol contains a set number of standardized units and each unit is assigned a particular code, inter-rater reliability represents the number of units for which the coders assigned the same code, as a proportion of the total number of units.

However, this is not the process used in the current analysis of the Delft individual design session. Rather, the process used to code the present design session relies on the division of the protocol primarily according to meaning, rather than standardized units, and so there is potential for variation between coders in the number and length of units identified, as well as differences in how these units were coded. Under these circumstances, the suitability of inter-rater reliability as a means of ensuring accurate coding is questionable. As a result, we adopted an alternative strategy to ensure that we achieved an accurate analysis of the design session. The strategy we adopted acknowledges differences in interpretation between coders and views these differences as a legitimate opportunity to explore the data in greater detail and to develop a coherent, consensus coding which reflects the structure in the data.

This strategy is based on the Delphi Method which may be characterized as a method of structuring group communication processes in order to achieve agreement[12]. A series of codings was conducted by two independent coders, with an arbitration process providing the means to resolve any differences in the codings in order to converge towards a coherent final coding. The strategy involved a four-stage process of coding. The two coders worked independently in the first three stages. In the first stage, each coder applied the coding scheme to the design episode. The process for applying the coding scheme is described in greater detail below. After a period of time (several days) the episode was coded a second time. The period of time allowed for a more independent second coding. After a further period of time, each coder carried out the first arbitration process, which was to compare their own two codings. Each coder worked independently to produce a third coding in which any differences between their first two codings were resolved. In the final stage, the arbitration between coders, the two coders worked together to arbitrate between the results they had each produced.

3.1 Some Background Information on the Two Coders

The first coder was experienced in the use of the coding scheme while the second coder was trained in the application of the scheme for this workshop. Training consisted of a number of discussions concerning the coding scheme as well as some practice. The second coder practised applying the coding scheme on 20-minute sections of two different videos. One video was of an electronic engineering design episode while the other was of the Delft individual design session. The results obtained by the experienced coder for the same sections of the videos were compared and used as a guideline for the revision of coding strategies.

3.2 The Coding Process for Individual Coders

This section describes in greater detail the process which each coder followed, while working independently, to apply the coding scheme to the Delft individual design session. The video, written transcript and the copies of the designer's sketches were all utilized by the two coders to guide their interpretation of the video. The video was viewed a number of times in order to increase familiarity with the experimental situation and the overall sequence of events. The actual coding was recorded on the written transcript. The transcript was formatted in a way which would provide a legible record of the identified segments and the codes which were applied to these segments. Specifically, the transcribed speech was formatted into a narrow column so that each line contained an average of six words. The format of the transcript is represented in Figure 11.1. Three blank columns, which correspond to the three broad classifications in the coding scheme, were positioned to the left of the text column. The codes were handwritten in these columns. Each coder, working independently, went through the whole episode and identified segments to which they applied a particular code from each of the three classifications. An exception to this procedure was that, where a segment was identified as an explicit strategy (the fourth group of codes in the classification *Strategy*), that segment was not coded according to the other two classifications. This was done because by nature, the explicit strategies do not have any bearing on the designer's consideration of the problem domain. The beginning and end of the coded segments were marked on the transcribed speech with a slash. When a certain code in any column is applied to a particular part of the episode, the speech or activity which follows is then examined in order to determine if the same code applies,

			00:56:00
1	S	Ps	side mounted alright /obviously
1	B	An	we might wanna put the
			backpack on this (mutter) like
			that oh let's see there's another
			way out here huh ah that's
			gonna give some stability issues
			here we have some stability issues
			here we have some stability
1	S	Ps	issues there em em /OK er we
1	B	An	might be able to em do this /I
			think I'd be worried about the
			pack sorta not really being well
			suited there (mutter) is facing in
		Ds	em top down /OK I have spent er
			fortyfive minutes or so now forty
			minutes doing that er so em
			... (mutter) going to check every
			possible location
			00:57:00

Figure 11.1
The transcribed speech has been formatted into a narrow column so that handwritten codes can be aligned with the corresponding segments. The beginning of each new segment is indicated by a slash.

or if it should be considered as a separate segment with a different code. A new event is associated with a change in code in at least one column. In some cases, new segments are associated with a change in all three columns, while in other cases, the code recorded in a particular column may simply be repeated from the preceding segment. Where a code is repeated, this indicates that the particular code applies for the duration of that series of segments.

3.3 Arbitration between Coders

In order to facilitate the arbitration process, both individual coders' results were added to the transcript, along with a third set of blank columns to record the arbitrated coding. As a result of the coding process used, in which each coder worked independently in dividing the design episode into segments, there can be multiple codes of a single segment of the design episode or the same code can persist over a number of segments. Therefore, for the purpose of arbitration between the coders, it was necessary to segment the design episode according to both codings. In doing so, wherever a segment had been identified by either coder, this became a new segment for the purpose of arbitration. This is demonstrated in Figure 11.2. A total of 575 segments were arbitrated between the two coders. The first step

FIRST CODER			SECOND CODER			ARBITRATION			TRANSCRIBED SPEECH
0	B	An	2	B	An		B	An	... alright so OK obviously if they cut loose the wheel would come loose too so we're not gonna worry about that too much
2	S	Cl	2	S	Ps	2	S		OK so we're gonna have to have these rear stays here go down to lugnuts here to lugnuts stay to lugnut alright alright and then this is this is going to ...
2	S	Re	2	S	Re	2	S	Re	em I don't like that 02:00:00
2	S	Ps	2	S	Ps	2	S	Ps	what we're going to do is we're going to run these stays to em to this er this fitting here
2	B	An	2	B	An	2	B	An	we wanna run this to here so that we don't have to rely on the lugnuts em unfortunately the angle at which this is gonna be is gonna be a function of the em bicycle so this is gonna have to be at different angles here too
2	S	Ps	2	S	Ps	2	S	Ps	so we're gonna have a pivot through there OK I got that figured out em we em er got to have a pivot here
			2	B	An				don't like that pivot here em don't like em that
			2	B	An				but unfortunately this is just this is just handling small loads like this tipping loads em

Figure 11.2
Transcript used for purpose of arbitration. Lines across the columns and transcribed speech indicate each segment. Identical codes were recorded in the arbitration columns.

in the arbitration process was to identify those segments where identical codes were given by the two coders. In identifying identical codes for each segment, each of the columns was treated separately, that is, an identical code could occur in any of the three columns and not necessarily all three. These identical codes were recorded into the arbitration columns; where the codes were not identical in any of the columns, that column was left blank and the final decision was added following arbitration.

A number of types of differences in codes were observed, all of which were amenable to discussion and arbitration. The first source of differences is similar to that of traditional qualitative analysis, in that the coders had applied different codes to the same segments. In these cases, each coder provided a justification for their choice with reference to the source material and an agreement was then reached. In some cases, it was decided to adopt the code applied by one of the coders; at other times a completely new code was applied. The other sources of differences involved only partial disagreement. On some occasions, although the two coders applied the same codes to a particular segment, the demarcation of that segment by the two coders differed slightly, usually only by a few words. Another source of differences in coding was when the two coders had segmented the transcript at different levels of detail. For example, one coder may have identified one large segment while the other broke this down into a number of smaller segments. Sometimes through discussion, the two coders decided that the greater level of detail proposed by one coder was not warranted. However, on other occasions, the finer detail was deemed appropriate and was added to the final code. This is an example of the way in which arbitration can be used to establish a more coherent and detailed final coding. In comparison, more traditional approaches may result in a loss of information from the original protocol. A single coder may overlook some of the finer detail. Where more than one coder is employed, the aim is usually to demonstrate that a high degree of similarity between codings can be achieved, rather than to synthesize the results into a richer interpretation of the protocol.

To achieve the final coding decision for each segment, both coders worked together to discuss and examine analytically the underlying reasons for the differences in codes, by referring either to particular parts of the transcribed speech or activity observed in the video or to previous coding decisions. The transcript was then reorganized according to the segments defined in the arbitration. The timer facility in the video was used to log the times associated with each event. An exemplary section of the final coding produced by the arbitration is presented in Appendix A.

4 Results and Example Analyses

Figures 11.3 and 11.4 illustrate how our coding scheme can represent events in a design protocol as a function of time over a full design session. The figures also demonstrate how a

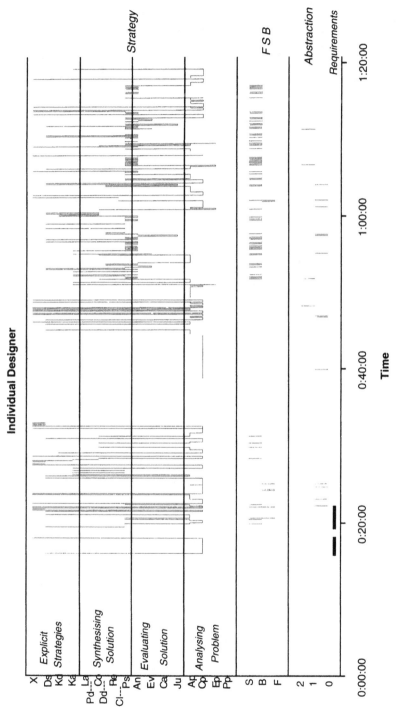

Figure 11.3 Coded events as a function of time for the first hour of the design episode.

Individual Designer

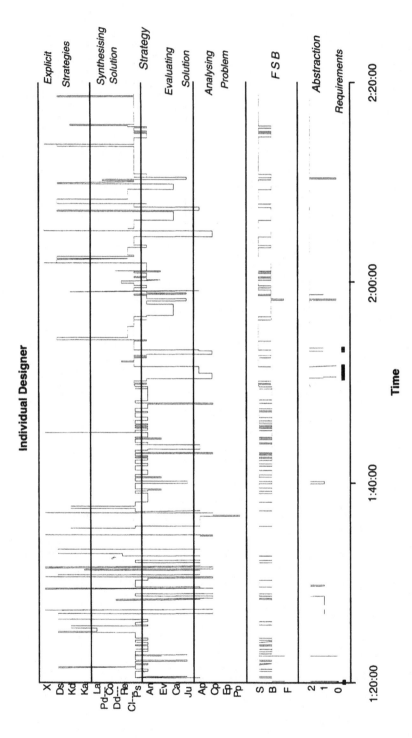

Figure 11.4 Coded events as a function of time for the second hour of the design episode.

complex set of coding categories can be visualized and how this facilitates examining the relationship between the categories. The design session has been divided into one-hour time segments. Figure 11.3 represents the first hour of the session and Figure 11.4 represents the second hour. Together, the two figures provide an overview of the designer's activity throughout the design session, in terms of the 27 categories used in the coding scheme.

For each of the classifications *Strategy, Problem Domain: Function, Structure and Behaviour* and *Problem Domain: Level of Abstraction*, the pattern of vertical and horizontal lines portrays the designer's activity at any particular time. The vertical position of the line corresponds to the relevant code and the horizontal position of the line indicates the amount of time associated with that code. For example, the upper portion of Figure 11.3 indicates that the first event in the design episode has been coded as an explicit reference to a design strategy (Ds), followed by a period of analysing the problem (Ap). By examining the lower part of the figure, it can be seen that this analysis of the problem corresponds to the designer being engaged in reasoning about Function (F) at the highest level of abstraction, the system level (0). The designer also considered the initial requirements at this time. In the figures, there are some quite noticeable breaks in the line. These breaks occur for two reasons. Firstly, some sections of the design episode pertain only to the experimental situation and not the design process and so these sections were not coded. The large break which occurs between 00:35:00 and 00:40:00 corresponds to the time when the designer made a phone call and was waiting to get through to the relevant person. The second reason is that explicit strategies are not associated with the two problem domain categories, and so, whenever an explicit strategy occurs, a break appears in the lines which represent the *Problem Domain: Function, Structure and Behaviour* and *Problem Domain: Level of Abstraction* categories.

As well as providing information about the designer's activity at any time, the figures also allow the reader to discern patterns of activity across the design episode and comparisons across the three classifications can also be made. A brief interpretation of the design episode derived from the figures is outlined below.

Overall, the designer utilizes four main strategies throughout the design episode, Proposing a solution (Ps), Analysing a proposed solution (An), Analysing the problem (Ap) and Consulting information about the problem (Cp). The pattern observed in the first half of the design session clearly demonstrates that the designer is building an understanding of the problem (see Figure 11.3). The longest period of time associated with any one strategy occurs between 00:39:54 and 00:44:28, when the designer uses a telephone conversation as a means of consulting information about the problem. Indeed, the designer spends a considerable amount of time consulting information about the problem (Cp) and analysing the problem (Ap) until around 1:40:00, after which this type of activity gradually decreases. This coincides with an increase in the amount of time engaged in synthesizing a solution (see Figure 11.4). In the latter two-thirds of the session, the designer proposes and then analyses a solution in a rapid series of cycles. In the final stages of the design session, the designer concentrates almost exclusively on proposing solutions.

In terms of *Problem Domain Function, Structure and Behaviour* the figures indicate that a greater amount of functional reasoning occurs at the beginning of the design session (until around 00:35:00) relative to the latter half. A return to function corresponds to the highest level of abstraction (0). A pattern which is repeated throughout the design episode is that of a period of rapid changes between reasoning about behaviour and structure followed by a period of slower changes. The periods of rapid changes appear to be associated with synthesising and evaluating solutions at the lowest level of abstraction while the slower periods seem to coincide with analysis of the problem.

Unlike the changes observed in *Strategy* and *Problem Domain Function, Structure and Behaviour,* the patterns observed in relation to *Problem Domain: Level of Abstraction* reflect a less rapid rate of change. The designer focuses on a particular level of abstraction for quite lengthy periods of time. The overall pattern suggests that the designer steadily decomposes the problem, beginning at the system level (0), then considering the subsystems (1), and finally addressing the detail level (2). Occasionally, the designer switches back from detail to the system level, but not from detail to subsystems. The designer mostly considers the requirements of the problem at the very beginning of the design session; however, he also addresses the requirements towards the end of the session, prior to an increase in reasoning about structure.

Figures 11.5 and 11.6 illustrate how theoretical predictions can be tested using this approach. Gero *et al*[7-91] have argued that design involves three types of reasoning – reasoning about function, about behaviour and about structure. Furthermore, this view proposes that design generally proceeds from a conceptual description of a problem (which involves reasoning about function and behaviour) to a description of an artefact as a solution to the problem (this involves reasoning about structure). A simple specification of this hypothesis is to examine the percentage of the time spent using these three types of reasoning over a design session. In order to explore this hypothesis, the design episode was divided into one-minute intervals. For each interval, the percentage of time spent by the designer in reasoning about Function and Behaviour (as compared to Structure) was calculated. Using a spreadsheet program, graphs were generated which indicate changes in the proportion of Function/Behaviour to Structure over the duration of the design episode. Using this method, it is possible to alter the time scale of the graph. Since the average event in the design session is around half a minute long and some events are longer than one minute, a graph which consists of one-minute intervals tends to oscillate from zero to 100% and this obscures any pattern in the data. To discover if patterns do occur and reduce the effect of noise in the data, moving averages were calculated over five- and ten-minute intervals. That is, at each one-minute interval, averages were calculated over the preceding five- or ten-minute period.

Figure 11.5, which presents the percentage of Function and Behaviour codings averaged over 10-minute time intervals, represents the overall trend in the data. This figure demonstrates a very systematic decline in these two types of reasoning over the full design session, indicating the increasing importance of structural reasoning. The design session begins with 100% Function/Behaviour and finishes with 100% Structure. Further it is apparent that there are two broad rates of decline in these two types of reasoning. The first is relatively slow and occurs throughout the majority of the design session. The second is very rapid and takes place at the end of the session. However, it is also apparent that within this overall trend there are systematic variations – that is periods where structural reasoning declines and the other two types of reasoning show a relative increase. The systematic nature of these variations

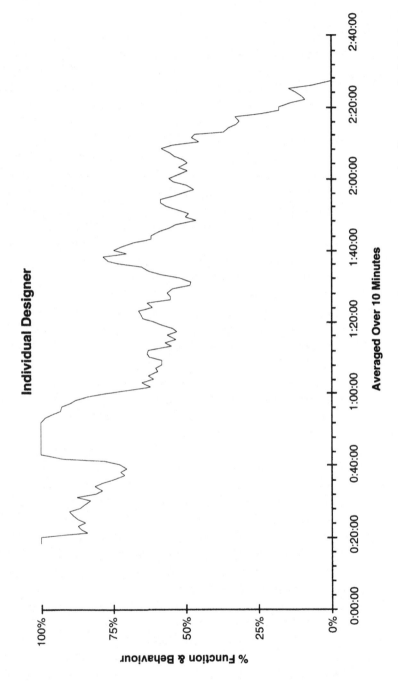

Figure 11.5 The percentage of time spent in reasoning about Function and Behaviour as compared to Structure, calculated at one minute intervals and averaged over periods of ten minutes.

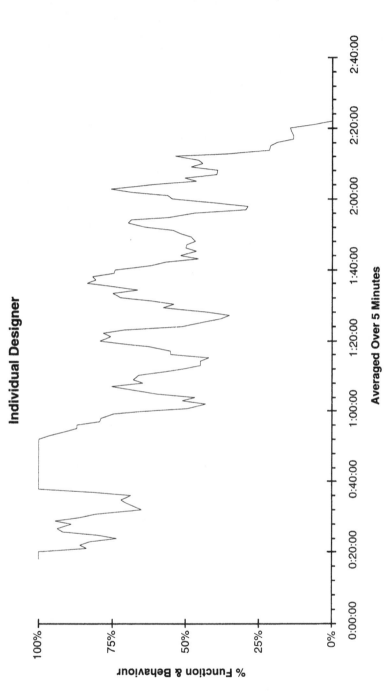

Figure 11.6 The percentage of time spent in reasoning about Function and Behaviour as compared to Structure, calculated at one minute intervals and averaged over periods of five minutes.

within the overall trend can be demonstrated through the finer-grained analysis provided by the five-minute moving averages shown in Figure 11.6. From this graph it is apparent that this variation within the overall trend is very systematic with a series of clear cyclic changes present, particularly in the second half of the session. It is also apparent that the rapid decline in these forms of reasoning at the end of the session is essentially unaltered in the finer-grained analysis.

If moving averages over a five-minute period are calculated separately for Function, Structure and Behaviour (Figure 11.7), another significant aspect of the design session becomes apparent. The amount of reasoning about Function is quite small overall and occurs as a very high percentage at the very beginning of the session with little of this type of reasoning subsequently. This is immediately followed by a rapid rise in reasoning about Behaviour which essentially dominates over the next 40-minute period. This is then followed by a rapid rise in the amount of reasoning about Structure with the succeeding period consisting of cycles between reasoning about Behaviour and Structure. Finally these cycles of different types of reasoning are quite abruptly replaced by a rapid decline in reasoning about Behaviour and a rapid increase in reasoning about Structure as the design is finalized.

In terms of the Gero et al. model[7-9], these analyses reveal that the first part of the design session does consist of conceptual reasoning as predicted. However the two types of conceptual reasoning effectively occur independently of each other with Function occurring at the very beginning of the session followed by Behaviour. Reasoning at the conceptual level also does not only occur at the beginning of a design session but a particular form of conceptual reasoning, reasoning about Behaviour, persists until very near the end of the session. While these patterns are based on one design session involving one designer, they are potentially theoretically significant and demonstrate how the approach developed here can be used to assess theoretical positions.

These patterns clearly pose the question of what is actually happening when these cycles are occurring. Figure 11.8 is a five-minute moving average based on the sets of items in the Analysis, Synthesis and Evaluation coding categories (see Table 11.1). Comparison of Figures 11.7 and 11.8 demonstrates that the patterns in the two graphs are closely synchronized.

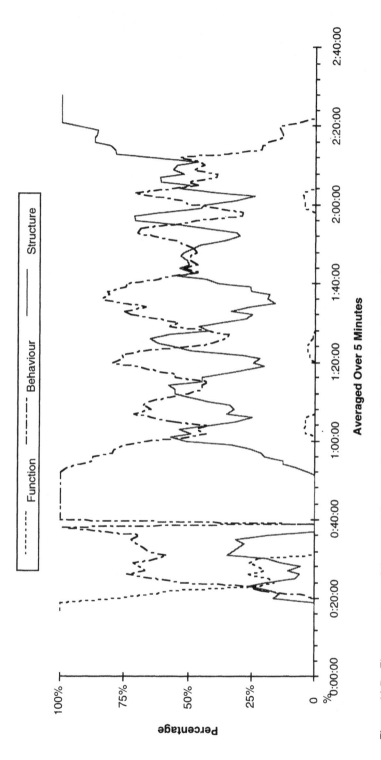

Figure 11.7 The percentage of time spent in reasoning about Function, Behaviour and Structure, calculated at one minute intervals and averaged over periods of five minutes.

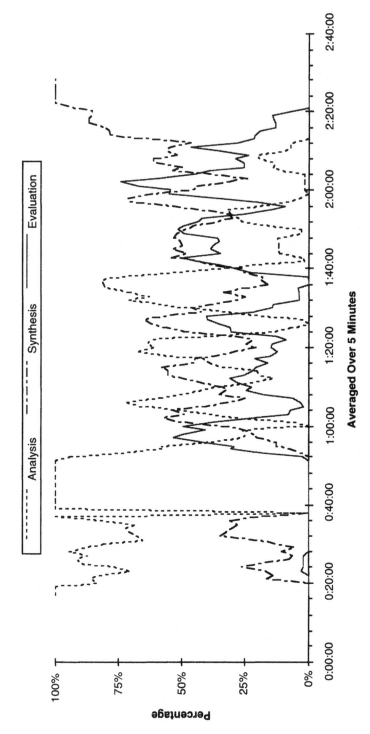

Figure 11.8 The percentage of time spent in Analysis, Synthesis and Evaluation, calculated at one-minute intervals and averaged over periods of five minutes

Analysis dominates the first part of the session when the designer is reasoning about Function and Behaviour. There is then a similar change to that which occurs in the Function. Behaviour, Structure graph. Increases in Analysis are followed by increases in Synthesis and Evaluation. These cycles correspond closely to the Behaviour, Structure cycles indicating that Behavioural reasoning is associated with Analysis and Evaluation while Structural reasoning is associated with Synthesis.

This more detailed analysis in conjunction with the Function, Behaviour, Structure analysis also allows an evaluation of a number of the other theoretical positions that can be found in the literature. For example, Lloyd and Scott[5] identify an engineering model of design based on analysis of problems followed by the synthesis of solutions and an architectural model of design where solutions need to be generated so that the problem can be analysed and understood. The patterns demonstrated in this design session indicate that, if they occur with other designers and with different problems, a more complex model is needed. Analysis clearly plays a central role but that role does not simply involve a period of analysis at the beginning of a design session. Rather analysis dominates at the beginning of the session with the analysis involving both reasoning about Function at the System level and about Behaviour at the Subsystem level. Analysis about Behaviour, however, continues after a marked drop in reasoning about Function and later in the session enters a systematic pattern of relationships with reasoning about Structure at the level of Details about Subsystems which is also associated with Synthesis and Evaluation. This part of the overall record of the design session is therefore consistent with the architectural model of the design process. Examination of the details of the protocol also reveals that the cycles of Analysis and Synthesis/Evaluation do not appear to be consistent with a view that argues that the designer breaks down the initially ill-defined problem into a series of well-defined problems which are then solved through an analysis followed by a synthesis process. Rather the cycles appear to governed by shifts in attention rather than a systematic reasoning process. The approach to the analysis of protocol data we have developed therefore seems to provide a way of developing more detailed theoretical models of the design process and would appear to be sufficiently sensitive to allow the examination of some of the particularly significant issues which have been appearing in the literature. Such issues as the differences between students and experts, between experts with differing types of training, and between problems which are typical and atypical for a discipline, can all be assessed by detailed comparisons of protocols obtained with the appropriate combinations of different groups of designers and different types of problems.

References

1   Glaser, B.G. and A.L. Strauss, *Discovery of Grounded Theory: Strategies for Qualitative Research*, Aldine, Chicago (1967)

2   Akin, Ö. A structure and function based theory for design reasoning, in N. Cross, K. Dorst and N. Roozenburg, (Eds), *Research in Design Thinking*, Delft University Press, Delft (1992)

3   Christiaans, H. and K. Venselaar, Practical implications of knowledge-based design approach, in N. Cross, K. Dorst and N. Roozenburg (Eds), *Research in Design Thinking*, Delft University Press, Delft (1992)

4   Eckersley, M., The forms of design processes: a protocol analysis, *Design Studies* 9(2) (April 1988) 86–94

5   Lloyd, P. and P. Scott, Discovering the design problem, *Design Studies*, 15(2) (April 1994) 125–140

6   Tzonis, A., Huts, ships and bottleracks: design by analogy for architects and/or machines, in N. Cross, K. Dorst and N. Roozenburg, (Eds), *Research in Design Thinking*, Delft University Press, Delft (1992)

7   Gero, J.S., Design prototypes: a knowledge representation schema for design, *AI Magazine*, 11(4) (1990) 26–36

8   Gero, J.S., K.W. Tham and H.S. Lee, Behaviour: a link between function and structure in design, in *Preprints IntCAD '91*, IFIP, Ohio State University, Columbus (1991)

9   Rosenman, M.A. and J.S. Gero, The what, the how and the why in design, *Journal of Applied Artificial Intelligence* (1994) (in press)

10 Gero, J.S. and T. McNeill, A method for the protocol analysis of design sessions submitted to *Research in Engineering Design* (1995).

11 McNeill, T. and E.A. Edmonds, Empirical study of conceptual electronic design, *Revue Sciences et Techniques de la Conception* 3(1) (1994) 61–86.

12 Linstone, H.A. and M. Turoff, (Eds), *The Delphi Method: Techniques and Applications*, Addison-Wesley, Reading, MA (1975)

Appendix A   Excerpt from the Final Coding of the Design Session

Time event begins	0–2	FBS	Strategy	Transcribed Speech
00:16:06			Ds	You bet. So I've been given the assignment and I'm reading the assignment I'll read it out loud I guess in part
00:16:06	0	R F	Cp	(reads brief) That's not a Batavus Buster over there..but..OK
00:18:00			X	*X No that one.. I could have.. I was going to say this at the end of your reading the assignment, that is not the Batavus Buster, that is a typical bike, but that is the backpack which is referred to.*
00:18:10	0	R F	Cp	OK. (continues reading brief aloud)
00:19:14			Ds	OK so I am going to... make a concept design for the device and this is a ...I will write this down so that I will not forget it
00:19:27	0	R S	Ps	..carring fastening....device.. alright.. and it is to attach to a bicycle, a mountain bike and to me that makes it different. Em, mountain bike....
00:19:51	0	R B	Ap	and this happens to be a ..em..
00:20:01	0	R B	Cp	OK let me just verify that this a..the um.. em alright and um..alright so it's a.. the HiStar backpack is a framed backpack it's an external framepack external.....
00:20:24	0	R B	Ap	OK so we know that it's actually an existing frame backpack it's an existing bike
00:20:35	0	R S	Ps	and we're making a device that is going to attach one to the other.
00:20:40	0	R F	Ap	Em and I have to focus on ease of use...good looks.....technical issues....price..
00:21:01			Ds	OK...Em OK em and the first thing I would do as a quick way of getting started is to ask you for the picture of a prototype and the test report.
00:21:27	0	R F	Cp	Em and er and you have the preliminary design for the device em but anyhow that'll show me that in this em..
00:21:37			X	*X So you would like the preliminary design, the design of the prototype which has been...*
00:21:42			Ds	Yeah right right there's no sense in designing something..er..if it's already.. there's no sense in starting from scratch if you can start at square two instead ofsquare one or square zero and um I (inaudible)
00:21:54			X	*X These are the sheets*
00:21:56	0	R F	Cp	get some ideas of what's not acceptable ha ha
00:22:00	0	S	Co	Alright and er these we know are um designs that the chap had worked out OK.
00:22:07			X	*X You also asked for a report. Which report was that?*
00:22:10	0	B	Cp	Em it says something about the em test report,

Time event begins	0–2	FBS	Strategy	Transcribed Speech
00:22:14	1	B	Ap	OK and it probably right off the bat says the backpack's too high or something like that and that bicycle stability's an issue . .
00:22:22	1	S	Ps	em off bat immediately my impression is hey it's nice to have putting it on the front handlebars
00:22:29	1	B	Ju	because you uh like low inertia on the handlebars
00:22:34			X	X *This is the er User Trials report.*[Experimenter adjusts window blind]
00:22:36	0	B	Cp	OK Look through this thing here it's em OK so it is a.......OK let's just see so I've got an idea of what this backpack design looks like em eh
00:23:05	1	B	Ap	interestingly enough em..... let me see... I see how the backpackinterfaces interestingly enough em... doesn't directly take advantage of the eh frame .. nor the fact that there is a frame. The prototype interfaces (inaudible)
00:23:40			X	X *One thing to say Dan, I'm sorry to interrupt you but if these sheets are, we'd like you not to mark on them but if you want to draw on them we can get you photocopies made straight away.*
00:23:48	0	B	Cp	OK no that's OK. Em most important conclusion to our (reads aloud from report) OK (continues reading aloud) OK
00:24:20	0	F	Cp	The product was considered ugly .. alright Takes a while to get used to cycling with this weight..
00:24:29	0	B	Cp	OK now em the ugliness is related to the frame itself OK Takes a while to get used to the way the attachment device itself because thats what the device is is.
00:24:46	1	B	Cp	Er the weight of the frame of the fastening device was er enough to cause problems,
00:24:56	1	B	Ap	well that's interesting it means that it was made too heavy probably
00:25:00	1	B	Cp	and no location on the centre of gravity attaching the fastening device to the pack tended to put the rear fixing points too low

12 *Comparing Paradigms for Describing Design Activity*

Kees Dorst and **Judith Dijkhuis**
Delft University of Technology, The Netherlands

This collection of papers will help to show that there are many ways of describing design processes. Each researcher will have attacked the design process in his or her own way, based on a unique choice of assumptions and goals. In this paper, two basic and fundamentally different ways of approaching the design process will be discussed, and evaluated on their descriptive value.

1 Two Paradigms for Describing Design Activity

Over the years, many systems for describing design processes have been developed. The 'first generation' methods of design methodology in the early 1960s were heavily influenced by the theories of technical systems. The positivist background of these theories made for design being seen as a rational (or rationalizable) process. Criticism of these models raised interest in the fundamentals of design theory, the logical form and status of design. It also fostered a need for more detailed descriptions of the design activity, leading to more attention for designers and design problems, rather than just for the design process.

Problem solving theories introduced by Simon[1] provided a framework for this extension in the scope of design studies by allowing the study of designers and design problems within the paradigm of technical rationality. Simon also provided a sound, rigorous basis for much of the existing knowledge in design methodology. This paradigm, in which design is seen as a rational problem solving process, has been the dominant influence shaping prescripitve and descriptive design methodology ever since. Most of the work done in design methodology today still follows the assumptions, view of science and goals of this school of thought (see Section 1.2).

A radically different paradigm was only proposed some 15 years later, by Schön[2], describing design as a process of reflection-in-action. This constructionist theory can be seen as a reaction to the problem solving approach, specifically made to address some of the blind spots and shortcomings Schön perceived in mainstream methodology (see Section 1.3).

The two paradigms for design methodology represent two fundamentally different ways of looking at the world, positivism and constructionism. These two ways have been with us literally since Plato disagreed with Aristotle. We do not aim to solve that disagreement in this paper, but we do hope to shed some light on the properties and limitations of these two modes of looking for the study of design.

1.1 Design as a Rational Problem Solving Process

Seeing design as a rational problem solving process means staying within the logic-positivistic framework of science, taking 'classical sciences' like physics as the model for a science of design. There is much stress on the rigour of the analysis of design processes, 'objective' observation and direct generalizability of the findings. Logical analysis and contemplation of design are the main ways of producing knowledge about the design process. Simon quotes optimization theory as a prime example of what he believes a science of design could and should be[1].

The problem solving approach means looking at design as a search process, in which the scope of the steps taken towards a solution is limited by the information processing capacity of the acting subject. The problem definition is supposed to be stable, and defines the 'solution space' that has to be surveyed. The view of design as a rational problem solving process has helped to give a much-needed stable basis to design methodology, and has informed much of our knowledge about design today.

1.2 Design as a Process of Reflection in Action

In *The Reflective Practitioner*, Schön[2] has developed what he calls a 'primer' for a 'new theory of design'. He argues that the prevailing positivist paradigm is hampering the training of practitioners in the professions. He sees the training programmes as being defined in terms of generalities about the design problems and design processes, without any attention to the crucial and

Item	'Simon'	'Schön'
designer	= information processor (in an objective reality)	= person constructing his/her reality
design problem	= ill defined, unstructured	= essentially unique
design process	= a rational search process	= a reflective conversation
design knowledge	= knowledge of design procedures and 'scientific' laws	= the artistry of design: when to apply which procedure/piece of knowledge
example/model	= optimization theory, the natural sciences	= art/the social sciences

Figure 12.1
The rational problem solving paradigm and the reflection-in-action paradigm summarized

difficult problems of the linking of these two in a concrete instance. Any design problem is unique, a 'universe of one', and a core skill of designers lies in determining how every single problem should be tackled. This has always been left to the 'professional knowledge' of experienced designers, and not considered describable or generalizable in any meaningful way.

Schön calls this the essence, 'the artistry' of design practice. Thus he finds it unacceptable that these problems cannot be described in the prevalent analytical framework, and that their solving therefore cannot really be taught in the professional schools. To describe the tackling of fundamentally unique problems, Schön proposes an alternative epistemology of practice, based on a constructionist view of human perception- and thought-processes. He sees design as a 'reflective conversation with the situation'. Problems are actively set or 'framed' by designers, who take action (make 'moves') improving the (perceived) current situation.

A summary of the two paradigms is shown in Figure 12.1.

2 Criterion: Describing the Design Situation

The aim of this paper is to compare the different paradigms for describing design processes on the closeness of their descriptions to the design activity. 'Closeness' will now be defined as the measure in which these methods capture design in the way designers themselves *experience* it. Design is not just a process or a profession, it is *experienced as a situation* in which a designer finds him/herself.

2.1 The Design Situation   Until now design methodology has failed to take into account this situational aspect of design. But if the academic field of design methodology wants to influence design practice and education, it should address the problems designers have, and do that in a way that designers recognize (experience them).

A more fundamental reason for dwelling on the designer's experience of design situations is that the multi-step process of designing is 'controlled' by the designer's *decisions*. These decisions are based on the *perceptions* of the designer at work in his or her design situation. This makes the understanding of (at least this perceptual aspect of) the design experience a prerequisite for any real understanding of the design activity itself. But what kinds of things do designers experience whilst in the design situation? This has been touched upon for instance by Winograd and Flores[3]. A situation is defined by the subjects' perception of the current state, goals and possibilities for action, and his or her way of dealing with 'thrownness'.

Winograd and Flores use a situation comparable to design (chairing a meeting) as an illustration of Heidegger's notion of 'thrownness'. When designing, you are in a situation in which

- you cannot avoid acting
- you cannot step back and reflect on your actions
- the effects of action cannot be predicted
- you do not have a stable representation of the situation
- every representation is an interpretation
- you cannot handle facts neutrally; you are creating the situation you are in.

The philosophical theory behind this approach to reality is phenomenology, created by Husserl, and later extended by Heidegger[4] and Merleau Ponty[5].

As a designer, you are in a situation in which you are continually faced with the very concrete challenge of your perceived design problem, and you have to decide on the *kind* and *content* of the action to take in this situation. 'What does this situation mean?' and 'What action can or should I take in this situation?' are eternally recurring questions. In most cases, considerations linked to the *content* of the design situation (the perceived design problem, the designer's goals and the perceived possibilities for the next step) will determine the 'kind of action' (*process*-component).

Note that designers also make process-driven decisions, in particular when they are making a planning or checking their progress. But this requires them to 'step out of their design situation'. These 'jumps' into a wholly different way of thinking can easily be seen in any protocol of a designer at work. The conclusion must be that these process-driven decisions – the object of much current design methodology – are not really part of the *core* design activity itself.

2.2 The Limits of Design Methodology

In studying design as a process, one is looking at the process-component of largely content-based decisions. This severely limits the power of a process-oriented methodology to *understand* what is going on in the design activity, and to help designers that are trying to work their way through the design situation towards a solution. Because of this process-focus very little knowledge and hardly any theory has been built up about the kinds and content of design problems, or the kinds of goals designers have (such as coherence and integration). We are strongly convinced that in order to get to a deeper understanding of the design activity, design methodology should now start to address at least some more aspects of the design situation.

2.3 The Layout of This Study

This paper will focus on finding out how much we know and could come to know about the design situation, using the two main paradigms of design methodology as the best 'tools' we have. This will be done by performing a protocol study, and trying out two different data processing systems that correspond with (are distilled from) the two paradigms that lie at the basis of design methods. The performance of these data processing systems in capturing the design situation will be seen as a measure for the 'descriptive value' of the paradigms themselves. Attention will focus on the ability of both description methods to preserve the process-content link in design decisions, and the perception of the design problem.

Please note that this paper is a rather informal presentation of a bigger study, in which the two paradigms were compared much more thorougly. The notion and mechanism of 'integration' was defined in both paradigms, and protocol data were processed with the two data processing systems. Space here does not allow the full treatment of that approach, and in the end we decided to omit it altogether. But some of the conclusions at the end of this paper have been drawn on the basis of this much bigger and more thorough analysis. That analysis will be reported fully in the thesis by Dorst[6].

3 Applying the Two Paradigms of Design Methodology

The encoding systems based on the two paradigms are introduced, demonstrated and discussed in Sections 3 and 4, respectively. Here we will draw some more general, theoretical conclusions on their behaviour. Overall conclusions will be drawn in Section 5.

3.1 Describing the Design Activity as a Process of Rational Problem Solving

There is an uneasy relation of the paradigm of rational problem solving with empirical research. Logical reflection has always been the main productive method for the researchers working within this paradigm. Protocol studies are seen as worthwhile, although the value of the results is constantly under fire; case studies are definitely seen as irrelevant, because they do not lead directly to generalizable knowledge of the design process. But there have been a large number of empirical studies that clearly operate within the paradigm as outlined above. One of the empirical studies that is most interesting to us is reported in the paper by McGinnis and Ullman[7]. It is a detailed study that gets very close to a description of the content as well as the process of design decisions while staying within the rational problem solving paradigm. This analysis has been extended and repeated here.

Every 15 seconds of the design process was scored with a data processing system containing five main categories:

1. Acts – what does the designer do: write, think, sketch, take a break, etc.?
2. Goals – with which goal does the designer perform this action: determining the problem, making a performance specification, building the concept, plan, etc.?
3. Contexts – from which perspective does the designer look at the problem: the user, the bicycle company.?

4. Topics – which topic is the designer dealing with: the bike, company policy, the maximum size of the product, materials, etc.?
5. Auxiliary topics – is the designer working in comparison with other products, referring to earlier projects, or in any other way reflecting on his or her own way of working?

A score for a 15-second stretch of a design process then looks like this:

03 02 05 35 00

which translates back into:

The designer is writing (03) the performance specification (02), looking from the viewpoint of the users (05) at the location of the backpack (35) (without reflection on design or comparing to other products (00)).

Note that there is a direct correspondence to Simon's problem solving theories[8]. The 'knowledge state' of the designer can be deduced from combining the 'context' and 'topic' categories. A 'problem–behaviour graph' could be constructed by combining these categories, and mapping them cumulatively.

3.2 An Overview of the Protocol

This overview of the individual protocol takes the form of four graphs, reflecting the four main categories of the encoding system. For each main category, the scores used for encoding are listed, followed by the corresponding graph of the workshop protocol of the individual designer.

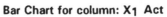

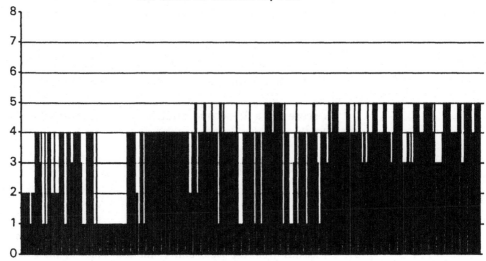

Observations

Category 1:   ACT
ask info   01
read   02
write   03
think   04
sketch/draw   05
model   06
take a break   07
sort/get things   08

Bar Chart for column: X$_2$ Goal

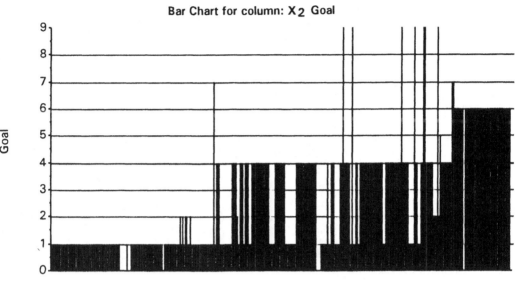

Observations

Category 2:   GOAL
none   00
determine the problem   01
make performance specification   02
generate concepts   03
build concept   04
make an evaluation, decide   05
prepare presentation   06
manage process (policy & strategy)   07
make an overview   08
decide   09

Bar Chart for column: X_1 Context-5

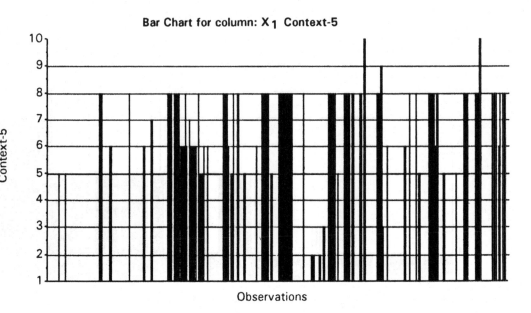

Observations

Category 3:   CONTEXT
none   00
neutral   01
by stakeholders – the company HiAdventure   02
                 – the company Batavus   03
                 – the designer   05
                 – the users   06
by aspects   – ergonomic   07
                 – technical   08
                 – form   09
                 – business   10

Bar Chart for column: X₄ Topic

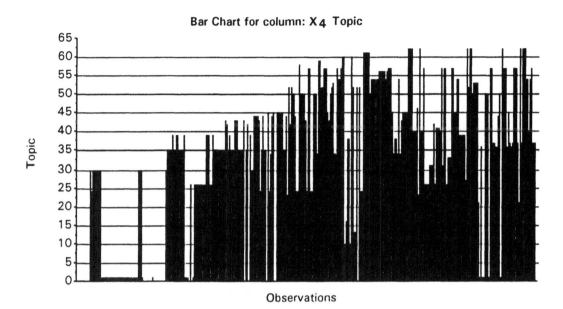

Observations

Category 4:	TOPIC		
none			00
the subject:	brief, the problem as such		01
the project	HiAdventure:	– policy	10
		– resources (time & tools)	11
		– position (role in the project)	12
Batavus:		– policy	13
		– resources (time & tools)	14
		– position (role in the project)	15
users:		– policy	16
		– resources	17
		– position (role in the project)	18
the designer		– goals (ideals)	20
		– resources (planning)	21
		– position (role in the project)	22
the environment	physical environment of the product		23
the bicycle		– all	24
		– the Buster	25
	the backpack		26
	user behaviour		27
	competition/patents		28
	norms/rules/laws		29
the product in	– general principle/layout		30
terms of basic		– use – put on/off backpack	31
problems/solutions		– put on fastening device	32
		– technical – put on/off backpack	33
		–put on fastening device	34
		– location of backpack	35
		– materials	36
		– production method	37

	– costs & price	38
	– bicycle stability	39
	– quick joining/accessory	40
	– fit to most mountain bikes	41
	– wobbling of content	42
	– looks	43
	– sizes	44
	– stiffness	45
	– safety	46
the product in terms of physical	– fastening device	50
parts:	– joining mechanism backpack	51
	– joining mechanism interface	52
	– stack-away mechanism	53
	possible physical parts of the system:	
	– pins	54
	– snaps	55
	– braze-ons	56
	– tubes	57
	– crank	58
	– fenders	59
	– brackets	60
	– mounting point	61
	– lugnut	62

3.3 Discussion

The space allotted to us here does not allow a complete discussion of the results. A few remarks will have to give an impression of what struck us most in this analysis.

The encoding

The 15-second interval time for scoring the protocols was just about right. (See also the paper by Akin and Lin, Chapter 2 in this volume.) The designers seldom change subject or approach twice in such an interval.

The subjects that the designer is dealing with, 'topics' in this encoding system, were surprisingly hard to score. Although the encoding system does provide a large number (over 60) of *a priori* categories to make scoring easier, these were by no means enough to capture the vagueness and subtleties of concept manipulation. Varying degrees of interpretation were required.

The encoding system

The encoding system was based on a simple classical analysis model, which means that observations are classed in *a priori* categories and that connections between subsequent observations are severed. This makes the encoded data hard to 'read'.

In order to find out what is happening, one really has to

refer to the written-out text of the protocol. Apparently, some vital information is lost in the severing of the links between observations, and in the use of the *a priori categories*. Just reading the encoded data gives you no idea what the concept looks like.

The written-out protocols give the impression of the designer being pretty consistent in his concept-building process, but the graphs show no pattern at all in the concept-building stage. This must mean that the encoding demolishes or at least fails to capture the pattern that is there.

Statistical analysis of these nominal data can give some information on how often this or that category has popped up. But this analysis only allows us to compare stretches of time, and tells us nothing of the possible importance of a category for the design activity.

The persistent recurrence of a certain category (in this case, topics 35 and 39, 'location' and 'stability', for instance) can give some inkling of its possible importance.

The graphs could lead to remarks on the time scale of some of the design activities: many people comment that the designer starts drawing much later than they expected (but the one-hour mark is pretty consistent with our own earlier research). The lack of basic design theory precludes any firm conclusions.

Nevertheless, the alternation between the different activities, contexts, and topics does provide information about the general nature of design processes.

3.4 Conclusions on the Descriptive Value of Seeing Design as a Process of Rational Problem Solving

The 'topics' category, combined with the acts and goals, gives some idea of the reasons for the different steps and the eventual course of the design process. But this way of looking at the design process has no way of dealing with the logical links of the one to the other. Links can be reconstructed, but textual analysis (concentrated on the content of the design problem, and consequently outside this paradigm) remains necessary to forge them solidly.

Patterns can be found in the scored protocol data, echoing strategies or heuristics in the design activity. These patterns can be most clearly seen in the information and embodiment phases, when you could expect more routine-like behaviour. The conceptual phase shows an erratic jumping between activities, with hardly any pattern at all.

The paradigm of rational problem solving does not provide a basis for the study of design problems and their structures, and is very much focused on the process component of design decisions. That limits the understanding one can get from analyses like these of the design situation. The lack of theory on design problems makes this way of looking at the protocols little more than a 'bookkeeping' of the design process[7].

The rational problem solving paradigm does not provide us with some detailed theory on what would be a 'good' or 'healthy' design process. Some general principles like concentric development, and working from the abstract to the concrete can be found in every protocol hitherto collected by the authors; they cannot explain the differences in quality of the end result.

This kind of data processing system can be very valuable in comparing design processes. (This is the way it is used in the thesis that we mentioned before. There the results of this analysis are also linked to quality of the end result, which gives a refreshing insight in what could be good and bad in design processes. Overviews like these can be instrumental in the formulation of more detailed hypotheses on the process-component of the design activity.)

4 Describing the Design Activity as a Process of Reflection in Action

Schön's well-written description of his architectural protocol[2] sparks immediate, intuitive recognition by designers. It inherently combines the content and process components of a designer's actions. The essence of Schön's theory is that designers are active in structuring the problem, and that they do not evaluate concepts, but that they evaluate their own actions in structuring and solving the problem. The unit of 'doing design' is not a design concept, but an action.

Designers work by *framing* a problem in a certain way, *making moves* towards a solution and *evaluating* these moves on the criteria of

- coherence (am I following a line of reasoning?)
- accordance with the specifications (am I on the right track?)
- the problem solving value (have I made things worse?).

The frames are based on an underlying *background theory*, corresponding with the personal view of the designer on design problems and his or her personal goals.

The first column gives the time in minutes and the designer is quoted in the second. The third column gives the kind of statement: MV (move), FR (frame), BTH ('underlying background theory'). 'Blackburn' refers to Blackburn, Inc., a large bike-accessories manufacturer which was telephoned by the subject when he was 24 minutes into the design process.

005	there's no use starting from scratch if you can start at square two	BTH
006	stability's an issue	FR
007	doesn't directly take advantage of the frame (efficiency in construction)	MV BTH
010	centre of gravity is very high	FR
010	you want to keep that as low as possible	MV
012	don't try and reinvent the state of the art	BTH
014	stability (connected to Blackburn)	FR
019	don't redesign things	BTH
023	... wanted to know what's the tradeoff between carrying panniers on the front versus the rear	FR
025	centre of gravity is very high (question to Blackburn)	FR
025	did it in the front (question to Blackburn)	MV
025	not much space between the frame (question to Blackburn)	MV
026	pack on one side (question to Blackburn)	MV
028	push it further back (question to Blackburn)	MV

We can summarize further by taking into account only the 'succesful' moves and frames (the ones that the designer stuck by). This gives a clear picture of the what, how and why of the design concept.

The design problem was *framed* as a stability problem. This rules the positioning of the backpack, and generally the technical context for the concept design (designed for maximum stiffness).

The most succesful *moves* were:

- to position the backpack as low as possible lying above the rear wheel of the bike
- to adopt a triangular structure for the product.

The *background theories* informing these moves and frames were:

- the efficiency of a process: don't reinvent the state of the art
- the efficiency of a construction: triangular supports give the most *stability* for the least material

4.2 Discussion

The encoding

This data processing system is reasonably easy to score. Most mistakes come from a confusion between frames and moves; better definitions could help the encoder 'getting his head straight'.

The encoding system

This way of looking at the protocols is very much content-focussed; something of the process–content link is preserved, but the process is not very well described. The links between consecutive statements are preserved. But a study of those links from these data still requires a lot of reconstruction and interpretation on the part of the researcher.

This description of the design process is staunchly problem-dependent. Research along these lines would result in studies that would be very hard to compare, and thus hard to draw general conclusions from – case-studies, in fact.

The formation of the concept can be followed very closely by looking along the lines of this paradigm. Information and embodiment phases are only reported very sketchily, because the more or less routine sequence of moves in these phases requires only a few decisions to get started, and is only evaluated at the very end.

The consistency of the design activity is much clearer in this description than in the other one. The frames provide the glue that tie together the subsequent design statements.

4.3 Conclusions on the Descriptive Value of Seeing Design as a Process of Reflection in Action

The link this paradigm provides (conserves) between design process and the content of the design problem is most valuable. But the treatment of design as a reflective conversation lacks the clarity and rigour achieved by the rational problem solving paradigm.

This paradigms gets us closer to describing design as experienced than looking at design as a rational problem process does. The process–content link in design decisions is preserved, and so is the perception of the design problem.

The weakness of the underlying theory makes it very hard to draw any general conclusions from this description of design.

For example, because there is no theory on the structure of design problems, there is no basis for judging the appropriateness of a certain frame. This limits the usefulness of this theory of design as reflection-in-action to providing a very structured way of making case-studies (for the time being, that is).

5 Overall Conclusions

Describing design as a rational problem solving process is particularly apt in situations where the problem is fairly clear-cut, and the designer has strategies that he or she can follow while solving them (as was the case in the information and embodiment phases of this workshop protocol).

Describing design as a process of reflection-in-action works particularly well in the conceptual stage of the design process, where the designer has no standard strategies to follow and is proposing and trying out problem–solution structures.

Seeing design as reflecton-in-action manages to decribe the design activity without totally severing the close link between the content and process components of design decisions. Taking the action (move) as the 'unit for studying design' also gets us much closer to the activity of design as experienced by designers. This would put a very extended and systematized version of Schön's theory in a very good position for possible application in design practice and education. The theoretical base of this theory should be developed further (e.g. through building a taxonomy of design problems, and of frames) so that more rigorous and generalizable conclusions can be drawn from this. There is no theoretical reason why this could not be done, and it has to some extent already been done by some builders of expert systems[9].

References

1   Simon, H.A., *The Sciences of the Artificial*, MIT Press, Cambridge, MA (1992)
2   Schön, D.A., *The Reflective Practitioner*, Harper Collins, New York (1983)
3   Winograd, T. and F. Flores, Understanding and being, in *Understanding Computers and Cognition; A New Foundation for Design*, Ablex, Norwood, NJ (1990)
4   Heidegger, M., *Being and Time* (1962)
5   Merleau-Ponty, M., *Phenomenology of Perception*, Routledge, London (1992)
6   Dorst, C.H., PhD thesis, Faculty of Industrial Design Engineering, Delft University of Technology (1996)

7   McGinnis, B.D. and D.G. Ullman, The evolution of commitments in
     the design of a component, *Journal of Mechanical Design*, **144** (March
     1992) 1–7
8   Newell, A. and H.A. Simon, *Human Problem Solving*, Prentice Hall,
     Englewood Cliffs, NJ (1972)
9   Tzonis, A., Huts, ships and bottleracks: design by analogy for ar-
     chitects and/or machines, in N. Cross, K. Dorst and N. Roozenburg
     (Eds), *Research in Design Thinking*, Delft University Press, Delft
     (1992)

13 Use of Episodic Knowledge and Information in Design Problem Solving

Willemien Visser
INRIA, Le Chesnay, France

Problem solving based on 'reuse' of problem-solving elements, i.e. particular solutions to particular problems, rather than on the use of general problem-solving knowledge ('problem-solving reuse', or 'design reuse' in this text), is considered to play an important role in design[1-3] (*cf.* Case-Based Reasoning[4]). This often-made statement is generally based on introspection by authors who are design methodologists, A.I. researchers or designers themselves. The conclusion which may be drawn from their papers is that designers may have good reasons to proceed to design reuse. A question which may be asked, however, is: do designers indeed proceed to reuse and, if they do, why, and how do they proceed? Only a few empirical studies have been conducted on this question, almost exclusively in the domain of software design[5,6].

The study presented in this paper originally intended to examine the question of design reuse in order to identify requirements for reuse support tools through a characterization of the actual reuse activities. The analysis conducted and the first results obtained led to a change of perspective: I discovered that, in addition to the reuse of problem-solving elements, the use – or reuse – of other types of 'episodic' data, particularly knowledge, also played an important role in the design problem-solving examined. This paper will present the questions and hypotheses formulated at the start of the study and the elements which led to modifying the original perspective

(Section 1), the results on actual problem-solving reuse (Section 2), and the results concerning the role played in design by episodic data not specifically linked to problem solving (Section 3). In Section 4 I will review and discuss these results and present some suggestions for design assistance.

1 Protocol Analysis in Search of Problem-solving Reuse: Identification of Episodic Data

At the start of the study, I intended to analyse the protocol of an individual designer (Dan) with respect to problem-solving reuse. The paper does not present the technique adopted for the protocol analysis. Its rather informal quality seems justified given its exploratory character and its aim, which is to examine if designers proceed to reuse, and if so, to identify possibly different reuse modes and/or types of knowledge reused.

The analysis was guided by several questions based on results obtained in three previous empirical design studies[7]. Although these previous studies on designers working in different domains (software[8], mechanical[9], and aerospace-structure[10] design) were not especially conducted in order to examine reuse, Visser[3,6] identified several characteristics of design reuse. Two of them are related to aspects of reuse identified in the present study and will be discussed in the final section of this paper.

Target–source relationships First, the association between the design problem in hand (the 'target' problem) and a reusable problem-solution association (a 'source') is often one of similarity (analogy or another kind of similarity), but also other relationships may be exploited by designers, especially opposition[7]. Second, the source can come from a more or less remote domain (even if the 'distance' in question is difficult to measure).

Authors of the sources Designers refer both to sources developed by themselves and to solutions designed by colleagues or anonymous authors (e.g. designs presented in the literature, such as famous architectural projects).

1.1 When Does One Call Use of Problem-solving Knowledge 'Reuse'?

Knowledge, i.e. data collected, processed and/or elaborated in the past, and integrated into memory, always plays an important role in problem solving. The knowledge used in problem solving which has been studied most is abstract knowledge (problem-solving schemas or rules) referring to types, or categories, of problems and solutions. Recently, researchers have started to discover the importance of problem-solving reuse: the use of 'episodic', i.e. *particular*, experience-linked sources which are at the *same abstraction level* as the target problem ('cases'), rather than general knowledge structures at a more abstract level. Notice that both can be applied in one and the same problem-solving situation[11].

1.2 Different Types of Episodic Data

I use the term 'episodic data' as an extension of Tulving's[12] notion of 'episodic knowledge'. Next to knowledge, i.e. data from an internal source, a problem solver may indeed use information on particular experiences coming from external sources.

In order to identify elements reused by the designer, I analysed the protocol by searching for references to the use of episodic data exploited in order to solve the design problem. Looking for this kind of data, the analysis led to the discovery of episodic problem-solving elements. But I also identified examples of episodic data not especially linked to a problem-solving context, that were still very important for the resolution of the design problem.

When referring to these data, often the designer verbalizes a link relating them to a particular episode which was their 'experiential' source. For example, the knowledge that, 'in pack-to-bike attachments, the centre of gravity must be kept as low as possible' or 'the pack should be attached in the centre' can be considered general rules applicable to designs involving bike and pack combinations. The way in which Dan (the individual designer from the Delft Protocol Data) evokes this knowledge as related to (based on) his personal biking experience, leads to qualifying the knowledge he refers to as 'episodic':

> 00:55   'That sorta looks classy, having a backpack in the centre; in fact when I biked around Hawaii as a kid, that's how I mounted my backpack. It was a framepack and I have to admit that, if there's any weight up here, this thing does a bit of wobbling and I remember that as an issue.

In addition to the reuse of problem-solving elements, I therefore analysed the use of such other episodic elements. This 'use' could have been called 'reuse', and could not have been distinguished from the 'pure' problem-solving reuse. I will come back to this point in the Discussion.

2 Reuse of Problem-solving Elements

I first present briefly a global analysis of the activity and the ideas Dan has about reuse. I then characterize the problem-solving sources reused and the different types of their use.

Global analysis of the activity   In order to relate the use of data to the main components of the design activity, two 'stages' were distinguished in the design process analysed, according to the activity which played the major role in the stage:

00:16–00:52 (36 minutes)   Stage 1. Construction of a problem representation

00:52–02:18 (86 minutes)   Stage 2. Solution development and evaluation

Two remarks should be made concerning this decomposition:

- I have many reservations concerning decompositions of cognitive activity into separate, independent, consecutive stages[13].
- The distinction between problem-representation construction, solution development and evaluation is particularly ambiguous in design where the progressive construction of a problem representation is a very important component of solution development.

Dan's ideas about reuse   On several occasions, Dan's *remarks on* his activity, translating his metaknowledge – which I (try to) distinguish from the *verbalizations of* particular traces of his activity – may make us think that reuse is playing an important role in his activity. Five minutes after he has begun the design session, Dan asserts:

> 00:21   'There's no sense in starting from scratch if you can start at square two instead of square one or square zero'.

A few minutes later, he states:

> 00:27   'My general philosophy is: don't try to reinvent the state of the art if it already exists'.

At 00:35, he restates this philosophy:

> 00:35   'Don't redesign things'.

Sources may be qualified with respect to many features. Five have been selected to be used here, because of their interest in the study.

Two families: Batavus and Blackburn   Data on particular designs and devices play an important role in Dan's construction of his problem representation. These data mainly come from two families. One is the design of a carrying/fastening device and a corresponding prototype elaborated by the client, Batavus. The other comprises various carrying devices developed and commercialized by a competitor, Blackburn.

Two aspects: descriptive and evaluative   Dan examines these sources from two angles: a descriptive one (technical–commercial presentations) and an evaluative one (users' evaluations, manufacturer's tests and judgements by an expert, a Blackburn employee).

Different status: from preliminary design to validated product   The solutions have different status in the design–validation cycle: Batavus' solutions are a preliminary design and a prototype; Blackburn's devices are final products, i.e. validated solutions.

Different degrees of 'satisfactoriness'   The Assignment mentions that 'the user's test performed on [the] prototype ... showed some serious shortcomings'. Thus, the prototype design is an existing, unsatisfactory solution to the present design problem. The Blackburn solutions are more or less satisfactory: the main object of Dan's phone call to Blackburn is to identify the degree of 'satisfactoriness' of different solutions and the underlying factors.

Availability   The existence – and thus possible availability – of the Batavus solutions and evaluations had been referred to explicitly in the design assignment; that of the devices by Blackburn or other companies had not.

The different ways in which Dan uses the sources will be detailed under two headings translating the two global objectives corresponding to the two 'stages' mentioned above.

Three examples of the way in which sources are clearly used to construct a problem representation are given in this subsection, but a more equivocal example may be found in the next section ('Solution shift or problem-representation shift?').

Examine data: first descriptive, then evaluative In order to construct his problem representations, Dan analyses the sources, going from preliminary design to final product, and from descriptive to evaluative data. After he has read the Assignment, he asserts his general goal, translating one of his first problem representations:

> 00:19   'OK, so I am going to make a concept design for the device and this is a ... 'carrying/fastening' 'device' ... and it is to attach to a ... mountain bike.'

Examine 'what exists already': a 'quick way of getting started' An examination of the Batavus prototype and its users' evaluation are considered a 'quick way' of getting some ideas of 'what exists already' (descriptive data) and 'what's not acceptable' (evaluative data).

> 00:27   'OK. This is the part of the process where I'm sort of just getting some familiarity with the design and with the design issues and state of the art.'

Examine similar, existing solutions After a quick analysis of the 'home-made' solutions, Dan asks for information on 'any other backpack, framepack/backpack or carrier devices [of which] the Company [has] done any market surveys'. The experimenter presents him with 'comparable products available in the US', from Blackburn and another company, Winchester. Dan examines the Blackburn documentation, then skips through the Winchester material, which he judges as not relevant for his present task.

After these descriptive data, Dan asks for 'user evaluations of the different types of any of these products' (see above, 'Examine data: first descriptive, then evaluative'). When the experimenter cannot provide these, Dan tries to obtain them from another source, i.e. a Blackburn employee, whom he is going to question by telephone, a normal information-collection mode to Dan.

The first three uses of the sources described below are presented as mainly contributing to solution development, but the second one illustrates the arbitrary character of the distinction between problem-representation construction and solution development and evaluation.

Solution proposal Source analysis and evaluation, mediated by a more or less elaborated problem representation, lead to solution formulation. As soon as he has looked through the three drawings of the prototype, Dan evaluates, negatively, the device:

00:23 '[It] doesn't directly take advantage of the frame.'

One of the two key features of his own solution is exactly going to take advantage of the frame:

01:54 'That is going to be *the* feature: ... [four] clips holding the backpack.'

His quick examination of the prototype drawings also leads Dan to formulate his first solution idea:

00:22 'Put it on the front handlebars.'

This solution proposal may be explained as the result of a particularly strong expectation-driven processing of the source: the Batavus carrier is a frame mounted *on the rear* of the bike.

Solution shift or problem-representation shift? Mainly because of information gathered during the phone call with Blackburn, the 'front-handlebars' idea (see above) is abandoned. Dan opens the solution-development stage asserting:

00:51 'My first thought is: "Hey, the place to put it is back here".'

One might analyse the observed shift as leading to either a 'new' solution to 'the' problem or a solution to a 'new' version, i.e. another representation, of the problem.

Solution-attribute assessment The importance attributed to the solution constraint 'heel and thigh clearance' (as Dan reports during debriefing) is based on data gathered on the Blackburn solutions during the phone call. This importance leads Dan to calculate, approximately, the distance required between the pedal and the backpack.

The following three uses of the sources contribute to solution evaluation.

Anticipation of possible problems and solutions   Dan questions the Blackburn employee, consecutively, on a great number of characteristics of the carriers and their evaluation (00:39–00:44), in order to anticipate the things 'people can complain about' and to try to find out 'what problems I am going to have and pick the right option'. So he asks very precise questions about the possible positions of the carriers: front vs. rear; above the wheel (top-mounted) vs. alongside the wheel (side-mounted); if side-mounted, on one or on both sides; if high above the bike, front vs. rear.

Proposal of two solution options, but elaboration of only one   The side-mounted location is considered by the Blackburn employee as one of two possible options; the other is the top-mounted location. Dan does not, however, elaborate both options. Mainly for reasons of making what – he thinks – 'everybody likes', he privileges the central top position:

> 00:57   'There is that issue of it being off the side: you know, from the aesthetics standpoint, everybody likes things symmetric.'

During debriefing, Dan notices that, due to bad scheduling, he did not have time enough to elaborate both options:

> 'I probably woulda wanted to make two designs or something like that, try the side-mounted one and the back, [but] scheduling my time on these things like this, that's the hardest part'.

Confirmation of the possibility and/or validity of a solution proposal   Based on the use of both problem-solving sources and personal experience, Dan develops a solution to the 'parallelogram' problem in the form of a triangular structure. In order to evaluate his idea, he examines if Blackburn had implemented such an option, i.e. he uses previous problem-solving elements to confirm the possibility and/or validity of his idea. When he discovers that Blackburn indeed did, Dan elaborates the idea, which is going to become one of the key features of his final solution, 'a triangular rigid structure with no bends in it', 'providing the lateral stability for the rear of the bike'.

3 Use of Other Episodic Data

As noted above, not all episodic data used in order to solve a problem come in the form of problem-solving elements. The analysis of the protocol shows various interesting examples of other episodic data used in order to solve the design problem. Episodic data can come from an internal source – a person's own memory – or from external, other sources. Both were exploited by the designer.

3.1 Exploiting One's Own Experience

Personal experience in cycling (with or without a backpack, on a mountain bike or any other bike) plays an important role in the construction of a problem representation, in addition to the previous problem-solving elements whose reuse has been presented above. The personal episodic knowledge based on this experience can be used in different ways.

Solution proposal In order to determine a distance of the backpack to the frame that he wants to base on foot size, Dan takes his own body as the reference:

> 00:52–00:53   'I have about a 12 inch foot: I would assume that you probably don't wanna have a foot which is any more.'

(Note that this 'bodily' 'objects-at-hand'-based knowledge might be considered as different from episodic knowledge. For the 'objects-at-hand' and its pervasive use in Dan's design problem solving, see Harrison and Minneman, Chapter 19, in this volume.)

Solution evaluation   Dan also uses his biking experience to corroborate users' evaluations from the test report:

> 00:26 'I've read the Users' trials/evaluation. I have some agreements with that: the fact that the backpack is handled in a vertical position puts its centre of gravity very high: having used a backpack on a bike in the past and having ridden over many mountains...'

Solution-attribute assessment The importance attributed to a solution attribute may be based on episodic data. An example is the experience-based knowledge that the rigidity of the backpack–bike attachment is very important:

> 01:00 'The biggest thing that I remember in backpack mounting is that it's gotta be rigid, very rigid!'

The great importance of the rigidity attribute contributes to the choice of one of the design's key features, a 'triangular rigid structure'.

Attribute-value choice Personal experience may be a unique source for the adjustment of a theoretically developed value to reality. A possibility which 'holds' 'in theory' may be considered not 'realistic' because of what may happen. Personal experience may have shown that what 'might happen' indeed happens in real life – or that the risk that it happens is great. This may lead a designer to adapt the theoretical value:

> 01:58 'OK now the reason I'm also gonna do that is because I'm sure that people are gonna try and misuse it; in fact kids are gonna sit on this thing so it's probably gonna have to be able to handle a little bit more than that.'

Signal possible problems This function is more general than those presented above. The analysis shows that an important, possibly unique role played by episodic-data use is to point to possible problems (*cf.* the examples presented above, and especially the use of personal experience to adjust a theoretical value). Personal experience may make designers become sceptical concerning certain characteristics of a 'theoretical' design, and not base their final design on it:

> 01:35 '[Being] in line is something that I never believe fully and not only that: if you got into your first little bike accident and you ever bent this frame, you might very well not have them in line, so...'

3.2 Exploiting the Experience of Other People

The experience of other people also is an information source used in problem solving (*cf.* also the use of the phone call with the Blackburn employee whose expertise was exploited to gather design problem-solving elements). This experience can be accessed and exploited in several ways. The interest of data gathering from human informants rather than from non-human information sources will be discussed below.

Using other people's experience in problem solving

Other people's experience may be referred to explicitly as coming from them. The reference may also be implicit – or there may be no reference at all – when the experience is considered a shared one.

Using friends and colleagues as informants Dan often uses – or under normal conditions would use – friends as informants.

> 00:33 'What I would do at this time if I was [at my office]: I would get on the phone, call up some friends. ...
> I have a friend who happens to have worked in the bicycle business for many years and I would probably call him and pick his brain a bit.'

Designers often use colleagues as informants – or colleagues present themselves as such without being asked to do so[14,15].

Using other people's experience as 'shared' experience One can also use other people's experience without perceiving the need to refer to them as source. This may be due to the fact that one also possesses the knowledge in question on the basis of one's own experience. For example, without having asked, Dan knows how packing is done by 'many people':

> 00:51 'Framepacks: many people don't carry their sleeping bags inside: they would carry it outside.'

Thus, designers use other people's experience as 'shared' experience when they attribute to these other people knowledge, beliefs, appreciations or other experiences that they (also) have had themselves. In these cases, there is no distinction between one's own personal episodic knowledge and that of other people. An example is Dan's decision that the weight 'one' would carry on a mountain bike is going to constrain the weight one would put into one's backpack:

> 00:51 'You wouldn't carry that much on a mountain bike. I would say probably in the order of 20 pounds...; I think that that's more like it of what a person might have.'

(In the above extract, 'that much' refers to the 'theoretically' possible weight, which is going to be adjusted – see above.)

Interest of human compared to non-human information sources

Human informants may often provide information which could also be found in other, non-human sources (books and other technical documents or data-bases). There may be several reasons to prefer human to other information sources.

Access is 'direct' This point may be related to something Dan notes twice concerning calling up informants: it is 'a way of getting up to speed quickly':

> 00:46 'In fact I might very well call up a few local bicycle stores and get some information too if I were doing this and I had to really get up to speed high.'

Access to the underlying history and justifications Informants, if they have been involved personally in the experience on which the data they provide are based, may be able to provide, next to these 'pure' data, the history and/or reasons concerning the adoption or rejection of certain choices underlying these data.

Access to other expertises If the informant is an expert in a domain in which oneself is not, they can present or combine information in an expert way. The telephone call with the Blackburn employee is a good example: this person knows the Blackburn devices, and the information that he can present Dan with is incomparable with the information which Dan alone could get from the technical–commercial presentations received. Dan exploits this possibility with great ability: he is also an expert in design, even if not of bike and/or backpack devices (see below Berlin's[14] remarks concerning knowledge on information sources as an aspect of expertise).

Access to other viewpoints   Other people may have had the same information sources, and/or possess information on the same objects as me, but they probably have processed this information from another viewpoint – or they simply have processed it, whereas I did not (yet).

Tradeoff between the gain and the cost of using human informants   The fact that referring to human informants has a particular interest does not mean that Dan decides to exploit all human informants available: the interest of referring to people is generally a tradeoff between the gain of the information which can be collected from them and the cost of what has to be given to them in return, e.g. personal attention and interest in the other, which also take time.

> 00:45   'I had considered calling another friend. I'm not gonna do that right now because in the interests of time here...; I'm gonna have to ask about his wife and his kids and all these other things and also how the bicycle business is. ... If I were normally designing this thing, rather than calling him during the day, I might call in the evening.'

4 Discussion

A discussion of the results will be followed by a reflection on their repercussions for design assistance.

4.1 Episodic Data and Reuse

After a paragraph on the frequency of the use of episodic data in the design process, the results will be related to two important aspects of reuse identified in previous design studies (presented in Section 1 of this text).

Are episodic data often used in design problem solving?

Explicit references in the protocol to problem-solving reuse and other use of episodic data do not necessarily exhaust the effective use of these data, so these (re)use activities are probably more frequent than protocol analysis may suggest. I did not measure the proportion of the various design solutions elaborated by Dan based on episodic data compared to solutions based on general knowledge, 'solutions developed "from scratch".' Given these restrictions, the only conclusion on this point can be that this study has presented evidence for a multi-faceted contribution of episodic-data use to design. One may formulate the hypothesis that the use of other than strictly problem-solving linked episodic data has had such a great importance in the present design activity because Dan had never designed bike and/or backpack devices.

Target–source relationships: reuse indeed, but what about analogical reasoning?

The relationships between the sources exploited and the target problem which they contribute to solve, can all be considered to be relationships of rather 'close' similarity. The two families of previous designs reused were the solution to a first version of the target problem (Batavus' prototype), and the solutions to a very close problem (Blackburn's devices designed for carrying packs, without a specific fastening device and not especially for mountain bikes). The other episodic knowledge exploited came from the domain of biking, with and without a backpack. One might thus conclude that no analogies from 'completely different' or 'remote' domains were evoked.

Two remarks, at least, may be formulated concerning this conclusion. First, as noticed already, 'distances' between domains are difficult to measure. Second, and more importantly, what delimits a domain? The target problem, i.e. the design of a bike–backpack carrying/fastening device, may circumscribe as the relevant target domains those of 'bikes', 'backpacks', and 'carrying/fastening devices'. Many questions might be raised concerning this circumscription. The only one I want to evoke briefly is the following: is the domain of 'biking', from which emanate several instances of episodic data evoked by Dan, this domain of 'bikes' that I considered one of the three target domains? Can the knowledge concerning the everyday activity of biking, with or without a backpack, on a mountain bike or not ('general' or 'common-sense' knowledge), be considered as pertaining to the same 'domain' as the knowledge involved in designing devices for bikes ('technical' or 'scientific' knowl-edge)? Does the knowledge on the everyday use of an object pertain to the domain containing the technical–scientific knowl-edge on this object? Does a designer of an artefact proceed to an inter-domain shift when he or she refers to his or her everyday knowledge on the use of this artifact, and does she or he thus use analogical reasoning?

This question is related to the following: which kind of knowledge is necessary, or only relevant, for design? Is it – only or especially – technical–scientific knowledge? As far as I know, the research literature on design does not seem to consider that the role of common-sense knowledge is important – even if it may mention this role[16]. Secondly, even if simulation in the 'application domain' is often attributed an important role in design, the contribution of everyday knowledge in this simulation is never underlined. Still, in order to simulate mentally the use of a device, often the knowledge required is everyday knowledge. What about the present design if Dan had never ridden a bike?

Among the three other design projects I have analysed until now[8–10], an aerospace design study[10] was the only one showing a frequent exploitation of what I considered 'analogies' from rather 'remote' domains: for the design of an unfurling antenna, the designer referred to studied conceptual solutions such as 'umbrella' and other 'folding' objects. The aerospace project was, of all three analysed, the one requiring most innovation: 'unfurling' was a new concept for antennas. This may perhaps explain the importance of the use of analogues: researchers have often suggested that there is a close link between analogical reasoning and creativity, and that creativity is crucial in innovative design[17]. The present design project may be qualified as rather innovative, so the question concerning the possible dependence of creativity on analogical reasoning remains open to further examination.

Authors of the sources

The previous problem-solving elements which Dan reuses come from various authors, and from the designer himself. The episodic data Dan uses are based on his own, and on other people's experience.

4.2 Design Assistance

Do the results of this study have repercussions for design tools, other than the ideas already often suggested in the literature on design, and on design reuse[3]? Which were the specific problems encountered by the observed designer in proceeding – or trying to proceed – to reuse? After some remarks about repercussions of the results which are not specific to reuse, I will discuss the difficulty of assisting reuse.

General design assistance

Three problematic aspects of Dan's design activity are discussed, all three having been discussed already in other design studies, but all three not (yet) having received a real 'solution'.

Visualization aids When the experimenter asks Dan, in the debriefing, if there were any 'things which were particularly difficult', Dan refers to two problems. The first one is visualization, which he considers as 'one of the biggest problems' he had, 'at least in the preliminary design'. The problem is known from design literature, but this does not necessarily mean that tools providing effective assistance on this point are available[18,19].

Scheduling aids The other problem Dan refers to is 'scheduling my time on these things like this'. Visser[20] has discussed the problems involved in planning during design and in assisting the activity of organizing one's design activity. Most existing design support tools are (still!) backed up by hierarchical-planning models of the design task which do not correspond at all to the actual activity of design organization[7,13,20]; and they do not support conceptual design.

Assisting alternative-solution development Dan says that he would have followed up longer than he did with two options, were it not for scheduling and time problems. Several design studies have shown that designers generally stick to their first solution idea[21,22], amending and patching it, and that they neither change it for another, nor develop it in parallel with other solution candidates. Tools assisting working-memory management may be an aid on this point, but other factors than memory limitations certainly play a role and might require specific support modalities.

Design-reuse assistance: providing access to the state of the art

As long as we have so little data on the activities implemented in reuse, it will be difficult to specify assistance tools. Given the results of the present study, what can be said about needs for assistance?

Dan considers that 'there's no sense in starting from scratch if you can start at square two instead of square one or square zero': one should take advantage of 'what exists already'. Traces of previous designs exist, on paper and in computerized databases; books, documentation and catalogues are manifold; colleagues possess information on 'what exists already' – but they are not always there, especially in technical domains where there is a considerable turnover. However, providing designers with enormous databases, containing all possible information on 'what exists already', is not a solution. The difficulty is indeed the access to the data (for the support of analogical reasoning, in particular in the form of access to analogues and other candidates for reuse: see Falzon and Visser[23]).

Berlin[14] points out that an aspect of expertise is knowledge about information sources: if these are human, experts know their areas of expertise and their 'approachability'; concerning documentation, experts know 'its reliability, location and tricks of use' (p. 14). As Berlin notices, 'this type of information is rarely documented, yet is essential for proficiency' (*ibid.*).

Do designers need assistance in use of, and access to, 'other' episodic data?

This study has shown the use of other than problem-solving linked data, but not its absolute contribution to the solution process. Concerning real design situations, we still know less about this contribution – only some anecdotal information that may be glimpsed through introspection. And if the use of episodic data indeed plays an important role in design problem solving, do designers need support in this use? One may suppose that the difficulties involved are similar to those of problem-solving elements reuse, but they are perhaps greater, or more complex, because of the 'private' character of episodic data.

Educators worry about the difficulty of providing students with 'reference information': how can they manage to familiarize beginning designers with the enormous richness of experience in their field? Even if they possess (elements of) 'libraries of past designs', these are not documented with respect to important questions for novices in a domain, such as solution procedures and alternatives which have been taken into consideration, and choices made and their underlying justifications. And the question of access and exploitation is an enormous problem (see above).

These considerations lead me to conclude that, in the present state of knowledge on use of episodic data, design-reuse assistance specification remains a difficult question. As defended already, here and elsewhere[7], I suggest that the main way to advance on this point is to analyse empirical data collected on actual reuse activities, in experimental contexts, but definitely also in real, professional work situations.

Acknowledgments

The author wishes to thank Laurence Perron-Bouvier for several embarrassing comments on the first version of this paper: I did not manage to take all of them into account. Thanks also to Jean-Marie Burkhardt and Françoise Détienne.

References

1   Pu, P., Introduction: Issues in case-based design systems, *AI EDAM*, 7(2) (1993) 79–85, and the other papers in this special journal issue

2   Trousse, B. and W. Visser, Use of case-based reasoning techniques for intelligent computer-aided-design systems, *Proc. IEEE/SME'93 Int. Conf. on Systems, Man and Cybernetics – Systems Engineering in the Service of Humans*, Le Touquet, France 17–20 October (1993)

3   Visser, W. and B. Trousse, Reuse of designs: desperately seeking an interdisciplinary approach, in W. Visser (Ed.), *Proc. Workshop 13th Int. Joint Conf. on Artificial Intelligence 'Reuse of Designs: an Interdisciplinary Cognitive Approach'*, Chambéry, France, 29 August 1993: INRIA, Rocquencourt, France (1993)

4   Kolodner, J.L., *Case-based Reasoning*, Morgan Kaufmann, San Mateo, CA (1993)

5   Visser, W. (Ed.), *Proc. Workshop 13th Int. Joint Conf. on Artificial Intelligence, 'Reuse of Designs: an Interdisciplinary Cognitive Approach'*, Chambéry, France, 29 August 1993: INRIA, Rocquencourt, France (1993)

6   Visser, W., Raisonnement basé sur des cas: une thématique transversale en psychologie et ergonomie cognitives, in S. Rougegrez (Ed.), *Séminaire 'Raisonnement à Partir de Cas'* (Rapport interne du LAFORIA, no. 93/42), Institut Blaise Pascal, LAFORIA, Paris (1993)

7   Visser, W., Designers' activities examined at three levels: organization, strategies and problem-solving, *Knowledge-based Systems*, 5(1) (1992) 92–104

8   Visser, W., Strategies in programming programmable controllers: a field study on a professional programmer, in G. Olson, S. Sheppard and E. Soloway (Eds), *Empirical Studies of Programmers: Second Workshop*, Ablex, Norwood, NJ (1987)

9   Visser, W., More or less following a plan during design: opportunistic deviations in specification, *International Journal of Man–Machine Studies. Special issue: What Programmers Know*, 33 (1990) 247–278

10 Visser, W., Evocation and elaboration of solutions: different types of problem-solving actions. An empirical study on the design of an aerospace artifact, in T. Kohonen and F. Fogelman-Soulié (Eds), *COGNITIVA 90. At the crossroads of Artificial Intelligence, Cognitive Science, and Neuroscience. Proc. Third COGNITIVA Symposium*, Elsevier, Amsterdam (1991)

11 Détienne, F., Reasoning from a schema and from an analog in software code reuse, *Fourth Workshop on Empirical Studies of Programmers*, New Brunswick, NJ (6–8 December 1991)

12 Tulving, E., Episodic and semantic memory, in E. Tulving and W. Donaldson (Eds), *Organization of Memory*, Academic Press, New York (1972)

13 Visser, W., *Giving up a hierarchical plan in a design activity* (Research Report No. 814), INRIA, Rocquencourt, France (1988)

14 Berlin, L. M., Beyond program understanding: a look at programming expertise in industry, in C.R. Cook, J.C. Scholtz and J.C.

Spohrer (Eds), *Empirical Studies of Programmers: Fifth Workshop*, Ablex, Norwood, NJ (1993)

15 Visser, W., Collective design: a cognitive analysis of cooperation in practice, in N.F.M. Roozenburg (Ed.), *Proc. ICED 93. 9th Int. Conf. on Engineering Design* (Vol. 1), The Hague, 17–19 August 1993, Heurista, Zürich (1993)

16 Eder, W., Creativity demands knowledge – types for purposes, in W. Eder, V. Hubka, A. Melezinek and S. Hosnedl (Eds), *Engineering Design Education*, Heurista, Zürich (1992)

17 Visser, W., Use of analogical relationships between design problem–solution representations: Exploitation at the action–execution and action–management levels of the activity, *Studia Psychologica*, **34**(4–5) (1992) 351–357

18 Candy, L. and E. Edmonds, Artefacts and the designer's process: implications for computer support to design, *Revue Sciences et Techniques de la Conception*, **3**(1) (1994) 11–32

19 Guindon, R., Requirements and design of DesignVision, an object-oriented graphical interface to an intelligent software design assistant, *CHI'92* (1992) 499–506

20 Visser, W., The organisation of design activities: opportunistic, with hierarchical episodes, *Interacting with Computers*, **6**(3) (1994) 235–274

21 Kant, E., Understanding and automating algorithm design, *IEEE Transactions on Software Engineering*, **SE-11** (1985) 1361–1374

22 Ullman, D., T.G. Dietterich, and L.A. Stauffer, A model of the mechanical design process based on empirical data, *AI EDAM*, **2** (1988) 33–52

23 Falzon, P. and W. Visser, Variations in expertise: implications for the design of assistance systems, in G. Salvendy and M. Smith (Eds), *Designing and Using Human–Computer Interfaces and Knowledge Based Systems*, Elsevier, Amsterdam (1989)

14 Observations of Teamwork and Social Processes in Design

Nigel Cross and **Anita Clayburn Cross**
The Open University, Milton Keynes, UK

Most of what is known about design activity and the design process comes from studies of individual designers[1]. Teamwork in design has been studied relatively little. However, teamwork is of considerable importance in normal professional design activity, and is becoming of even greater importance in product design as it becomes a more integrated activity. There has been a growing number of studies of teamwork, particularly in the context of CSCW – computer supported cooperative work[2]. In this paper, we draw observations from the recorded three-person team design session.

Working as a member of a team introduces different problems and possibilities for the designer, in comparison with working alone. Some of the areas of difference can be surmised from the practical necessities of the situation – such as the need to communicate with other members of the team – and some others we have noticed from observation of the particular team recorded in this experiment. We have selected the following aspects for observation:

Roles and relationships
Planning and acting
Information gathering and sharing
Problem analysing and understanding
Concept generating and adopting
Conflict avoiding and resolving

In a team there will be various *roles and relationships* to be acted out, or which will affect the work of the team in some way. It has been conventional in considering design teamwork to ignore these social or psychological factors of group dynamics. However, the social dimension of teamwork in design has been acknowledged more recently[3,4]. There are some aspects of roles and relationships within this particular team that we feel are important to comment upon.

Whether working alone or in a team, it would seem necessary to have to plan one's activities to fit within the available time, but in fact overt planning of activities is not always evident in either individual or teamwork. Furthermore, it seems to be necessary in design work for unplanned or 'opportunistic' activities to be pursued when they are perceived as relevant by the designer[5]. This particular team does overtly plan its activities, but if opportunistic activities are to occur, it should be interesting to investigate how *planning and acting* are handled by the team.

In any design task, information relevant to the task has to be gathered from a variety of sources. Information search strategies for designers are poorly understood[6]. A particular feature of the experimental design of this study is that information on aspects of the problem was kept in a file by the experimenter, to be given to the designers if and when they ask for particular items. This formalizes and makes explicit and observable some aspects of the necessary *gathering and sharing of information* that any team would have to undertake.

In design, it is not normal to have a clear and immediately apparent problem given as the task, in the way that is normal in other problem-solving studies. The ill-defined nature of design problems means that *analysing and understanding the problem* is an influential part of the design process[7]. Individual designers can form their own, possibly idiosyncratic understanding of the problem, but a team has to reach some shared or commonly held understanding of the problem.

Since a design task also means that the goal is to produce a design proposal for some artifact, it is necessary to generate some ideas or concepts for what that artifact might be. An advantage of teamwork over individual work should be that a greater number and variety of concepts are generated[8]. Again, in teamwork it will be necessary to communicate and share such concepts and ideas. It should be interesting to see how *proposing and developing design concepts* are handled by the team.

A disadvantage of teamwork is likely to be that conflicts will arise between team members[9]. Different interpretations or understandings of the problem may become evident; different design concepts may be favoured by different members of the team. An inevitable part of design teamwork would therefore seem to be *identifying, avoiding and resolving conflicts*.

The aspects selected and identified above are built around a framework of studying essential or key processes of design – planning, information gathering, problem analysing and concept generating – within a context of teamwork and social processes. There are several other aspects of teamwork which we do not have space to address here – a significant one would be the using and sharing of the available work-media (drawing pad, whiteboard, etc.).

1 Roles and Relationships

An obvious difference from single-person work is that the team members have roles and relationships within the team, relative to each other. In a normal work situation, some of these roles and relationships may be formally established; for instance, there may be seniorities of position established within the team, there may be a team leader appointed by a higher authority within the organization, or there may be particular job roles.

We do not know the normal working background of the team members of this experiment (I – 'Ivan', J – 'John' and K – 'Kerry'). We do know that they all work for the same design consultancy firm, have approximately equal previous design experience, and have very similar job roles within their firm. We assume that they are all approximately equal in the hierarchy of their normal work situation, and that there were no pre-determined roles that they brought with them to the experimental session.

However, we observe from the video recording that different roles within the team were adopted. Some of this role-adoption was formalized within the team. Some other potential role-adoption behaviour was not acknowledged and formalized within the team. Informal role-adoption is evident through repeated patterns of behaviour or types of comments by an individual.

Let us illustrate this with just a few examples of the ways we saw roles and relationships being established and played within the team, and influencing what happened.

Example 1 Immediately after reading the brief, Kerry suggests that they begin by reviewing the design of the existing prototype:

> K   what do we need? I guess we should look at their existing prototype, huh?

John suggests a different activity – checking that they all share the same understanding of the problem:

> J   yeah, em, let me think; we could also just sort of like try to quantify the problem, because – what's your understanding of the problem first of all?

This 'problem clarifying' activity is then adopted as the first shared activity of the team; Kerry's suggestion has been overridden.

Example 2 *During the 'problem clarifying' activity, Kerry suggests that gathering information from the user evaluation report would be a useful activity:*

> K   they're not pleased with it so far, and the users' tests have some – in in fact it would be nice if we could see those users' tests to em see what the shortcomings were

This suggestion is ignored by the others. Shortly after, during the scheduling activity, Ivan mentions use of 'information' in the context of refining initial concepts. At this, Kerry again suggests that the user evaluation report might be a source of useful information. Again, this suggestion is not acted upon, and is dismissed as irrelevant to the task in hand:

> I   information or
> K   yeah we wanna look at the em customer feedback or the users' testing
> J   oh-yeah, so maybe, yeah, wherever that comes in in this list...

A little later still (and after meanwhile requesting from the experimenter the information on the target selling price of the product), Kerry eventually gets to ask for the user evaluation report – but note that now with the addition of Ivan's intervention:

K	I think I'd also like to get the information on em
I	the user testing
K	the user testing

Example 3 After the 'problem clarifying' discussion, Ivan suggests that they should prepare a schedule, and John and Ivan proceed to do this. Later, Ivan begins to sort out the various documents on the table top. John takes the opportunity to suggest that Ivan should adopt the role of being in charge of scheduling:

I	... let's get this stuff sorted out
J	OK you you were talking about schedule stuff before, do you wanna
I	yeah, I think we should uh just figure out
J	just set some time limits for ourselves

Later again, when Ivan is planning the schedule, this role is confirmed for him by John:

I	five-thirty we'll move on to the final cost and presentation ... let's leave ourselves a little bit of time
K	mm mm
J	Ivan's gonna be Mister Schedule
I	yeah [...] on time, under budget

Ivan adopts the scheduler/timekeeper role, and plays it throughout the session.

In these examples, we see that Kerry apparently experiences difficulty in getting the team to proceed in a way she would prefer; that Ivan apparently accepts quite happily a facilitator role as timekeeper; and that John apparently has a strong influence on what happens in the team. We believe that these examples demonstrate some of the patterns of roles and relationships within the team that are evident throughout the session. However, these roles and relationships are not simple. For instance, each member at times may take a leadership role, although playing that role in a personal style.

We also observe patterns of 'paired' behaviour. For instance, Ivan and John working together tend to undertake formalized activities, especially listing activities; Ivan and Kerry working together tend to undertake more informal activities such as open-ended discussion.

Much more analysis could be made of roles and relationships within the team than we have been able to do here. For instance, it would be relevant to take into account what members of the team are doing whilst others are actively 'centre stage'. Such apparently temporarily non-active members could be 'doing nothing'; or they could be working independently – whether on the current main activity or some other; or reflecting constructively – perhaps drawing; or pursuing another line of thought; or being attentive and tacitly supportive; or being distractive and unhelpful. Study of body language, for example, would need to be included in this kind of more in-depth analysis. Other forms of action and expression also appear to be very relevant, such as the use of laughter and jokes as ways of covering behaviour and avoiding conflict.

2 Planning and Acting

Within this team there is a consciousness of planning – members of the team are particularly aware of planning their activities and of keeping their activities to a schedule. This may seem like normal procedure for a team, but in fact not all teams in similar situations construct such an overt procedure as this team does.

Conventionally, much design activity – particularly in the conceptual design stage – is unplanned, intuitive and *ad hoc*. Other protocol studies of designer behaviour have made clear the 'opportunistic' behaviour of designers, which occurs when they deviate from current or planned activities in order to pursue ideas as they occur. It has been argued that opportunistic behaviour is appropriate behaviour for designers. However, this must create difficulties in teamwork, where activities need to be coordinated, and an opportunistic deviation initiated by one member may be seen as irrelevant by another. Our analysis here attempts to observe how this team deals with these aspects of planning and acting.

2.1 There is Explicit Planning of Activities and Scheduling of Time

Planning is initiated when Ivan suggests that they should prepare a schedule of activities, rather than just doing *ad hoc* activities:

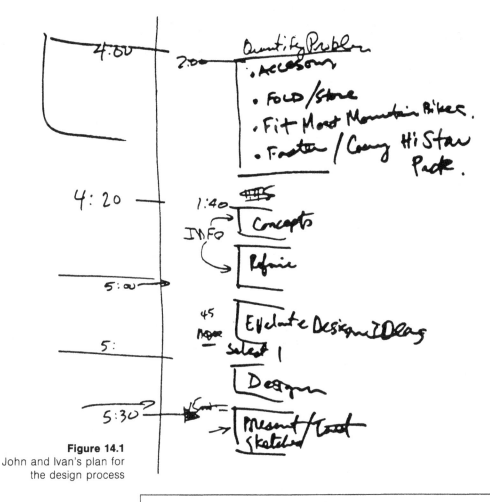

Figure 14.1
John and Ivan's plan for
the design process

> I   should we uh prepare a schedule and then just sorta
> stick to it, or should we uh just start working?
>
> J   no, it's probably a good idea to try to quantify our
> amount of time; the kinda time we have left (laugh)

Ivan and John proceed to list a design procedure (Figure 14.1):

Quantify the problem
Generate concepts
Refine concepts
Select a concept
Design
Present
Test

Figure 14.2
Kerry's plan?

This procedure seems to be derived from a conventional model of the engineering design process. Ivan and John seem to share a view that this is an appropriate design process to adopt, and this process is in fact broadly followed through the rest of the session.

However, whilst Ivan and John are listing this procedure, Kerry writes on the drawing pad a list which appears to be an alternative procedure (Figure 14.2):

Understand
Observe
Evaluate

Kerry does not draw any attention to this list, it seems to be something personal, like the personal diagram of the problem she has made previously. (This other diagram will be referred to later, under 'Problem Analysing and Understanding'.) This alternative procedure (if that is what it is), together with Kerry's apparent desire to start with gathering information on the existing prototype design and the user evaluation of this, suggests that she might have preferred to pursue this design task from the starting point of evaluation of an existing design. This is also a conventional approach to a design problem, often adopted by experienced designers.

2.2 A Role of 'Scheduler/ Timekeeper' is Assigned to One Member

In making the first suggestion that a schedule of activities should be prepared, Ivan seems to 'volunteer' as the scheduler. At a later point, Ivan begins to sort out the various documents on the table top and John takes the opportunity to suggest that Ivan should be in charge of scheduling:

I	. . . let's get this stuff sorted out
J	OK you you were talking about schedule stuff before, do you wanna
I	yeah, I think we should uh just figure out
J	just set some time limits for ourselves
I	how how much we wanna spend on each thing yeah so we can just move on

Later, Ivan is timetabling the planned design process, and this role is confirmed for him by John:

> I   five-thirty we'll move on to the final cost and pre-
>   sentation ... let's leave ourselves a little bit of time
> K   mm mm
> J   Ivan's gonna be Mister Schedule

At important points, Ivan remembers to draw attention to schedule and time:

> I   we have to start making decisions, we're already at
>   five-fifteen

John again confirms the scheduler/timekeeper role for Ivan:

> J   yeah, I don't have my watch on; you'll have to be
>   official timekeeper

Towards the end of the session Ivan keeps the time pressure on the others; for example:

> I   OK um keep moving along, we have er fifteen minutes
>   to finish our design

2.3 At Times Someone will Draw Attention to What to Do Next or What They Should be Doing According to the Plan

During the listing of requirements (John using the whiteboard), Ivan and Kerry get involved in discussing some structural design implications arising from considering the existing proto-type design. Kerry makes a suggestion about the design of the mounting forks. John suggests that they are 'moving on to ideation', and brings attention back to the scheduled task of listing requirements:

> K   it'd be nice if you could have the forks coming more
>   like that, so that
> I   right
> J   mm mm
> K   the bouncing
> J   it sounds like
> K   these'd have to be self-attached
> J   it sounds like in a way we're starting to move on to
>   ideation already, but uh have have we kinda fleshed
>   these major things out?

After listing design requirements, Ivan suggests it is time to move on to generating ideas. John agrees, but decides to check through the information again:

> I   OK shall we move into uh idea- ideation?
> J   yeah; I think we've, have we covered the uh all the stuff that (inaudible)? I'll just read some of this out loud 'cos I (reads from marketing report)

2.4 There are Unplanned Discontinuities of Activities

Despite the conscious planning that the team undertakes, not all activities proceed strictly according to the plan, and there are some curious discontinuities of activities. These discontinuities are usually initiated by one member of the team, who draws one or both of the others into this unplanned activity.

During the period of information gathering and listing of requirements (John listing requirements on the board), Kerry suddenly shifts attention to the design of the existing prototype.

> K   (referring to drawing) now this looks like a snap-in feature

This observation is then taken up by Ivan, and Ivan and Kerry spend a little time discussing structural design implications of the jolting and vibrating motion of the backpack on the rack. This leads to Kerry making a suggestion about the design of the mounting forks (this is an example of an opportunistic deviation leading to a suggestion for a design feature):

> K   it'd be nice if you could have the forks coming more like that

During discussion about alternative materials, John asks for a tape measure and begins to take measurements of the backpack frame. This activity displaces the previous listing activity, and Ivan and Kerry get drawn into it (e.g. John asks for the measurements he is reading off to be recorded):

> J   do you have a ruler or a measuring tape anywhere?
> Ex Yes there is a ruler
> J   just wondering how big this rack is; oooh... this looks like

> I   eighteen
> J   em eyeballing it looks like seventeen
> I   OK
> J   seventeen there, and em shall we record some of this
>   stuff?

Later, John acknowledges that his measuring activity had interrupted the previous activity, and gives a justification for it – his
train of thought had been triggered by earlier comments about
strength and stability:

> J   I kinda disrupted our materials discussion but I was
> I   that's OK
> J   when Kerry started talking about sorta the strength
>   issue I was thinking well how big are we looking at?

Kerry then immediately returns the discussion to strength and
stability.

A little later, discussion falters, during generation of ideas for
the kind of rack that might be designed. After a pause, Kerry
resumes discussion with something that concerns her – how to
make a 'nice connection' between the rack and the bike frame.
John turns away to ask the experimenter for information on
dimensions of the backpack; the experimenter provides a
drawing. Kerry and Ivan continue to discuss the issue of a
bracket design. John inspects the drawing but then abandons it
when he realizes it has Dutch writing on it, and interrupts the
others:

> J   OK who reads Flemish or Dutch or whatever this is?
> I   where is it?
> J   there's not too much on there (laugh)

John rejoins the discussion about the connector design. The
drawing is abandoned; no information about dimensions of the
backpack is gained.

*2.5 Activities may be
Initiated Tacitly, Rather
than there being a
Formal Decision to
Undertake the Activity*

When Ivan suggests it is time to move on to the so-called ideation phase, John agrees, but begins reading aloud from the brief in order to check that the listing of requirements is complete. Whilst he is reading, the other two become restless: Ivan gets up and goes to look at the bike; Kerry finishes her coffee, gets up to put the cup in the bin, then picks up the backpack and goes to the bike with it; Ivan lifts the bike down from its stand; Kerry positions the backpack behind the saddle. Nothing prior is said: there appears to be a tacit agreement between Ivan and Kerry that they will work directly with the bike and the backpack at this point. They are ignoring John, who concludes reading and then goes to join the others at the bike, and immediately enters into the activity, suggesting that they fill the backpack:

> J   so I think we've covered that ...
>   is there anything we could put like some weight into
>   this? to kinda give us the right

Although they had earlier agreed that it was time to move to ideation, there was no overt decision about how to do this.

John now straddles the bike and begins to talk about placing the backpack within the central diamond of the bike frame. Kerry points out the impracticality of this; she holds the backpack in the position behind the saddle that the prototype design has adopted. John suggests positioning it in front of the handlebars. Ivan records suggestions on the board. The team has now entered upon an activity of considering alternative mounting positions for the backpack, but there was no overt decision made to adopt that activity.

*2.6 There is
Opportunistic Drifting
from the Agreed Plan*

One form of opportunistic deviation from a plan is 'drifting'. For example, during discussion of weights (of the backpack; of the product that is to be designed), Kerry asks for information on existing bike racks (presumably to check the weight of comparable products). When this information is made available, it becomes more interesting as a source of design ideas:

> K   this looks a lot like the little backpack frame doesn't it?
> I   yeah ... you see we've been, it seems like mentally
>   trying to just, because of a similarity in size and shape
>   between the two, thinking of ways to er use the same

> product for the same thing but I dunno that we neces-
> sarily – I mean we're on a target for fiftyfive dollars, I
> mean if they're able to make that for fortytwo nine-
> tyfive ... and if we just add a plastic part

This discussion further develops into talk based on Ivan's experience with a child seat on his own bike. After a while, John brings the discussion back to the consideration of the weight of the product:

> J   can I interject here?
> I   yeah
> J   on our weight spec, if you just wanna sorta look at
>   these they have weights; yeah they're between four
>   hundred and thirty grams to er six hundred and thirty
>   grams

These tacit and unplanned, drifting and discontinuous changes of activity mean that it is not always easy to track what is actually happening in teamwork. This has implications for the construction of design 'rationales' and for the design of support systems which must tolerate such implicitly-understood shifts of activity.

3 Information Gathering and Sharing

The difficulties experienced by Kerry in persuading the others to collect items of information early in the session have already been referred to. The experiment design, with its controlled access to the available information, means that gathering information is a more overt activity than it might be otherwise. Relevant information not only has to be gathered, as in any design task, but also extracted from its source and somehow shared amongst the team.

3.1 Gathering and Sharing of Information is not Formalized

Kerry is the team member who first identifies that it might be useful to gather some of the specific information that is available, and which is mentioned in the brief. In doing this she is effectively 'volunteering' to be the information-gatherer for the team, in the way that Ivan 'volunteered' to be the scheduler/timekeeper. A formal role for an information gatherer/sharer might have been instituted by the team, in the same way that a scheduler/timekeeper role was instituted.

Despite the difficulties in 'persuading' the others to agree to gather information, within the first 15 minutes of the session Kerry asks for and receives information on the target selling price, the user evaluation report, and the prior prototype design. She interrupts the listing of the 'Functional Specification' on the board by Ivan and John to ask for information on the target selling price when this is mentioned as an item for the specification:

> J   ...cost target, we don't really know what that is (laugh)
> K   low
> J   low but
> K   maybe they have a – do we have that information, let's see do we ask for em – do we have any specificaton on what the uh reasonable price range is?

Having gathered several items of information in the form of reports the team tries to assimilate all this new information by simultaneous reading of the reports over each other's shoulders. Some parts (which apparently interest them or seem relevant to them) are read aloud by individuals. No formalized method for gathering and sharing information is instituted by the team, apart from the 'public' listing of requirements and concepts on the board.

At one point in discussing the removability of the rack, 'theft' is raised as an issue. Kerry refers to the marketing research report:

> I   was theft er an issue?
> K   er let's see – user marketing research?

Kerry then reads through the marketing research report, whilst the others continue to list design requirements, and 'theft-proof' is added to the list. In fact, there is no mention of theft as an issue in the marketing research report, but Kerry does not report this.

3.2 Errors in Understanding the Design Requirements, Misinterpretations of the Information and Forgetfulness of Requirements are Evident

There is no suggestion in the design brief that anything other than the specific, 'HiStar' external-frame backpack is the backpack for which the carrying/fastening device has to be designed, and the experimenter also makes this clear. But both Ivan and John are confused about this, and Kerry has to correct their misinterpretation:

J   OK I missed that

I   which part did you miss?

J   oh the fact that – I I thought I picked up that they were going to, that they were conceiving of making an internal frame pack but em I guess that's not what they're saying; you're saying that they make external frame packs currently?

K   mm hmm they make external

J   does it say that they want to stick with that?

I   well it doesn't say anything about going uh external or internal so that I think that you raised a good point

K   they just, yeah

I   yeah that we have that freedom right now

J   OK maybe we could get something that we're gonna propose to them that if it has any advantage in this application, right?

I   sure

J   OK

K   but they wanna use it with this external frame backpack it looks like

I   right with this, well let's see

K   because the HiStar, this this is a best-selling backpack – the mid-range HiStar

I   right and they have their best-selling bike, right

K   they've decided to develop an accessory for the HiStar

Late on in the session, as details of the concept design are being resolved over a drawing, Kerry and Ivan have forgotten that a requirement mentioned in the brief is that the device 'should fold down, or at any rate be stacked away easily':

K   what, the rack has to fold?

J   yes the rack has to fold

K   where does it say that?

J   it says that in our spec

I   where?

K   our spec?

J   says right here

K   (reads from brief) should fold down or – stacked away easily

3.3 Misunderstanding of Apparently Shared Concepts is Evident

Ivan and John make several references to the 'rooster tail problem'. Kerry does not query this until quite late in the session, when it becomes evident that she has not shared the same concept as the other two. She asks if they are referring to a particular strap on the backpack, whereas they are actually referring to the splashing of water/mud from the rear wheel of the bike onto the rider:

> K   we're calling this the rooster tail, this little tail?
> J   no, the rooster tail – when you, when you ride in the rain and it goes whoosh all over your

3.4 They use Personal Knowledge, and are Prepared to Rely on this Individual Experience or Knowledge

Kerry's experience with bike riding is referred to early on by John, as the backpack is being prepared by having things stuffed into it:

> J   you ride with a pack on occasionally don't you Kerry?
> K   I used to use my em bike courier bags 'cos those don't pull on my shoulder

Later, Kerry refers again to her own experience and offers an opinion on off-centre loading:

> K   well I've done a lot of lake touring and I've done front panniers and I've done rear panniers
>   [...]
> K   yeah front panniers, you could, you could set it up so you could have one of these on each side – there's no guarantee you'd always have two but it's actually not as bad as you'd think to have just one that's off

Further on, discussing whether braze-on mounting positions are standard on mountain bikes, Kerry claims that they are, and John defers to her 'expert knowledge':

> K   but these are pretty standard though
> J   the lower ones I would agree, but the uppers?
> K   that's pretty standard too
> J   the uppers are?
> K   it's getting to be, yeah; I mean it's not on this, but – actually some mountain bikes are pretty scoopy and weird, but
> J   we can assume Kerry has expert knowledge (laugh)

The errors and misunderstandings suggest that the team did not have a very effective strategy for gathering and sharing information. The fact that this was a short, experimental design session, and that there was relevant personal knowledge available within the team, probably significantly affected the team's strategy. However, the reliance on personal knowledge, rather than public and more formalized knowledge sources, could again create difficulties for support systems and rationales. Even when information is apparently shared, misinterpretations and misunderstandings are evident, which means that common, shared understanding cannot be assumed in collaborative work.

4 Problem Analysing and Understanding

The first shared activity of the team is to seek a common understanding of the problem. An attempt is made to externalize this understanding of the problem, through listing design requirements and specifications in a public form. Attempts are also made to create a more internalized understanding, through 'framing' the problem in some more intuitive or conceptualized form.

4.1 Listing and Framing are Used as Means of Understanding the Problem

During the initial discussion and clarification of the problem, John starts to list on the drawing pad a set of design requirements. Listing requirements becomes the principal activity by which the problem is summarized and shared through use of the large, public whiteboard. However, immediately after the initial problem-clarification discussion, whilst John re-reads aloud sections of the brief, Kerry makes a small, personal drawing on the drawing pad which is a diagram of the problem, showing in a graphic form the basic concept of what has to be designed – an accessory that links the backpack to the bike (Figure 14.3). This may be an example of 'framing' the

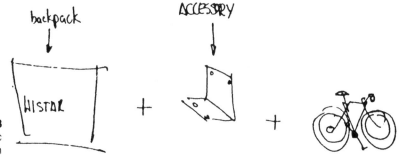

Figure 14.3
Kerry's diagrammatic 'framing' of the problem

problem in a way that helps the designer to internalize, as well
as externalize her understanding of the problem.

Problem framing may also develop through verbal exchanges.
For example, during the listing of the specification of design
requirements, Ivan and Kerry become involved in a discussion
of the problem, based on understanding how people are
expected to want to use the bike and backpack together:

I	do they talk about how the people wanna use it? they uh do these, do the vacations – they take long bicycle trips and then take short feet off uh short trips off by foot
K	mm mm
I	em so they use the bike to get where they're going and then do a little hiking; sounds like the bike becomes the
J	so you
I	it sounds like they wanta really ride the bicycle and just temporarily go to work or something but you wanna be able to ride the bicycle
K	right mm mm
J	does it sound like
K	ride it through the country and then you get to the base of the hill and you wanna take your backpack and summit the mountain or something
I	mm hmm
J	so
I	so you want like a real
K	and it's an off-road bike so you'd need a real rugged rugged attachment or a rigid attachment
I	mm mm

This conversation is probably instrumental in provoking a later
design concept proposal by Kerry for the rack to double as a
lock when the bike is left unattended. John interrupts this
problem-framing discussion to bring attention back to the listing
of a specification of the design requirements, introducing an
attempt to quantify a particular requirement:

J	so what's a reasonable time to like allow somebody to take this off their bike – should it take like under five seconds or under thirty seconds?

There is a contrast, and perhaps a conflict between attempts at understanding a problem through 'listing' and through 'framing'. Listing establishes an externalized specification, but it does not necessarily lead to an internalized conceptualizing or grasping of the problem in the way that attempts to 'frame' the problem do.

5 Concept Generating and Adopting

Clearly it is necessary for the team to generate design concepts, and jointly to build those concepts into a specific design proposal. The team therefore has to develop initial concepts into more detailed and robust versions, and it has to decide to adopt certain concepts from amongst the many that may be proposed. Two kinds of concepts are proposed by team members: general principles that should underlie or figure in the design solution, and specific ideas for the solution concept or a feature of the solution concept. An example of a general principle would be to make use of the existing frame of the backpack:

> K   you've already got that nice frame on the pack – it'd be nice if we can take advantage of that, it seems redundant like at this … it seems redundant to have that and the frame

An example of a specific idea would be to propose the concept of a tray:

> J   maybe it's like a little vacuum-formed tray kinda for it to sit in

There are many concepts proposed by the team members; the above are just two examples. We do not propose here to analyse further the proposing of concepts, but to focus on the ways concepts are developed together by the team and how team members persuade the others to adopt concepts.

5.1 Concepts are Built Cooperatively

A design proposal may begin life as a very sketchy concept that has to have a lot of development work put into it. Concepts need to be built up, with additions and variations being developed to turn the initial idea into something more robust. There are many examples of this concept building by the team members, cooperatively adding to and refining an initial concept.

Bike lock/bike stand:

K   maybe if you could flip it out and it becomes a bike lock, 'cos you know – lock up your bike while you go on a hike, that'd be kindof a neat feature so you could justify some extra cost maybe

I   right, right

J   kick-stand alternative (laugh)

I   pull it around your tyre and now you can stand the bike up

Shoulders/child/manikin:

K   what's what's kinda neat about their thing – it's not really a bag that this rack goes up inside those um webbing details in the back

I   right, that are already there

K   that kinda envelops it – it doesn't sit cinch down on this, which we could add, but it's kind of a nice nesting feature

J   maybe the rack wears the backpack straps just like we wear the backpack straps

K   sure

J   (laugh)

K   why not, see, like you mount shoulders back here

J   yeah, yeah, just maybe maybe you just mount a child seat back there and you give them a child (laugh) and make him wear the backpack

I   or a manikin

K   a manikin

I   with the top towards the

K   Harry the backpack holder

Net/retractable net:

I   uh uh what if your bag were big er what if your your on er if this tray were not plastic but like a big net you just sorta like pulled it around and zipped there, I dunno

J   maybe it could be part maybe it could be a tray with a with a net and a drawstring on the top of it, I like that

I   yeah I mean em

> J   that's a cool idea
> I   a tray with sort of just hanging down net you can pull
>   it around and and zip it closed
> J   (inaudible)
> K   it could be like a a a window shade so you can kinda,
>   it sinks back in so it just
> J   oh yeah
> I   it retracts yeah
> K   you pull down it retracts in

5.2 Persuasive Tactics are Used to get Favoured Concepts Adopted

As well as cooperating in the building and refining of concepts, team members may find it necessary to persuade the others of the value of a concept they particularly favour (usually a concept they generated themselves). It is common for designers to become committed to particular concepts, even to the extent of becoming emotionally attached to a concept. There are two clear examples within this team of the need to persuade others to accept a preferred concept. In the first, John persuades the others that his 'tray' concept is the best to adopt; in the second, Ivan and Kerry persuade John to accept the use of the fixed, brazed-on mounting points on the bike frame.

The tray concept initially just pops up as an idea:

> J   so it's either a bag or maybe it's like a little vacuum-
>   formed tray kinda for it to sit in

The concept is quickly adopted and discussed, with ideas being added to it, but John takes care to ensure that it is added to the public list of solution concepts:

> J   I think tray is sorta a new one on the list it's not a
>   subset of bag

John also confirms that he is emotionally attached to the concept:

> J   yeah I, I really like that tray idea (laugh)

This emotional commitment is reinforced by a claim that selecting a preferred concept is a 'popularity contest', which is passed off as a joke by Kerry:

> J   I think all design eventually comes down to a popu-
>   larity contest
> I   (writes on board) tray
> K   I hate that idea
> J   (laugh)

When it becomes time to proceed with one preferred concept,
John is quick to nominate the tray idea:

> J   well OK well we know we we like this tray idea,
>   right?

The second example of persuasion is based around using the
brazed-on mounting points. This is strongly supported by Ivan
and Kerry; John has some reservations, but acknowledges
Kerry's 'expert knowledge':

> J   so I I guess my point is I think if you designed it
>   specifically around mounting points – known mount-
>   ing points on this bike – you might get yourself into
>   trouble by limiting your market a lot
> K   but these are pretty standard though
> J   the lower ones I would agree, but the uppers?
> K   that's pretty standard too
> J   the uppers are?
> K   it's getting to be yeah, I mean it's not on this, but
>   actually some mountain bikes are pretty scoopy and
>   weird, but
> J   we can assume Kerry has expert knowledge (laugh)

When it comes to making a decision, Kerry conveys her commitment and attachment to the braze-ons with an enthusiastic
response:

> I   we're gonna go with the rack, let's go with er talk
>   about braze-ons – these braze-ons?
> K   yeahhh!

John still has doubts, and wonders why the designers of the
existing prototype design did not use the braze-on mounting
points, but Ivan closes the argument by denigrating the other
designers:

> J   see here's something that just surprises me is – why if
>   the braze-ons are available, why wouldn't they have
>   used them?
>
> I   'cos they're not hot designers!

6 Avoiding and Resolving Conflicts

It is probably inevitable that disagreement will arise between members of a design team. We have already seen that disagreement arose within this team over whether to use the braze-on mounting points. More serious disagreement might have arisen if there had been competing design concepts to which different members of the team were committed. However, provided that a team collectively desires to reach an acceptable conclusion to their design task, it will have to find ways of resolving, or perhaps avoiding conflicts.

In this team, we observe instances where the team members acquiesce in a kind of non-committal 'agreement' until one of them finds an argument that closes the disagreement, and where they postpone agreement and seem to 'agree to disagree'.

6.1 Non-committal 'Agreements' are Reached

A disagreement becomes evident over designing for adjustability in the rack. John proposes that the support legs of the rack should be adjustable, but Kerry feels sure that this is not necessary:

> J   you know one of the things that seems problematic,
>   and it would be great from a manufacturing stand-
>   point if you could get around it, is this distance is
>   going to vary with frame size all over the map
>
> I   right
>
> J   but so, y'know, we were talking about maybe those
>   legs could extend before so that you could get some
>   adjustability on your rack, maybe you need that
>   anyway just so that you can adjust to different rack
>   styles – like a telescoping tube here
>
> I   mm mm
>
> [...]
>
> K   I don't think you really need it
>
> I   (laughter) OK
>
> K   because this is a twenty-six inch wheel or whatever,
>   it's pretty standard and so if this distance – you're

> right it just does vary a lot, but what's gonna change is maybe your angle on your on your rack is gonna change er er what really is gonna happen is is this is gonna be a fixed distance because if we go onto a braze-on or something down here and and you want to make sure that there's clearance here and then as the bike grows it might pivot up a little bit more

Later, John returns again to the adjustability issue, appealing to the authority of 'good human factors'. Both Kerry and Ivan make rather non-committal 'agreement' responses; Kerry just shrugs her shoulders, Ivan says 'OK'. Their doubt is evident to John, who admits that what he is suggesting is 'opinion not fact'. Ivan resolves the issue for the time being by suggesting that they can 'look at ways of making it adjustable' when they are finalizing their design:

> J   I think good human factors says it should be adjustable so that people can find the position they like
>   (K shrugs)
> I   OK
> J   em that's my opinion
> I   whatever idea we come up with I think we can
> J   opinion not fact (laugh)
> I   we can look at ways of making it adjustable

6.2 An Argument is Found that Closes a Disagreement

The non-committal 'agreement' over adjustability is not permanent. At a later point John proposes a way of incorporating adjustability into the design, but Kerry comes up with an argument that resolves the disagreement in her favour – if adjustability was necessary then it would feature in the commercially available rack designs of the Blackburn company:

> J   one way to get that adjustability for the seat post height and all that stuff is if this, say this was a single bar and it went like this
> K   mm mm
> J   and it could slide along here, that way if you need to come up more, y'know pivots around the braze-ons if it needs to come up more for a taller person or for better wheel clearance or whatever, you just kinda slide it forward and put little lock-downs on it

> K yeah yeah, I don't think you need to change this length 'cos the wheel is fixed enough that you can rotate about the braze-on, and I mean if if you really need adjustment I think all these Blackburn racks would have adjustments

Interestingly, the work of other designers can be used either to support an argument (as here in Kerry's reference to the Blackburn designers) or to refute an argument (as in Ivan's earlier reference to the 'non-hot' designers of the prototype device).

6.3 Disagreement may Remain Unresolved for the Sake of Expediency

John disagrees with Ivan and Kerry over their proposal to include with the rack an Allen key for fastening/unfastening the rack to the mounting points. Because time is running out, Ivan suggests that they keep going without resolving this issue for the time being:

> J I think 'no tools' – they should come off with no tools, except for
> I maybe it should be an option
> J well except for the thing that you wanna lock
> I well see, OK, OK
> J (laugh)
> I well let's keep going – we can we can add features later

When the question of attaching the rack to the bike in a theft-proof manner is raised a little later, Kerry returns to the proposal to bolt the rack to the mounting points, using an Allen key. The disagreement on this issue is not resolved, but is defused by everyone becoming jokey about it, and Ivan and Kerry poking fun at John's alternative (time is running out and they want to reach a conclusion):

> I I'm talking lock in terms of theft
> K yeah
> J right um and I don't know how we address
> K that doesn't address that
> J yeah this doesn't address that yet
> K that's why a nice Allen wrench and bolt is nice
> I we'll just throw it in, have this and the other thing, yeah

> K   mm mm for the Johns of the world
> I   mm mm
> K   they can use these expensive
> I   big, ugly
> K   ugly
> J   (inaudible)
> K   aerodynamic-drag, heavy
> J   I'll just evaluate my idea now!
> K   (laugh)
> I   no, I think that's a good idea
> J   oh yeah!? (laugh) uh uh
> I   OK
> J   they're ganging up on me! (laugh)
> K   help I want out of this design exercise!

7 Conclusions

It is perhaps important to point out that this particular group of designers worked productively as a team and reached a relatively successful conclusion to the set task, within the prescribed time. In the debriefing after the working session, they reported that they were reasonably happy with what they had achieved in the available time, and that 'it was fun'. Despite some of the observations we have made about the roles, relationships and social interactions within the team, there were no overt signs of frustration or dissatisfaction within individual members of the team.

However, it is clear that teamwork is a social process, and therefore social interactions, roles and relationships cannot be ignored in the analysis of design activity performed by teams. From our analysis, many aspects of team design activity can be seen to be influenced by social process factors. For example, we saw it immediately in the very first planning activity of the team, when Kerry's alternative approach was ignored or overridden. We saw it in the ways the team shifted amongst planned and unplanned activities. We saw it in the way personal commitments to particular concepts lead to social process actions such as expressing commitment and persuading others. We saw it in the socially skilful ways in which conflicts were resolved or avoided.

We suggest that these observations are relevant to the analysis of design activity, and important to the design methodology of teamwork. Design methodology, particularly in the engineering domain, has tended to treat the design process as a technical process – as a sequence of activities based on a rationalized approach to a purely technical problem. More recently, and more particularly in the architecture, product design and software design domains, attention has also been directed to designing as a cognitive process – to the cognitive skills and limitations of the individual designer. Just a few studies have begun to suggest that designing is also a social process, to point out how designers interact with others such as their clients or their professional colleagues, and to observe the social interactions that influence the activities of teamwork in design. Design methodology now has to address the design process as an integration of all three of these – as a technical process, as a cognitive process and as a social process.

References

1 Cross, N., Research in design thinking, in N. Cross, K. Dorst and N. Roozenburg, (Eds), *Research in Design Thinking*, Delft University Press, Delft (1992)

2 Olson, G.M., J.S. Olson, *et al.*, Small group design meetings: an analysis of collaboration, *Human–Computer Interaction*, **7** (1992) 347–374

3 Minneman, S. and L. Leifer, Group engineering design practice: the social construction of a technical reality, in N. Roozenburg (Ed.), *Proc. Int. Conf. Eng. Des. ICED93*, Heurista, Zürich (1993)

4 Branki, N., E. Edmonds and R.A. Jones, Study of socially shared cognition in design, *Environment and Planning B: Planning and Design*, **20**(3) (1993) 295–306

5 Guindon, R., Designing the design process: exploiting opportunistic thoughts, *Human–Computer Interaction*, **5** (1990) 305–344

6 Kuffner, T. and D. Ullman, The information requests of mechanical design engineers, *Design Studies*, **12**(1) (1991) 42–50

7 Fricke, G., Empirical investigations of successful approaches when dealing with differently précised design problems, in N. Roozenburg, (Ed.), *Proc. Int. Conf. Eng. Des. ICED93*, Heurista, Zürich (1993)

8 Visser, W., Collective design: a cognitive analysis of cooperation in practice, in N. Roozenburg, (Ed.), *Proc. Int. Conf. Eng. Des. ICED93*, Heurista, Zürich (1993)

9 Klein, M. and S.C.-Y., Lu, Conflict resolution in cooperative design, *Artificial Intelligence in Engineering*, **4**(4) (1989) 168–180

15 Collaboration in Design Teams: How Social Interaction Shapes the Product

Margot F. Brereton, David M. Cannon, Ade Mabogunje and **Larry J. Leifer**
Stanford University, Palo Alto, CA, USA

Modern interdisciplinary design demands that engineers learn to work well in teams. Teamwork requires individuals to express ideas and misgivings, listen, negotiate, etc., that is to collaborate. Engineers need to be aware of various characteristics of collaboration so that they can identify successful and poor strategies within their own work practice.

Engineering design students at Stanford University, commenting on the usefulness of design process models and prescriptive design methods in their seven- month design project, noted that while the models and methods told them *what* to do, they provided little insight into *how* to do it. In particular, students wanted more help with group dynamics, solution development and project management.

The Delft Protocol Analysis Workshop presented us the opportunity to examine a two hour long videotape of a team of practising designers developing a preliminary solution for a product that mounts luggage onto a mountain bike. Videotape allows us to take a close, careful look at how the moment-to-moment activity steers the course of the design solution. Through repeated observation, analysis and discussion we can develop a better understanding of how professional designers do such things as collaborate, develop design solutions and manage their work. This chapter explores the collaboration of the design team. Early in the session the designers reveal different solution agendas and different ways of working. They reconcile their differences through effective collaboration. We find that the content of the evolving design depends heavily on negotiation strategies, among other more subtle and ubiquitous social processes. We describe how the design evolves through designers' negotiation strategies and through topic shifts prompted by the design activity itself.

1 Video Analysis Method

We made a deliberate attempt to come to the tape with an open mind, trying to consider several facets of design activity, sometimes looking at long segments of tape to get a broader perspective, other times looking at short segments over and over again. We took a qualitative approach focusing on describing *designer interaction*, rather than a quantitative approach focusing on counting *design acts* or *design content*. In the quantitative approach, which is most common in design protocol analysis, the researcher develops a coding scheme that categorizes the design activity by topic and then spends the bulk of the research effort coding, quantifying and analysing the data looking for interesting patterns in graphs or informative statistics. We considered the design activity to be so rich that reducing the activity down to a set of categories without considerable qualitative analysis of the raw data would make considerable assumptions about what was important and run the risk of overlooking interesting aspects of the activity. We needed to immerse ourselves in the raw video data until interesting patterns emerged because we did not yet know what to look for. So we took the approach of Video Interaction Analysis[9], in which an interdisciplinary team observes tape segments looking for standard and interesting practices. The team stops the tape frequently to discuss hypotheses and it tests them by reviewing the tape segment. Only those practices confirmed by interdisciplinary scrutiny that occur repeatedly in different parts of the tape are admissible in the analysis. Using this method we did develop a scheme of categories. It is intended to illustrate our interpretation of collaboration in the design team. We illustrate the scheme against five minutes of raw transcript data so that the reader may evaluate its merits. However we do not attempt to quantify the scheme.

The analysis team consisted of four researchers with engineering design and design research backgrounds. Watching the tape for the first time to make general observations, all team members found it difficult not to engage in designing the product! We were drawn into judging the content of each proposal. With each viewing, more facets of the activity appeared. In particular, it helped us to watch the tape with an interdisciplinary team of social scientists, anthropologists, computer scientists and engineers. The tape served as a catalyst, provoking recall of hypotheses based on experience from outside the tape as well as from within. Repeated video watching determined which hypotheses were validated by several occurrences of supporting events in the tape.

The time limit and restricted setting of the protocol for the purpose of video taping emphasized certain aspects of design such as time pressure and decision making, whereas other aspects such as information gathering, organizational context, the ability to mull over ideas or engage in opportunistic solutions were restricted or removed from context. All participants felt out boundaries of what was within the 'rules' to a greater or lesser extent. Collaboration was chosen for analysis both because of our interest in this area and because it was considered relatively insensitive to the protocol design. The analysis explores how the group works together under the given conditions, whatever they might be. Also, it became evident that social interaction strongly influenced the content of the emerging product.

2 Analysis Overview

Section 3 introduces the members of the design team. As the session progresses the designers reveal quite different preferences towards ways of approaching the design problem. Furthermore, the designers reveal preferences for different types of solution. These differences in ways of working and solution preferences require the designers to negotiate skillfully a solution. Section 4 describes how the design discussion transitions from one topic to another. It describes the designers' negotiation strategies. It describes how they express their commitment to proposed solutions. Section 5 summarizes our analysis of the collaboration.

3 Roles and Solution Preferences

As if in a novel or play, the characters begin to reveal themselves in their opening lines (Transcript 1). Kerry, who has worked on products for bikes, bids to look at the hardware (comment no. 1). John, who often steps back to look at the process, counters by proposing that they develop a shared understanding of the problem (comment no. 2).

Transcript 1

[All sitting at table with problem statement]
00:07:00

1 K what do we need? I guess we should look at their existing prototype huh?

2 J yeah, em, let me think we could also just sort of like try to quantify the
problem because what's your understanding of the problem first of all?

Throughout the early stages of the session Kerry is engaged by the existing hardware and concepts. She seeks to ground the problem by examining the existing hardware and concepts in detail (Transcript 2; comments 1, 3, 6, 8). John observes this process and then bids to work back at a higher level to get a broad view of the process they will follow before diving in (comment 9).

Transcript 2

00:19:00

1 K yeah you have to put a lot of tension on to these em bolts 'cos that's the opening is in the direction of the

2 I right

3 K loading the force it'd be nice if you could have the forks coming more like that so
that

4 I right

5 J mm mm

  K the bouncing

7 J it sounds like

8 K these'd have to be self-attached

9 J it sounds like in a way we're starting to move on to ideation already but uh have have we kinda fleshed these major things out or

---

Transcript 3 illustrates how John begins to seek out ambiguity in the problem statement (comments 5, 9). He negotiates for flexibility in wanting to consider internal and external frame backpack designs as well as all types of mountain bikes (comment 22). Kerry seeks to ground the problem, countering that the problem statement suggests that they only consider the HiStar external frame backpack (comments 4, 12, 14, 16, 18). Ivan's role is emerging as one of arbitrator. He considers both positions and falls to either side. As we enter Transcript 3 they are discussing the backpack...

Transcript 3

00:09:00

1 J   OK I missed that

2 I   which part did you miss?

3 J   oh the fact that I thought I picked up that they were going to that they were conceiving of making an internal frame pack but em I guess that's not what they're saying you're saying that they make external frame packs currently?

4 K   mm hmm they make external

5 J   does it say that they want to stick with that or

6 I   well it doesn't say anything about going uh external or internal so that I think that you raised a good point

7 K   they just yeah

8 I   yeah that we have that freedom right now

9 J   OK maybe we could get something that we're gonna propose to them that if it has any advantage in this application right

10 I   sure

11 J   OK

12 K   but they wanna use it with this external frame backpack it looks like

13 I   right with this well let's see

14 K   because the HiStar this this is a best-selling backpack the mid-range HiStar

15 I   right and they have their best-selling bike right

16 K   they've decided to develop an excs accessory for the HiStar

17 I   yeah

18 K   this is the HiStar backpack the HiStar (holding sample backpack)

00:10:00
19 J   where do you see that?
20 I   at the top here
21 K   very beginning
22 J   yeah here it is on the basis of this marketing report
       HiAdventure has decided to develop an accessory
       for the HiStar and these are the two kinda func-
       tional criteria it says a special carrying fastening
       device that would enable you to fasten and carry
       the backpack on mountain bikes and then the
       device would have to fit on most touring and
       mountain bikes so it doesn't sound like it's specific
       to this one

Kerry has managed temporarily to gain an agreement from
John and Ivan that they design for the HiStar backpack. But in
the last statement of transcript 3, John, in agreeing to designing
for the HiStar opens up a bid to design for most touring bikes
(comment 22). This is counter to an earlier suggestion in
00:08:00 by Kerry to 'make it a special mountain bike so it could
have the stuff required attached[1]something] to it', tacitly agreed
to by the group.

Kerry's preference to try to pin down part of the solution is
repeated throughout the tape, as is John's preference for preser-
ving ambiguity. Table 15.1 illustrates their design solution preferences for key issues.

The group's roles are summarized in Table 15.2. Clips
throughout the paper will illustrate the development of these

Table 15.1 Design solution preferences. Two of the three designers
exhibited consistent preferences on how to approach the problem

Pin down solution (Kerry)	Preserve ambiguity (John)
Design for HiStar backpack and Batavus Buster bike	Design for various backpacks and mountain bikes
Focus on rear placement (most promising candidate)	Consider all possible placements
Design for a fixed position	Make device adjustable
Use emerging industry standard attachment method (braze-ons)	Use attachment method usable by all bikes

Table 15.2 Designer roles. These emerged during the first quarter of the design task

Ivan	John	Kerry
• Whiteboard manager • Arbitrator • Timekeeper/ keeps group on track	• Theorist: abstracts process from context • Uses process rationale as commentary to keep group on track	• Bike expert and user advocate • Seeks out context and detailed knowledge • Seeks to ground the design with specific solution alternatives

roles. Ivan gradually emerges as an organizer, timekeeper and occasional arbitrator between John and Kerry. He manages the whiteboard lists and generally keeps the group on track. Are these roles predetermined or adopted? From the tape we cannot know whether the characters play the same roles in other design situations. Nor can we tell how the emergence of one role depends on that of another. However, it is likely that a certain amount of compensating occurs, each designer seeking to fill in gaps they perceive in the group approach. Kerry exhibits behaviour noted by Guindon[2] to be consistent with that of domain experts, quickly pruning the search space to promising solutions. In Section 4 we characterize how the individuals reconcile their different positions to engage in effective collaboration.

4 Making Progress – Foci and Transitions Mediated through Social Interaction

4.1 Designing from Context

Many researchers have characterized design as being opportunistic or chaotic, moving fluidly between requirements and details in response to information or ideas uncovered that are worthy of immediate exploration[2]. The session is replete with designers abstracting from the context to gain perspective and seeking context to ground the design with affordable, manufacturable alternatives. They develop requirements by considering candidate solutions and user scenarios and tinkering with hardware. For example, requirements for easy attachment, low centre of gravity and strap containment all emerge from the context of the problem, by working with the backpack around the bike or examining the user specifications. (The structured design methods do not address how to seek out alternatives and generate solutions from the context of the problem.)

4.2 A Scheme to Describe Design Progress Mediated through Social Interaction

As the design team negotiates the problem space, each designer makes bids to have issues they think important discussed by the team. Having called focus on an issue, the other designers might engage in the focus, adding ideas towards a partial solution. The content of design then evolves through discussions adding incremental solution additions, use scenarios, justifications and information-seeking questions. The designers acknowledge other contributions with nods and short phrases or they call into question an aspect of a proposed solution. They align themselves with various aspects of the evolving solutions and approach and distance themselves from others. We looked for evidence presented in the videotape that designers were happy with the alignment of the team and, if not, how they sought to change it.

We characterize the design discussion as focusing in on issues and then transitioning topics. However, there is evidence that the designers are continuously engaged in multiple activities at different levels. Although they focus in on issues, they continuously monitor the progress of the solution from the point of view of various requirements and solution alternatives. They reflect on their course of action, monitoring and modifying their process. There is evidence that they monitor their team-mates' utterances and actions on a moment-to-moment basis and moderate their talk accordingly. It is difficult to represent such a rich process in a scheme of categories. However, we have chosen to do so to try to illustrate how the evolution of the design content is governed by the social interaction in the team.

A scheme to describe how the design discussion focuses and transitions are mediated through social interaction is proposed in Table 15.3. Before offering an example of the use of the categories in this scheme, a few notes regarding their status and purpose are appropriate. The categories draw upon those used by other researchers, such as Guindon's 'partial solution' term[2], but are not an exact copy of any one set[1,3-6]. Many different

Table 15.3 Classification Scheme for Focus and Transition. Categories are chosen as an aid in conveying the notion of focus in a group setting by enabling a more detailed examination of the video, particularly the utterances.

---

call-focus – Asks for attention, attempting to focus the group around an issue of concern

start-partial-solution – Begins to describe a solution or specification proposed for consideration

add/refer-to-partial-solution – Adds or refers to some detail or constraint in the current partial solution or specification

use-scenario – Illustrates a proposed detail, with a usage scenario

justify/refer-to-higher-principle/req.. – Gives rationale, referring to requirements, higher principles or standard practices.

~~call into question~~ – Asks for reconsideration of some part of a partial solution, justification, process etc.

acknowledge – Shows that one is paying at least partial attention, using, e.g., a short utterance or a sentence completion

needs-information – Expresses a need/request for some information or further filling-in of a detail

---

coding schemes are possible, each appropriate when taking a particular point of view on the subject material. We make no claim that these categories are entirely mutually exclusive; many of the phrases in the transcript carry some force in more than one of the categories. However, the ambiguities are few enough that classification and interpretation provide useful insight into the dynamics of the group and problem.

Our presentation focuses on a five-minute piece of transcript, 70 minutes into the design session. The team members are attempting to persuade each other of the merits of their suggestions on rack attachment. We begin by looking at how the content of the discussion evolves. Then we examine the social interaction at work.

4.3 Focus: Building a
Partial Solution

The segment begins with several calls-for-focus on the issue of where and how to attach the rack to the bike (Transcript 4a: comments 1, 3). Kerry there outlines a basic proposal for a partial solution (comment 4), with a use scenario to justify it; this is in part a restatement of one of the options that has already been identified. J and I offer acknowledgments (comments 5, 6) indicating that they now share her focus on this partial solution. Then, in a series of statements (comments 7, 9), K adds details and justifications to this proposed solution, which I and J acknowledge. J then calls a piece of the proposed partial solution into question (comment 10), offering some alternatives to a detail that is being considered; K and I bolster the initial proposal (comments 11, 12) with several references to both scenarios and broad principles such as strength and reliance on standards. Acknowledgments (comments 13, 14) indicate that each continues to be engaged by the focus.

Transcript 4a

[K and I have been looking at attachment to the bike. K leaves the bike to see if she can read Dutch on the drawing, noted by J. J and I are now around the bike and K is at the table]

[01:10:30]

1　I　**let's see we're just thinking of**

2　J　*of what (laugh)*

3　I　**we were thinking of ways** of er **put the bracket here**

[K moves toward bike]

4　K　**this is our idea of orientation** *so that we can get our bedroll without having without to move the* (inaudible) [gestures with hardware]

5　J　*the roll back I agree with that*

6　I　*right*

7　K　and then em I think it makes a lota sense to have a good compression member to hold this portion up

8　I　*mm mm*

9　K　and then then I think we can do that reliably kind of a la Blackburn rack use their kind of [points to standard attachment feature location on bike]

10　J　~~does it do we really wanna~~ use these lugs for speed of disassembly or does it make more sense to like just have something that like a plastic ferrule or something that goes around this that you

[01:11:00]

11 I   this is probably a bit stronger

12 K   this is strong and er *Allan wrenches are pretty* standard
          to be carrying around on
          a y'know

13 J   *bike anyway*

14 K   *on a back on a bike*

15 I   Blackburn, they include it with the (inaudible) just throw it in there (inaudible) Allan wrench is practically free *you're paying fifty dollars for that* so you just include it and these threads are standard or else you just include it in that tube

16 J   *yeah*

17 K   *mm mm*

4.4 Strategies of Persuasion

Notice that John and Kerry demonstrate commitment to their ideas and adopt persuasive strategies. The designers masterfully invoke the support of neutral parties such as common sense, higher principles or theories, and expert or standard practices to support their opinions. These serve to depersonalize the debate, in addition to being means of persuasion and explanation of rationale.

Common sense

Appealing to common sense is a tactic to build support for an idea from commonly held beliefs. Kerry's suggestion 'it makes a lotta sense' in Transcript 4a (comment 7) prefaced with 'I think', suggests she is embarking on a persuasive strategy but is open to negotiation. John counters with the same strategy in comment 10 suggesting of his idea: 'does it make more sense?'.

Higher principles/requirements/theories

Appealing to higher principles, theories or requirements also appears frequently in designers' efforts at persuasion. In comment 9 Kerry suggests 'I think *we* can do that *reliably*'; in comment 12 'this is *strong*, Allen wrenches are *standard*.' John counters with 'do we really wanna use these lugs for *speed of disassembly*' in comment 10 and 'it just doesn't seem real *elegant* to me' in comment 21. These higher principles or broad requirements serve as both explanation and means of persuasion.

A third persuasive strategy is to appeal to established methods by established experts. Kerry in comment 9 suggests 'I think we can do that reliably kind of *à la Blackburn rack* use their kind of [attachment]'.An even stronger appeal is made outside this section of transcript in the 62nd minute: 'I mean if you really need adjustment I think all these Blackburn racks would have adjustments'.

4.5 Commitment

Before continuing with Transcript 4, a brief transcript from earlier in the session during brainstorming illustrates how differently the designers behave before they develop a strong commitment to a solution. The speakers are much more committed to their positions in Transcript 4a than in Transcript 5.

In Transcript 5, J is seated on the bike and K is experimenting with backpack position, while I lists on the whiteboard. I repeats sentences as he writes on the whiteboard (comments 4, 6) explicitly communicating his interpretation of what J and K are doing. K places the backpack in several positions suggested by J, offering advice like 'see if you can steer' (comment 8). In Transcript 5, the designers offer little commitment to each statement, prefacing statements with 'maybe', 'what about', 'would it be too funky to'. They are proposing suggestions that don't necessarily constitute opinions and paying extensive attention to communicating their actions. Calls to question, accompanied by rationale, are quickly agreed to (comments 1, 2, 3). The designers complete each other's sentences, perhaps to communicate that they understand the other's concerns. There are no justifications or reiterations of position.

Once designers begin to preface statements with 'I think' or 'my opinion is', they are clearly in the realm of offering an opinion. However, depending on the context or intonation, these can be interpreted as either 'I think, but I don't know for sure' or 'I think and I don't care what anybody else thinks'. Thus there are several linguistic cues to determine the speaker's level of commitment to a proposition, as described in Schiffrin[7].

As they begin to embark on strategies of persuasion, parties may lessen commitment in the interests of negotiation, using phrases like 'it seems to me' and 'the way I see it'. 'It seems to me' rather than 'I think' somewhat lessens the control of the speaker, serving to depersonalize the debate. Skilful use of persuasive strategies, paying attention to communicating assumptions and appropriate moderation of commitment, maintains the negotiation process.

Transcript 5

[J is seated on the bike and K is experimenting with backpack position, while I lists on the white board]
00:27:00
1   K   so we want to put it in there ~~but I let's see if you got~~

[K places backpack near triangle]

2   J   ~~yeah you'd never really be able to~~
3   K   ~~you wouldn't be able to get your knees pedaling~~ ~~OK~~ **now what about maybe** we ought to have a prototype that kinda has it this way

[K places backpack at rear, behind J]

4   I   is it facing for ? yeah that's right facing forward
5   J   **would it be too funky to have it** on the like projecting from the front wheel?
6   I   handlebars? yeah try that
7   J   or off this handlebar stem even because that's fixed but if it's off the handlebars *you know it's like an old bike basket that way like the Wizard of Oz* (laugh)

[K places backpack on handlebars]

8   K   See if you can steer. It tends to
9   J   well, you could turn it long ways
10 I   **or if you could get it down low where the**

4.6 Opportunistic Strategies of Persuasion – Give and Take

We return to Transcript 4, where the group is still engaged in focusing on rack attachment. In Transcript 4a, Kerry backed by Ivan has tried to persuade John that they should use the standard hex bolt attachment method. John remains unconvinced (Transcript 4b: comments 18, 20), using the higher principle of elegance as a reason for his doubt, and Ivan and Kerry again offer justifications for the initial proposal (comments 21, 22). Each designer is maintaining their position. There is evidence of an impasse.

During this impasse, Kerry bids to shift the debate to more neutral territory (comment 23). She assumes tacit agreement on the issue of primary attachment with hex bolts (using Allen wrenches) and then seeks to address John's concern about elegance elsewhere in the design, offering several possible alternative solution details for secondary attachment. John joins the new focus with an alternative partial solution to secondary attachment (comment 24). Kerry has been successful in shifting the debate by using an opportunistic strategy of give and take. Seeing the discussion had reached an impasse she sought to move to another part of the design space, *taking* tacit agreement on primary attachment and *giving* ways to address the elegance concern through other means.

Transcript 4b

18 J   ~~just doesn't I dunno~~

19 K   *(inaudible)*

20 J   ~~it just doesn't seem~~ real elegant to me but

21 K   it's there you might as well use it and that reduces parts yeah

22 I   you could use it yeah like a one cent screw versus a couple of pieces or three plastic parts yeah

23 K   mm mm and then the issue that I think kinda remains is how to get this attached nicely to some portion of the bike back here and that's where you wanna get come up with your nice injection mould bracket that really works well or has a little bit of product identity maybe from a product standpoint from a marketing standpoint might be cool if there's some

24 J   maybe [01:12:00] maybe you put this down down to the er what d'you call them lugs [lifts backpack frame into vertical position] and then you er have a waist strap that goes around your waist and you don't need to attach to (inaudible)

25 I   oh yeah

26 K   yeah that could be cool

---

27 J ~~might get too much~~
28 I *yeah*
29 J ~~at your back~~
30 K *yeah (inaudible)*

4.7 Transitions

Topic shifts to avoid or postpone conflict

In addition to an opportunistic means of persuasion through give and take above, Kerry's topic shift away from the primary attachment to secondary attachment serves to avoid or postpone conflict in the group, since the group is at an impasse. Several other types of topic transition are identified in this section.

Ideas gracefully lose steam

The group encourages John's partial solution for secondary attachment, but he then calls it into question and no-one protests. This appears to be mutual recognition that the idea is not worth further consideration at this time. Rather than state as much, the idea simply loses steam. (Processes such as listing also tend to lose steam as ideas become exhausted.) This opens an opportunity for a bid for focus which Ivan takes and so a transition of focus occurs. In Transcript 4c, Ivan calls for focus (comments 31, 33) on a discussion of rack width, which has caught his attention as John gestures with the backpack frame. He develops a concern that the rack might be too wide for general riding, drawing upon the context of rack use.

Topic prompted transitions

John follows up with a question confirming Ivan's concern and Ivan bids to look for marketing information to answer their concern. He moves to the table to look for information on marketing research, and the group follows. As Ivan searches for the information, John also calls for focus on a current use scenario of bungee cords (comment 42) drawn from the comparison with the partial solution the group was working on; Ivan pursues his focus bid in parallel, rather than giving full attention to John's proposal. John laughs occasionally at his suggestions and prefaces them with 'maybe' indicating he has a low level of commitment. Kerry offers acknowledgments and supporting comments but does not engage in developing the idea. There is some sense of exploration in the conversation but nobody bids to change the topic except for John himself, indicating the group is not unhappy with the process or that perhaps they are relaxing. One topic seems to prompt another. The current usage of bungee cords prompts the concern that their product must compete with that, leading to the notion that they could purchase rather than manufacture a good solution, leading to a request for information on the manufactured rather than sales cost, leading to an assumption about the manufactured cost as a percentage of the sales price.

Halting to seek information

Ivan does not find information in the marketing survey relating to rack use to resolve the issue of rack width, but this is never explicitly stated within the group which is now following along with John's comments. The group does not return to the issue of rack width for another 25 minutes. It is worth noting that information seeking often halts a focus in the activity. Further, when the designers find information they often do not use it in ways observable to us, and if they do not find it, still they often return to a different topic. Information seeking serves to broaden the designers' knowledge of options, yet it rarely adds to the knowledge space in a predictable way, which perhaps explains why transitions of focus often occur during information seeking.

Returning to key issues

The group eventually and repeatedly returns to issues, such as positioning, attachment and materials, indicating that these are key issues to them in the design problem. This perhaps indicates their ability to shift topics effectively to stay productive. Hales[8] argues that design managers need to 'window out' and 'window in', 'concentrating effectively on the detail, while at the same time keeping the wider context in mind, a crucial aspect of managing engineering design'. In transitioning from topic to topic and yet returning to key issues, we see evidence of how a group manages itself in monitoring the broader problem yet focusing to define details, through social interaction.

Transcript 4c

31 I   **what what I was thinking is**

32 J   *(laugh)*

33 I   **if we have** like a <u>like a Blackburn rack</u> ~~the only problem that I see with it~~ <u>if you put that back up</u>

       here [J raises backpack frame back to vertical position so the width is visible] is that it it's fairly narrow I mean *you don't want a rack that's* this wide

34 J   *right*

35 I   *just for general riding*

36 K   *mm mm*

37 I   but if you had one that was close and *then you could just sorta*

38 J   did it say that people would want to use it as a regular rack or or is that a feature

39 I   *mm mm*

40 J   that we could incorporate

41 I   *I imagine* **let's see** where's the marketing research

42 J   **I mean I'm sure what** people do right now probably **is go buy a bike rack** and bungee cord this down on to the bike rack

43 I   *yeah*

44 K   *mm mm* that's definitely within the product target space

45 J   maybe we should sell them a Blackburn rack with two bungee cords for fiftyfive bucks

46 I   *(laugh)*

47 J   *so hey you're right in there they sell these for forty bucks* so

48 K   *mm mm you could buy some real nice bungee cords for that*

49 J   the purchased solution *(laugh)*

[01:13:00]

50 I   they say never make it if you can buy it out of a catalogue

51 J   no tooling (laugh) make back your tooling on your first order um but out out of fifty-five dollars **I'm wondering** if there's any sort of price breakdown that um *people want* like you know in other words what's the manufactured cost *if the sales cost is fifty-five dollars*

52 I   *yeah (inaudible)*

53 J   is there any sort of um cost specification for **we know** the sales price is fifty-five dollars but um

54 K   landed cost or

55 J   the manufactured cost of the product is there a target for that

56 X   I'm sorry say again Kees

57 J   is there any is there

58 X   No, I'm asking for the my assistant to say again

> have to estimate their own ratios OK yep you have
> to estimate your own
> 59 J   oh **we estimate our own ratios** OK
> 60 K   so we'll assume that is sell it to a retailer
> 61 J   <small>use the standard</small> one fourth model

4.8 Process Prompted
Transitions (Calls to
Order)

In the short phrases that follow (Transcript 4d: comments 62–74), the group seems to be mentally relaxing, exploring the topic of acronyms for their own amusement rather than exerting themselves to resolve issues raised recently. Interspersed in this, Ivan makes three repeated bids for a new focus (comments 75, 83, 85), indicating he is ready to move on and not happy with the current process of the group, one of relaxation. He calls to focus on an issue of process: what should we do next? The group continues to wander but he finally engages them by specifically proposing to strike issues he considers resolved from the whiteboard list, bidding for focus with comment 96 (not shown): 'positioning have we thrown out the folds in the middle?' The group responds by attending to the issues he raises from the whiteboard list.

Transcript 4d

62 I   *yeah*

63 K   *mm mm mm mm*

64 J   *manufactured costs will be one fourth the the MSRP*

65 I   *yeah*

[01:14:00]

66 K   *manufacturer's suggested retail*

67 J/I *suggested retail price*

68 I   *yeah*

69 K   from a I at an IBD ... independent bicycle dealer

70 J   *(laugh)*

71 I   **OK**

72 J   *I always wondered what that footnote meant IBID*

73 K   *IBID*

74 J   *and now I know it means independent independent bicycle dealer (laugh)*

75 I   **let's get a stock of where we are**

76 J   *OK ...* so what's a quarter of fifty-five bucks er twelve er twelve fourteen bucks

77	I	OK (inaudible) some nice round numbers
78	J	twelve fifty plus a dollar twenty-five fifteen seventy-five
79	I	OK fourteen
80	J	you're eating into the margin there (laugh)
81	K	those engineers always like to take the (inaudible)
82	J	the expensive way
83	I	OK um so where are we are we going to
84	J	I think we should goal (inaudible) (laugh)
85	I	let's talk about the decisions that we've made so far ... shall we do it based on the er based on the er brainstorming sort of

[01:15:00]

4.9 Breakdowns and Arbitration

In Transcript 6 Ivan senses an impasse and actively jumps in as an arbitrator to negotiate the conflict between John and Kerry, who are committed to opposing positions. Following an extended discussion in which Kerry has proposed the bike dimensions will not vary enough to merit an adjustable solution, John makes a strong bid for an adjustable solution (comment 1). John supports his opinion with appeals to theory of good human factors and the absent users' preference, indicating he is strongly committed to 'it should be adjustable'. Kerry shrugs, indicating she has stated her position and will not offer any more argument. The standoff is apparent to Ivan who takes upon the role of arbitrator and offers a compromise: 'whatever idea we come up with, we can look at ways of making it adjustable'. Since they are not making progress they defer the discussion of adjustability until there is a specific solution to consider and more progress can be made.

Having made a strong statement to which Kerry offers no response, John lessens his commitment with 'em, that's my opinion, opinion, not fact', in apparent effort to lower any tension that may have arisen. John then shifts the topic.

Transcript 6

[00:60:00]

| 1 | J | I think good human factors says it should be adjustable so that *people can find the position they like* |
| 2 | K | [shrugs shoulders] |

3	I	right OK
4	J	em that's my opinion
5	I	whatever we idea we come up with (inaudible)
6	J	opinion not fact (laugh)
7	I	we can look at ways of making it adjustable
8	J	**OK OK so well it's getting the materials**

4.10 Contextual Strategies of Persuasion: Stories

There are few stories in this session, perhaps because of the timeframe, but the designers use some techniques of storytelling. Schiffrin[7] offers a framework to describe how stories (or reporting) are used in persuasion, noting that story tellers use:

- Selective interpretation – recounting aspects of the event preferred by the narrator
- Evaluative devices – highlighting parts of the experience from the narrator's perspective, to show the narrator taking an orientation to what is being talked about
- Deictic shifts – shifting time, place and participants from the conversational world (storytelling world) to the story world
- Contextualization – framing an event within the story world.

John's ski pole story, in Transcript 7 (comment 2), though brief, appears to give a much more persuasive argument against selecting aluminium, than would be given by simply stating 'thin walled aluminium tube fractures easily in the cold'. The participants in the conversation are shifted to Denver and the ski pole fracture is highlighted in the context of skiing. There is no need to describe the wall thickness or pole length to anyone familiar with ski poles and skiing (although the temperature is open for debate).

Transcript 7

1	K	steel painting isn't that expensive is it
2	J	no but the only the only thing I know that's wrong with aluminum is *if you've ever skied I had my ski poles fracture on me in really cold temperatures and er I was skiing in Denver one time and my ski pole bent in half and not only did it bend in half it broke when it bent*

> 3 I   y e a h
> 4 J   *because it was so cold*
> 5 I   and you could fall on it yeah ... **OK so now**
> 6 K   and what material are you saying

Without even telling a complete story, selective interpretation and reframing of past design experiences or even of the problem statement (see Transcript 3), and creation of scenarios play an important role in developing, representing and communicating partial design alternatives.

5 Conclusion

The content of the evolving design depends heavily upon negotiation strategies and other more subtle and ubiquitous social processes that shape design work. Minneman[1] has also demonstrated that design emerges from social interaction. Team members' orientation to a solution or process is demonstrated by levels of commitment in utterances (and gestures). Depending on their level of commitment and other team members' alignment they adopt appropriate strategies of persuasion. They carefully moderate their commitment to their ideas to remain amenable to negotiation. They appeal to common sense, design theories, standard practices, expert practices, user preference and demonstrations with physical hardware in order to persuade.

Many solution proposals and interpretations of requirements clearly arise from designers' interacting with available hardware. They also emerge as part of the ongoing activity. We have focused on the designers as actors that interpret the hardware, examining how their utterances steer the activity. A compelling analysis would also result from examining how hardware acts as a negotiator to steer the activity.

The design progresses as the group focuses and transitions from topic to topic. Still there is evidence that team members are continuously engaged in monitoring multiple issues at multiple levels of attention. Transitions occur when:

- team members seek to shift the debate to another topic
- team members seek to change the process
- prompted by related topics
- topics lose steam

- processes lose steam
- team members stop to seek information.

We conjecture that the collaboration is successful because the group is well balanced in their roles and manages their negotiation well. Kerry seeks to pin down solution alternatives, John seeks to preserve ambiguity and characterize the ongoing process, and Ivan keeps the solution progress on track and acts as an arbitrator between John and Kerry.

This is a story of one group of designers that we have used to illustrate strategies of design collaboration. There are surely many other methods and interpretations. However, the tape provides a valuable means of introspection and reflection for the design student. Watching, discussing and reflecting upon such tapes provides a means for design students to become aware of the variety of productive and counterproductive strategies and processes available to them. The tape makes these processes available and identifiable. With this awareness it becomes easier to identify when oneself or members of one's own team are following counterproductive strategies. Videotapes offer the opportunity of process examples with the context necessary for the student to gain a fuller appreciation of strategies that work well in design.

Acknowledgments

The authors wish to thank the other members of the Xerox-PARC Design Studies Group (David Bell, Natalie Jeremijenko, Steve Harrison, Catherine Marshall, Scott Minneman, Susan Newman, Lucy Suchman and Randy Trigg) for reviewing the tapes with us and providing feedback on a draft paper. The authors take full responsibility for the analysis presented. We also wish to thank the University of Delft Faculty of Industrial Design Engineering and the participating designers for making this protocol analysis possible.

References

1   Minneman, S., *The social construction of a technical reality*, PhD thesis, Stanford University (1991)
2   Guindon, R., Designing the design process: exploiting opportunistic thoughts, *Human–Computer Interaction*, **5** (1990) 305–344
3   Baya, V., *et al.*, An experimental study of design information reuse, *4th Int. Conf. on Design Theory and Methodology – DTM '92*, ASME, Design Engineering Division, DE v. 42, New York (1992), pp. 141–147

4   Stauffer, L.A. and D.G. Ullman, Protocol analysis of mechanical engineering design. *Proc. 1987 Int. Conf. on Engineering Design*, Boston, MA; ASME, New York (1987)

5   Christiaans, H. and K. Dorst, Cognitive models in industrial design engineering: a protocol study, *4th International Conference on Design Theory and Methodology – DTM '92*,ASME, Design Engineering Division, DE v. 42, New York (1992) pp. 131–140

6   Tang, J.C. and L.J. Leifer, Observations from an empirical study of the workspace activity of design teams, *Proc. 1st Int. Conf. on Design Theory and Methodology*, Montreal, Quebec; ASME, New York (1989)

7   Schriffin, D., The management of a co-operative self: the role of opinions and stories, in A.D. Grimshaw (Ed.), *Conflict Talk: Sociolinguistic Investigations of Arguments in Conversations*, Cambridge University Press, Cambridge (1990)

8   Hales, C., *Managing Engineering Design*, Longman, London (1993)

9   Jordan, B. and Henderson, A., Interaction analysis: foundations and practice, *Journal of the Learning Sciences*, **4**(1) (1995).

16 *Concurrency of Actions, Ideas and Knowledge Displays within a Design Team*

David F. Radcliffe
University of Queensland, Brisbane, Australia

Researchers and the questions they pose are shaped by many factors including their *Weltanschauung*, the setting and context within which they work, the historical development of their group and its value system, their peer networks, the prevailing research cultures and serendipity through the opportunities that arise. This project presents us with the chance to explore some of those issues by asking different research groups to analyse a common design study. It attempts to overcome the relative isolation of design research studies one from another, in both their subject matter and their research methodologies. As such it is important that some of the influences that shape the ways in which each research group approached this task be declared at the outset, at least as far as such self-revelation is possible.

The analysis reported here was conducted in the context of the Engineering Practice Research Group (EPRG) within the Department of Mechanical Engineering at the University of Queensland, Australia. The dominant research culture in the Department, in common with most Engineering Schools, is reductionist, quantitative and laboratory-based; although there is a strong shift to computational work, it values the development of sophisticated experimental apparatus, instrumentation systems or computer analysis software, and is structured increasingly around domain specialist groupings. Research, defined in this way, is highly valued. In contrast, the EPRG places its emphasis on process[1], valuing the act of engineering as a topic worthy of scholarly enquiry. Its approach is evolving from notions of design as a learning process, an intellectual activity, a knowing activity, a social activity creating things of utility, the core activity of engineering, an exciting, intellectually challenging and rewarding activity, not simply problem-solving[2].

The research culture of EPRG is qualitative as well as quantitative, informed by methodologies from the social sciences, especially action research and grounded theory[3]. Its projects are located in a variety of product development contexts, including small manufacturing companies, whole agricultural sectors (e.g. the pineapple industry), cross-discipline rehabilitation teams in a hospital setting and the development of new teaching methods in design. EPRG is concerned with the development of engineering designers and design practices that acknowledge the design context as a learning environment. Practice, research and teaching are synergistically linked through common questions. Ideas, perspectives and insights circulate amongst these three.

This chapter presents an analysis of the team of three designers. It begins with an overview of the methodology used. Several initial impressions are then outlined. The body of the chapter deals with the relationship between the ways in which ideas enter and mutate in the team's design conversations, the notional design methodologies they profess and the moment-to-moment actions they take.

1 Methodology

The team experiment was chosen for analysis in keeping with the interest in the EPRG with team processes. The video and other documents were analysed in several overlapping ways. There was a period of initial familiarization with the material and preliminary appraisal by the author. This folded over into a series of two-hour analysis sessions in which members of the EPRG viewed and discussed the video. These group discussions fed back into a more intensive analysis of the team design tape by the author assisted by NUD.IST (Qualitative Solutions & Research), a software tool for qualitative analysis.

NUD.IST stands for Non-numerical Unstructured Data Indexing Searching and Theorizing. This package creates an environment to manage, explore and search rich text documents and non-textual sources, including diagrams, photos and video. Documents may be stored on-line or be in off-line media. Using multiple windows it enables the researcher to search documents, record ideas that emerge about the data and to cross-link these ideas. Theories about the data can be constructed and tested. The program produces summary and detailed reports including statistical analyses.

The transcripts were stored as on-line documents in NUD.IST. The size of the text unit, the fundamental unit of analysis, was chosen to be individual utterances by the designers, no matter how long or convoluted. The transcript was manually annotated to highlight issues that stood out during the group viewing and subsequent personal viewing of the tape. The scope and focus of these markings changed with successive passes through the tape. The annotations triggered searches of the transcript for occurrences of particular words. This helped to cross-link repeated occurrences of terms, issues and themes and to test to see if perceptions of the frequency and distribution of a word or theme were as they appeared to the viewer forced to approach the tape serially. Non-transcript data including the physical activities of the three designers at discrete time steps, were also included.

Parts of the transcript were indexed under multiple headings. These headings were an orthogonal set of categories that might enable the relationship between different dimensions of the work to be explored. The initial index heading used in this study was the typical roles adopted by members in a team (agenda fixer, summarizer, note taker, chair, monitor, time-keeper). To this were added the types of knowledge and the content of knowledge, based on the hierarchy presented by Christiaans and Venselaar[4]. They proposed four types of knowledge from a perspective of cognitive psychology; declarative, procedural, situational and strategic knowledge. For each type they distinguished between three content-related knowledge components; domain-specific basic knowledge, domain-specfic design knowledge and general process knowledge.

This study began by classifying the content knowledge components as the third dimension much more broadly, as everyday knowledge, domain experience (knowledge of bikes), abstract knowledge (of the type gained through formal study) and professional knowledge (of the type gained by experiences as a designer). The final index heading was the individual. This initial indexing system evolved throughout the analysis as the following account demonstrates.

2 Preliminary Observations

Two issues stood out during the initial familiarization and group viewing of the tape. As these inform the subsequent detailed analysis phase it is important that they be aired at this point.

2.1 Natural Work or a Game?

It was noted that the task was well matched to the experience of the designers being studied and the sort of task they might encounter working in a product design consultancy. Some concerns were raised, however, as to how seriously the three designers were taking the exercise. For instance, on several occasions when asking for information, the designers made several asides that suggest they may be seeing the exercise as a game:

```
00:47
X   We do have some facts on the use of the backpack
I   yes ... just hand over the book
J   (laugh)
K   (laugh)
J   here's your book it'll be back to you in a while ... OK
    ...
```

These concerns were countered by seeing the natural, free flowing design discussions that took place during the vast majority of the exercise. There was ample evidence of carry-over of style and traits from their everyday work environment to suggest they were being themselves in the main. This comes through in some of their interpersonal exchanges. The commitment, intensity and displays of humour and personal feelings were broadly similar to those observed in design work recorded in natural industry settings[3]. The movement of the designers in the work space, the ebb and flow of discussions in relation to the demands for documentation and the use of artefacts in discussions were very reminiscent of actions observed in the rehabilitation engineering team in its regular clinic sessions[5].

2.2 Adherence to Schedule and Due Process

The shared concerns and expectations expressed by the designers about the need to schedule the work were striking. Ivan sets the tone when he declares, 'Let's start' as soon as the opening instructions have been completed. He asks the others if 'we should prepare a schedule and just [sorta] stick to it or should we [uh] just start working?' (00:11), to which John replies 'no it's probably a good idea to try to quantify our amount of time ... we have left'. The pair take turns at preparing a schedule on the worksheet; John articulates the key steps (00:12), 'generate concepts', 'refine the concepts', 'evaluate design ideas' and 'select one', 'design', 'present, test', simultaneously listing them on the worksheet. Curiously, after mentioning 'design' he adds the gratuitous 'which everybody always forgets' and laughs self-consciously.

Later John draws attention back to the schedule list (00:20), 'you were talking about schedule stuff before do you wanna [just] set some time limits on ourselves?' and Ivan agrees 'we should ... figure out how much we wanna spend on each thing'. John proposes times to be allocated to each phase and then writes the agreed duration against each activity. Ivan confirms the schedule and reinforces it by adding the time of day to it. Kerry participates in both episodes but especially the latter, indicating at least tacit agreement for the whole scheduling thing. As John is listing the key steps, Kerry appears to annotate her sketch of the design task with the words 'understand, observe, evaluate'. The meaning and significance of these is not immediately clear but it could be her personal checklist of design stages. John confers on Ivan the title of 'Mr Schedule', to which he acknowledges with 'on time and under budget'. The ready agreement to this ritual by all three suggests that this may be part of their everyday work culture, a common habit in undertaking a design task.

There is evidence that certain actions should be performed at prescribed stages. For example:

01:38	
J	so *let's get some dimensions* on this turkey and er *detail drawing phase* do we know what this feature looks like it's just another lug feature right

Equally striking is the public attention given by the group, particularly John and Ivan, to adherence to the correct sequence of design stages; that due process should be observed. Between listing the stages (00:10) and settling on the timing of each (00:20), John is concerned that they are commencing to generate concepts prematurely, before completing the functional and feature specifications.

> 00:19
> J   it sounds like in a way *we're starting to move on to ideation* already but uh have *have we kinda fleshed these major things out* or

There is an incident in which Ivan raises an issue but then pulls back from it in a moment of self-recognition, perceiving that he might have violated the brainstorming protocol:

> 00:59
> I   then we have the foot kicking over problem *but we're on a brainstorm* so ...

Has he broken the rules of brainstorming; should new problems be raised during brainstorming? Later in the roles of agenda fixer and summarizer, Ivan calls the others back to the correct process sequence as listed in the schedule:

> 00:14
> I   *let's talk about the decisions that we've made so far*, shall we do it based on the er based on the er brainstorming sort of ...

Implicit in the schedule may be an unstated expectation that the rationale underlying each design decision should be recorded. The reaction to a call by John to fulfil this possible expectation draws a mixed reaction that is difficult to interpret; do they believe it is appropriate design practice to record the rationale or do they think that as they are undertaking a design research exercise that they should therefore provide such a rationale for our benefit?

> 01:17
> J   yeah and if you're gonna have dirt and stuff in it *should we like break down a rationale for our killing off some of these ideas too*
> K   *nah no*
> I   *we'll just say it and they'll record it*

In the closure when congratulated on being the first group to get to calculations, Ivan is clearly conscious of the role of his 'strict timing' in this achievement and this seems to be appreciated and understood by the other two. Without being prompted, they also expressed reasonable satisfaction that due process had been followed:

> K   and it was kinda *what we planned to do how because we kept pretty close to our schedule*, right?
> I   yeah
> J   yeah *we were about fifteen minutes behind I think*
> K   *concepts refine evaluate design present* a bit more present I think
>
> . . .
>
> J   *we actually got to the to the point of like going back and checking to see if we met our functional criteria which which we did more around the sides* (laugh)

Obeying the accepted rules of engagement in design techniques such as brainstorming and adherence to schedule seem to be important values in the group, sufficient for them to acknowledge deviations publicly. For example:

> J   *kinda disrupted our materials discussion* but I was
> 01:07I
>   that's OK
> J   *when Kerry started talking about sorta the strength issue I was thinking well how big*

2.3 Work Loci and Design Objects

The layout of the room created three basic work spaces; the table, used for reviewing documents and sketching, the whiteboard, used for listing and sketching, and the physical artefacts, bike and backpack. An immediate impression gained from the video was of movement of the group from one setting to another and back again throughout the task. In an attempt to characterize better this phenomenon the times during which each of four external design objects, documents, sketches, product artefacts (bike or pack) and the whiteboard, appeared to be a locus of work were recorded. Often two loci were involved simultaneously, for instance with one person sketching at the table while the other two worked at the whiteboard. These loci are plotted in Figure 16.1.

If nothing else this diagram captures something of the visible

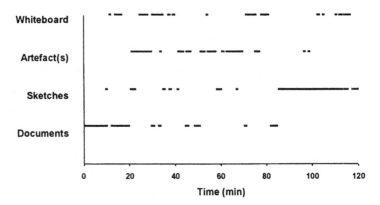

Figure 16.1

rhythms of the work. From an early focus on documents, attention swings to the whiteboard and then the artefacts, especially
the bike and then back with periodic gathering of new documents or reference to existing ones. Shifts in locus seem to arise
in the moment rather than be part of some planned work
pattern. The external objects predispose the designers to a particular mode of activity. They shape the type of exchanges that
are likely to occur. For example, within the total work area, the
table is the natural setting for sitting and reading, for spreading
out documents and drawings, and for sketching. When new
documents are asked for and received the locus of the work
usually shifted back to the table. The act of reading is an individual one, as is the gaining of insights and understanding from
new documents. Perhaps not surprisingly, the brief and other
documents commanded their attention as they struggled to
précis it and extract the key issues. As they shared their
emergent understanding of the task verbally within the group,
Kerry and John each sketched quite different summaries of the
task. Kerry's was visual, simply picturing it as an accessory in
the form of a simple L-shaped rack. John expressed the task as a
list, a series of dot points – accessory, fold/store, etc. – in the
form of a summary specification.

In contrast, the whiteboard usually evoked a didactic
response, where one scribe tended to control proceedings from
the board. In the main it was used as a summarizing tool. One
of the few exceptions to this pattern is when Kerry and Ivan
sketch together on the board. For the group this seems to be a
natural use of this design object, commencing with Ivan's offer
to 'work up there' (00:13). Many of the work products of
genuine group activity are created in the context of the white-

board. This is not surprising as the things recorded on it are available to all the group, more physically accessible than say a sketch whose orientation varies for each person at the desk. The upright orientation of the board allowed mobility between it and the main artefact, the bike. The bike was frequently a point of cross-reference for conversations held at the board.

The artefacts, bike and backpack, became a major source of interaction once the team initially got up from the table. They fill the pack and briefly wear it, and sit on the bike. To do this they bring the bike down from the table into their more immediate world. More significantly, however, they express ideas and even simulate their operation using the two artefacts. Given that they are designing some accessory that will combine the pack to the bike it is natural that they should gravitate to it. In the concluding phases, the focus shifts to the media for presentation of the concept, the sketch pad on the table and the whiteboard for listing.

3 Systematic Methods and Emergent Ideas

The record of work on the whiteboard reveals an apparent systematic listing of functional specifications, features and constraints. Concept generation appears decomposed into subproblems: position of the pack, joining methods (pack-to-rack and rack-to-bike) and materials selection. A summary of the design features of the chosen alternative and a subsequent Bill-of-Materials create the impression of a systematic design method at work. It has already been noted that there was considerable concern by the designers that due process be observed at each stage of the design and that the sequence of stages be adhered to. However, despite their best efforts the group could not control the appearance of design ideas.

3.1 Life Cycle of the External Frame as Rack Concept

The history of the concepts of using the external frame of the existing pack as a rack is illustrative of this reality. While the team are reading the brief, the experimenter makes the point (unsolicited) that the bike in the room is not the Batavus Buster but the back pack is the HiStar. Kerry responds by picking up the backpack and places it in the midst of the table. As they explore it she makes the following observation which amounts to two design proposals. The second will be referred to as the frame-as-rack concept.

00:08

K   they're getting busted by the internal frame folks but they think they think an advantage would be to *make this external frame also be mountable to a rack or become a rack*

I   yeah

K   and *that would be pretty cool* too

They return to their reading and sharing of their perceptions of the task and the idea is not pursued. Later while generating the list of possible concept areas in their 'ideation' phase, Ivan (at board) and John (at table) come up with joining as one of the areas. They seem intent on decomposing the ideation space into discrete areas; position, joining methods, materials, etc. At this point, John at first makes a strong point on how to systematically decompose the joining issue but immediately goes on to re-propose the frame as rack concept. This proposal seems to run counter to the decomposition agenda embodied in the listing on the whiteboard. Kerry supports the idea with no reference to her earlier, apparently identical proposal.

00:35

J   we have *two joining problems* we have *the frame to the bike* and then we have *the pack to the frame*

K   the frame (tearing a sketch sheet off)

J   the *the obvious solution just says really its very its like add something to your er internal* (external?) *frame* of the pack and increase the cost of the pack

I   how shall we do this

J   I mean *it's an option you can buy with the pack* or something but

K   *you've already got that nice frame on the pack it'd be nice if we can take advantage of that it seems redundant like at this*

J   (mutter) hardware stuff

K   *it seems redundant to have that and the frame it' be nice if we can take advantage of that* it seem redundant like at this

---

As the discussion on joining continues with the focus on the list on the board, Ivan confirms and records the pack-to-rack and rack-to-bike dichotomy. In the midst of this John comes back to the rack-as-frame concept and in so doing indicates that perhaps this falls outside the neat framework for decomposing the task that they have constructed on the whiteboard. Kerry does not take part in this exchange as she returns to developing a sketch of a concept on the top left of the work sheet.

00:40

I   OK *you wanna split it in half* and then look at them

J   yeah yeah I mean *there might be common solutions among*

I   OK this is gonna be pack to rack

J   (laugh)

I   *pack to rack* ... and what's the other one *rack to bike*

J   rack to bike yeah

I   OK oh go ahead

J   I just think another there's *a kind of other class of solutions outside of our design problem* and that's that you could *somehow use the external frame* and *wouldn't need the rack* maybe it's some sorta like

I   mm

J   I don't know maybe

I   (inaudible)

J   yeah

I   *we're assuming what the rack is right?*

In contrast to the clarity of the earlier assertion that we have two joining problems, John stumbles for words to articulate his proposal. The emergence of this concept is a much more brittle and uncertain process than the clearly understood and shared agenda of decomposing the task and listing it on the board. He moves to the bike and simulates the cantilevered pack over the rear wheel but his words and other utterances suggest he is still lacking confidence about the concept. Ivan (at the board) shifts the emphasis and John concedes ('I'll buy that') that the fundamental issue is that there will be 'another part', which will effectively be the rack in the rack-to-pack and rack-to-bike dichotomy. This allows the agenda to be firmly switched back to Ivan and brainstorming on joining methods for these two cases. Kerry joins the others in a rapid-fire generation of joining methods: Velcro, snaps, straps, bungee cord, quarter-turn fasteners, etc. Within minutes, however, Kerry and John propose design concepts that fall outside the isolated joining techniques demanded by the whiteboard agenda.

Kerry closes a concept episode that culminates in the light-hearted 'Harry the backpack holder' and then she gets serious. The frame-as-rack concept returns, via Kerry, some 35 minutes after she first proposed it, but with a twist. It emerges as she is simulating with the pack in the vicinity of the back of the bike.

00:43

K   now *we're kind of assuming that there's some rack to attach this to but what if the rack was really um something that attaches to this* (the frame of the pack) and just flips down so maybe you hook it on to a bracket up here but you just flip down and it clips in here

---

John rejoins with an ownership claim on the idea, but acknowledges that it goes beyond his original proposal. Ivan shows no sign of recognizing the restated or mutated concept.

J   oh I like that *I was kinda suggesting something similar*
K   something something that
J   but *I didn't have that extension in mind*
I   say that again
K   so maybe rather than
J   (inaudible) off the internal frame
K   *maybe the attachment is kind of a a leg that attaches right to the external frame*

Ivan records the concept on the whiteboard as the 'legs on internal frame' under the 'pack to rack' heading of 'concepts – joining'. There is no explicit acknowledgment that this concept falls outside the pack-to-rack/rack-to-bike dichotomy. In terms of the recorded work of the team, listing concepts under suitably decomposed headings takes precedence over capturing less structured or controllable emergence of design ideas.

John shifts the focus from Kerry and the artefacts to the table, creating a simple sketch of the concept. It is interesting to speculate on the motivations for this action. Is it to reinforce ownership of the idea? Does he wish to demonstrate that he follows the enlarged concept (frame plus folding legs)? Is it because he does not have access to the pack (in Kerry's hands) to simulate his ideas as Kerry has just done? Is the paper simply the medium at hand? When this episode commences he is at the table watching Kerry explain her idea with the aid of the pack and the bike. He reaches for a pen momentarily, then puts it down and stands up and motions to move across to the bike. As if having a second thought he returns to his chair and clears the sketch pad of extraneous paper and begins to sketch. The others are drawn in:

J	like um if you um *if this is the external frame of the pack maybe there's these like little sets of legs that you can fold out*
K	yeah
I	right
J	and they *some sort of clip details and maybe when you're hiking then you can then you use them to stand your pack up* if you're er
I	oh yeah

John's sketch is drawn without reference to the bike. Kerry clarifies how the frame might be attached to the bike by making reference back to the artifact, to the rear axle of the bike.

K	but then they *they mount into kind of like what they've drawn* there *there's a some some bracket that you attach to your rear forks* and then
I	yeah
K	*it clips in to there*

John seems to acknowledge Kerry's suggestions on attachment by adding the annotation 'clips to forks' to his sketch. Ivan then leads a physical investigation of how the pack is put together, apparently to test out exactly how the proposed attachment to the bike might work in practice. Such a detailed assessment of the implementation of the concept is far removed from the rapid-fire brainstorm of a few minutes earlier; it is in clear violation of the purported methodology.

Although the frame-as-rack concept has been at the centre of discussions and idea articulation for the past several minutes, a concept apparently now understood by all, Kerry and John both express some surprise (or recognition) of the fact as Ivan liberates the external frame from the pack.

J   yeah *look at that internal frame*
00:45
J   *it looks exactly like it looks like a big version of* um
K   *a rack*
J   *a mountain bike rack*
I   *right it goes like this*

Their reaction raises the question as to how deeply they appreciated the similarity between the external frame and a bike rack previously, or how confident they were of it. The frame now forms a focus for exploring the location and orientation of a rack on the bike. It seems that the artefact has seduced them away from their brainstorming/listing agenda. There is still further reinforcement of this recognition of the similarities between the pack frame and the commercial rack a little later when they gain access to Blackburn product brochures on carrying frames for bikes.

00:52
K   *this looks a lot like the little backpack frame doesn't it*

The backpack frame-as-rack idea seems to have struck a chord with the team for its simplicity and use:

00:54
K   *this looks so much like our em backpack frame*
I   yeah
K   this kind of frame up here it'd be really cool if all we had to do was em add y'know two pieces that look like these three members
I   right
K   and so somehow like clip it on to there and then take advantage of this braze-on most I mean this is getting it's pretty standard on mountain bikes to have a rack braze-on
I   right

The discussion then branches off onto weight (an omission from earlier considerations) and the realization that it is necessary to include a bedroll with the pack. Ivan calls them back to the schedule (1:15) to 'talk about the decisions we've made', to go through the list of decomposed concepts and integrate them into an overall product concept. In the process, the frame-as-rack concept was summarily dismissed. It is not clear from the preceding discussion on weight that they had decided to take weight off the pack. Almost simultaneously, the final form of the pack–bike interface, the tray, emerges serendipitously.

3.2 Tray Concept

In contravention of the external display of systematic design, the tray concept arises only during the evaluation phase (at 01:18). The plan is that concepts listed under the decomposed headings of position, joining and materials are being selected or eliminated. In reality new ideas are also being added, in violation of the normative rules of the process. In contrast to the frame-as-rack concept, the tray idea appears and is adopted as part of the final concept in the space of a couple of minutes. It comes about as a consequence of a discussion as to how straps can be kept out of the way;

> J   so *it's either a bag* or *maybe it's like a little vacuum formed tray* kinda for it to sit in
>
> 01:19
>
> I   yeah a tray that's right OK
>
> J   'cos it would be nice I think *I mean just from a positioning standpoint if we've got this frame outline* and we know that they're gonna stick with that you can vacuum form a a tray or a (inaudible)
>
> I   right or even just a small part of the tray or I guess they have these
>
> K   (inaudible) so something to dress this in
>
> J   yeah
>
> I   or even just em
>
> J   maybe *the tray could have plastic snap features* in it so *you just like kkkkkk snap your backpack down in it*

Its apparent attraction was that it seemed to fit with other parts of the design and solve some problems;

> J   it takes care of the easy *it takes care of the rooster tail problem* on your pack
>
> I   uh uh what if your bag were big er what if you're you're on er in this tray were not plastic but like *a big*

> *net you just sorta like pulled it around and zipped* there I
> dunno
> J   maybe it could be part maybe it could be *a tray with a*
> *with a net and a drawstring on the top of it I like that*
> I   yeah I mean em
> J   *that's a cool idea*
> I   a tray with sort of just hanging down net
> 01:20
> I   *you can pull it around and and zip it* closed

4 Knowledge Displays

The designers displayed the full spectrum of knowledge types – declarative, procedural, situational and strategic. These types take various forms that can be classified on two orthogonal axes; concrete to abstract and generalized to domain specific. These are related to the axes of Kolb's learning styles[6]. Classifying knowledge displays in this manner, however, may convey a false impression of discrete pieces of knowledge with simple relationships between them. The striking thing is that types and forms are displayed in no particular order, cascading out as they will, often concurrently. This is illustrated in the extracts from the preceding section on the external frame episode and in the following section.

4.1 Knowledge Cascade

The form of a solution concept is frequently expressed initially in terms of the manufacturing processes that will be used, with the materials implied with the process, e.g. injection moulding, wire form, vacuum form, die-cut pieces. How it will be made and possibly the relative cost of that process comes before function and form. For instance, injection moulding was introduced into the design conversation just as they were beginning to prepare their functional specification on the whiteboard (00:16).

Heuristics are intertwined with general knowledge of bikes; intuition with knowledge of manufacturing processes; abstract knowledge of loading with design values, as the following extract illustrates:

> I   yes OK well maybe em OK let's look at *materials* you
> were talking *injection moulding*
> J   *injection moulding* em *wire form* what else comes to mind

on top of these maybe like er *cloth with some sewn in em diecut pieces* or *something for plastic reinforcement*

01:05

I   oh yeah

J   like sorta like *backpack construction technology* emmm come on come up with some ideas (laugh)

I   er rubber

K   *wire form plastics* and *injection moulded bracket* things the *small things to injection mould the the small bits so the tooling isn't too light*

J   why *why does that make sense to you what intuitively makes sense about it*

K   em cos that's where you need some maybe *tricky bracket* or maybe you want some custom

J   but the rest you're just going for *minimum*

K   *minimum cost* and *maximum strength* wireform rack seems like

J   we're not

I   yeah I mean the only

K   *you don't need to reinvent the wheel* it seems like

I   we already have some *stability here* so the rack *doesn't have to be all that strong* just strong enough to

K   that's true

I   *keep it from wobbling* I mean *in this in this plane* or whatever if we're if we're gonna go in this direction this plane is already very stiff so

4.2 Mimicry and Simulation

Many ideas were expressed with supporting knowledge displayed through the agency of the physical artifacts. In particular the frame of the backpack was used repeatedly for impromptu prototyping of concepts. This is similar to that previously observed in a cross-discipline rehabilitation team[5]. A mixture of artefact location and gesture is used to express a concept. An additional phenomenon was previously observed here: the use of sounds to mimic operations or function. For example to emphasize the simplicity of detachment of a rack Kerry (00:55) says 'yeah that's easy', simultaneously gestures with both hands in the vicinity of the attachment point and goes '*bzzzzzt*', apparently simulating the detachment process. As an extension of the tray concept, John (01:19) suggests that maybe 'the tray could have plastic snap features in it so you just like ...', and he motions with both hands the act of positioning the pack in an imaginary tray (adjacent to the bike) and mimics the sound of a snap fit being closed: 'kkkkkkk ... snap your backpack down in it'.

4.3 Design Values

Many of the knowledge displays embody design values. These values range from examples of 'best practice' through to matters of personal taste. Examples of the former used by the three designers include simplicity, the notion of standard parts, using minimum number of parts, multifunction parts, reducing cost, and 'never make it if you can buy it out of a catalogue'. There are recurring references to the importance of human factors in product design and concern expressed that all the solutions are 'like all these things are human factors issues and they're not in our spec.' (01:23). In addition to universal design principles, domain specific variants occur, e.g. 'lower is better regardless' (K 00:46).

Aesthetics are also important to some members of the team. They use words like cool, zappy and snappy to signify approval of design suggestions by others or underline their own contributions. Individuals have their own interpretation of what is cool; one goes for 'cosmic colours' while another doesn't. They exhibit an awareness of product appearance.

> K   that's where you wanna get come up with *your nice injection mould bracket that really works well or has a little bit of product identity maybe from a product standpoint from a marketing standpoint* might be cool if there's some

But there are inevitable compromises between values and design heuristics, for example:

> 01:11
> J   *it just doesn't seem real elegant to me* but
> K   it's there *you might as well use it* and that *reduces parts* yeah

The team demonstrates that design is not a value-free activity nor is it a process of rational decision making; for example:

01:18

J   depends on what kind if it's like *snaps on cloth* em you
     know like those little Stimtex em press in
I   mm mm
J   *like rivet kind of snaps I don't particularly care for them but
     if it's like Fastex kind of snaps those are reasonable*

Yet a somewhat self-conscious comment by John (01:00) reveals
an underlying value system that admonishes non-rational behaviour in design. Even when apparently presenting a generally
held design value or belief he feels obliged to declare:

J   *I think good human factors says it should be adjustable so
     that people can find the position they like*
I   right OK
J   em *that's my opinion*
I   whatever we idea we come up with (inaudible)
J   *opinion not fact* (laugh)

A more direct assertion of design values comes from Ivan;

I   unless we can cut a little *we make some sort of pro-
     prietary er mounting thing but I'm sort of like morally
     opposed to that*
J   so am I
01:26
I   *because its not design*
J   (laugh)
K   should be universal, go with any bike

That these are stated and not left implicit is interesting. Are
these displays for the benefit of the individual or for their
partners? Are these remnants of an acquired heuristic, public or
self-recognition which is confirming or reassuring? Are they
used to establish or confirm membership of the group?

01:22

J   I;I think all design eventually comes down to a popu-
     larity contest

5 Concluding Observation

5.1 Protocol Studies and Natural Settings

The question of the authenticity of the design processes observed must be raised for exercises performed outside a natural work setting. The advantages of controlling the time, the scope or the space in which the action takes place, inherent in a protocol such as this one, have to be balanced against the realism of work that is displayed by the designers. The participation of practising designers does not of itself guarantee the authenticity of action. Assessing how well this balance has been struck is not straightforward. For example, fun exchanges that grow out of zany suggestions by one member are part of the natural social intercourse of group design work. There are several such exchanges in this study. The degree of levity on each occasion could be interpreted either as indicative of a natural working style or as a sign that the group were taking the task less seriously than might be expected in their daily work. Without seeing the group interacting in their normal work environment it is not possible to say which of these two interpretations is the more probable.

5.2 Concurrency: the Natural Order of Design

Team design is inherently concurrent in nature: parallel processes exploring the design space informed by individual knowledge histories. Each individual is an independent cognitive and social agent, whose actions are intertwined through the necessity to produce some group work product. They form an intellectual federation, constituted around a set of agreed understandings and rituals that frame their collective work. These understandings are presumably negotiated over time, restated and reinterpreted, as with any confederation.

This study demonstrates that the emergence of design ideas cannot be constrained to a particular place or sequence in a systematic design methodology. It is clear from the tape that design ideas emerge where they will in the continuum of the design conversation; they are not subject to the rules or discipline of a work schedule of the type proposed by the group. Ideation cannot be constrained to occur only during the prescribed time for this activity as dictated by notions of due process and proper sequence of phases in design. Design conversations are refreshed by a well-spring of ideas that bubble to the surface, the result of cognitive processes that are not available to the observer, and possibly not even to the designers at a conscious level. These design ideas have a life beyond the external ritual of design; quantify problem, generate concepts, refine, evaluate, design, test and present. Ideas don't simply present themselves, are recognized and recorded. Rather they appear, they may only be partly appreciated at that instant, then reappear subsequently, sometimes with a hint of recognition as ideas that have passed this way before and sometimes without.

Knowledge of vastly different types and forms is displayed and interacts simultaneously in the acts of idea presentation, sharing, mutation and acceptance. Consideration of the manufacturing processes to be used to make the product appear concurrently with issues of loads, stability and strength. Aesthetics and design values mingle with an exploration of the design context of the product, for instance how the frame might be connected to the bike.

The non-obvious relationship between the external process rituals in this case and the sub-text of emergent design ideas raises several questions. What are the roles of ritual in design activity? Is it that design be seen to be done, satisfying professional honour? Are they used to bind the team, and to publicly acknowledge their shared story of how to design? How does the decomposition agenda played out on the whiteboard influence the design outcomes in this case?

The challenge for the design team is how to synchronize their activities to achieve a suitable outcome in a reasonable time while working with the natural processes at work rather than fighting them. At a more general level, the advocates of concurrent engineering need to consider how the concurrency inherent in natural design processes can be usefully channelled to meet their vision. How do we harness the concurrency of actions, ideas and knowledge displayed by designers at work to make the process as effective as possible while reconciling these natural attributes with our normative models for organizing product development in an industrial setting?

Acknowledgments   The author acknowledges the contribution of the members of the EPRG who participated in the group watching sessions:

Sandra Munoz, Michael Buchanan, Glen Horton, Jinglu Qi, Fiona Solomon, Philip Teakle, Gil Logan and Pam Swepson.

References

1   Holt, J.E., D.F. Radcliffe and D. Schoorl, Design or problem solving – a critical choice for the engineering profession, *Design Studies*, **6**(2) (1985) 107–110

2   Holt, J.E. and D.F. Radcliffe, Learning in the organisational and personal design domains, *Design Studies*, **12**(3) (1991) 142–150

3   Radcliffe, D.F and P. Harrison, Transforming design practice in a small manufacturing enterprise, *Proc. ASME Design Theory and Methodology Conf.* (1994)

4   Christiaans, H. and K. Venselaar, Practical implications of a knowledge-based design approach, in N. Cross, K. Dorst and N. Roozenburg, (Eds), *Research in Design Thinking*, Delft University Press, Delft (1992), pp. 111–118

5   Radcliffe, D.F. and P. Slattery, Emergent learning and interaction in a cross discipline design environment, *Proc. ASME Design Theory and Methodology Conf.* (1992)

6   Kolb, D.A., *Experiential Learning*, Prentice-Hall, Englewood Cliffs, NJ (1984)

17 *Design as a Topos-based Argumentative Activity: a Protocol Analysis Study*

Brigitte Trousse[1] and **Henri Christiaans**[2]
[1]*INRIA, Sophia-Antipolis, France*
[2]*Delft University of Technology, The Netherlands*

Empirical studies of the design process have divergent motives, varying from the improvement of design education to the development of supporting tools for the designer. In the study discussed here the aim is to understand the nature of the design process in order to develop knowledge based expert systems for design. Therefore the general issue is related to knowledge modelling of the designer or the designers in a team. One of the limitations of the current systems in complex problem solving is that hardly any attention is paid to the discursive activities performed either in isolation or in cooperation between the group members. In general, every agent solving a problem starts a communication process and thus a process of argumentation. The word 'agent' refers to a designer in this case. For a better understanding of problem solving activities it seems valuable to analyse the process within an argumentation linguistics framework. On the basis of such a framework more valid and reliable specifications for the development of real cooperative knowledge-based design support systems could be derived.

Hence the main question in this study is what role argumentation plays in a design activity. With designing, as is the domain in this study, we primarily mean the field of industrial and architectural design. The data are directly derived from an industrial design task. The design activity is understood as an information-processing activity. It consists of the mental formulation of future states of affairs, having a claim on generating creative and innovative artifacts. By definition design problems are ill-structured in that design tasks involve unspecified goals and operators[1]. The essential difference between design and nondesign problem-solving seems to lie in the task environment[2] with aspects like the distribution of information, the nature of constraints, the feedback loop, etc. For an overview of design studies see Reference 3.

1 Theory

This section presents our protocol-analysis approach which is based on two theoretical viewpoints. The first considers design as a discursive and a meaning-construction activity and proposes a semio-linguistic and action-based analysis of the problem-solving activity. The second viewpoint considers the discursive activity as a 'topos'-based argumentative activity. Both will be explained here.

1.1 A Semio-linguistic and Action-based Framework

The analysis of the role of argumentative knowledge is based on a theoretical framework derived from previous work, influenced by three main disciplines, Semio-linguistics, Pragmatics and the Theory of Complex-Systems. This framework is used as a tool for analysing a complex problem solving activity as a *discursive* activity and thus as a subject matter for semio-linguistic analysis. Design is viewed here in a new perspective as a *meaning-construction* activity. Within design as a development project the argumentative activity aims at *effectiveness*, which means that it aims at producing certain determined effects. It is a kind of action and as such it is pragmatic in nature. Argumentation derives the meaning of this effectiveness from the resources of language; as such it is pragmatic in its objective and linguistic in its mechanism.

Hence, the main assumption in this study is that every problem solving activity integrates in its execution an argumentative activity. According to this assumption the design activity is considered as a discursive activity and thus provides subject-matter for a semio-linguistic and action-based analysis.

Within the framework proposed here three skills and accordingly three spaces are distinguished. According to this framework skills are postulated for both the agents who perform the problem-solving task and the analyst who describes the task, while spaces are related to the analyst's viewpoint: see Figure 17.1.

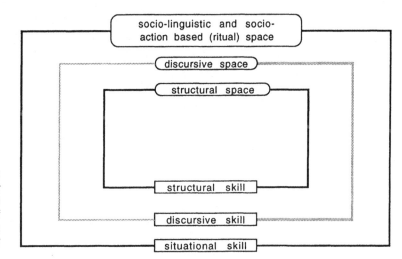

Figure 17.1
Proposed framework for
the semio-linguistic and
action-based analysis of
a problem-solving
activity

The skills can be described as follows.

1. The problem solving activity makes sense as far as it responds to a prescribed task, the design task. By this we implicitly hypothesize the existence of a *discursive skill*.
2. The activity makes sense 'within a human organization'. We implicitly hypothesize the existence of a *situational skill*. This skill pertains to the knowledge of situations or environment which is called 'socio-linguistic and socio-action' and refers to a more general cultural context. In another context Christiaans and Venselaar[8] use the concept 'situational knowledge'. This has a different meaning while it is synonymous with 'conditional'. In their model it is the specific application of a rule: 'If on this condition..., then...'.
3. Finally the activity makes sense at the execution level – being practicable – which means that it expresses itself in terms of gestures or operations identifiable both through existing rules and through a practical entity. Besides, the entity must respect semantico-logical and physical constraints. Here the existence of a *linguistic and action-based skill* is postulated.

Each of the three skills is performed in a specific space. According to the adopted semio-linguistic viewpoint for the analyst it is called a 'meaning-construction space'. The three spaces are:

1. A *structural space* (called *s-space*). In this s-space the action-based and linguistic skill manifests itself. In our study all

objects are coded in the s-space which relates to the technical aspects of the design: for example, the geometry of the artifact.

2. A *socio-linguistic and socio-action based (ritual) space* (called *S-space* for Situational space). In this S-space the situational skill manifests itself. In our study all contextual objects related to the design and to the task are coded in the S-space: for example, the costs of the artefact and the situation in which the artefact is used.

3. A *discursive space* of the activity (called *D-space*). In this D-space the discursive skill manifests itself. But this space also plays a specific role itself. In this space the designer implicitly pre-constructs his cognitive devices that will be used for design purposes in the two other 'meaning-construction spaces', previously mentioned as the structural and the situational space. This discursive space is then based on the ability to propose and interpret constraints linking some attributes of objects from each of the two other spaces. The more the discursive space is based on relations between the two other spaces, the higher is the discursive skill of the agent.

The agent's knowledge in the 'socio-linguistic and socio-action' validates and guides his argumentation. If agents do not share the same space, then a valid statement of one agent can be of no meaning for another agent who has to interpret it. Hence, the criterion here is not objectivity (or efficiency) but *efficacy*.

For an agent performing an activity and using the three skills – in a variable way – the argumentation will be the way to orient and modify the real world in order to adapt it at each skill. The argumentation then in this perspective is an action of refinement, useful for each activity.

1.2 The Topical Viewpoint

A second part of the analysis is focused on the notion of 'topos'. We consider in our study the discursive activity in design as a topos-based argumentative activity. For this, we use a theory of argumentation in Linguistics, inspired from the topoi of Aristotle[9,10]. Let us introduce and define briefly the notions of *topos* and *topical model*. For Anscombre *et al.*[9], topoi are a kind of gradual relations or argumentative rules such as:

> 'The more an object O has the property P, the more the object O' has the property P'.'
> 'The less an object O has the property P, the less the object O' has the property P'.'

Symbolically, '+P[O], + P′[O′]' or '–P[O], –P′[O′]'.

A topos links the sense of variation of two topical fields (P[O] and P′[O′]). A topical field is a field (for example, slide-up/ down [bike]) with an ordered set of possible values. The value domain could be continuous such as [0,250 kg] for the mass of an equipment, or discrete such as (horizontal, vertical) for the orientation of, for example, the backpack. Usually two types of topoi are distinguished: positive and negative topoi. The positive topoi corresponds to a variation in the same direction (see T1). A positive topos can generate two topical forms (see below T1a and T1b generated by topos T1).

For example, in the Delft protocol used in this study we can find the next statement (with a topical form) uttered by one of the designers (agent 'John'):

The more there are potholes in the ground,
the more the backpack will slide up and down. (T1a)
(time: 00:15:00)

This topical form refers to the positive topos T1 which we code as follows:

+ *number-potholes[ground]* → + → *slide-up/down[backpack]* (T1)

We could construct another topical form from this topos, like:

The less there are potholes in the ground,
the less the backpack will slide up and down. (T1b)

Actually, because of the orientation of the utterance (time 00:15:00 in the Delft protocol) we identify that agent John uses the optical form T1a in the text.

We can also find the following topical form in the protocol (time 00:15:00, agent John):

The more the backpack will slide up and down, the less is the stability on the bike (T2a)

which can be coded as

+ *slide-up-down[backpack], –stability-on-bike[user] (T2)*

T2 is a negative topos as is the following example in which agent John uses implicitly this topical form:

'The less time it takes to remove the backpack from the bike, the easier it is for the user'

–time-to-remove[user], + easy-to-use[user]

The use of topoi is usually limited by conditions of applicability and by the validity of the meaning:

- the conditions of application define the situations when the topoi are valid,
- the validity of the meaning indicates the allowed values: for example, *the higher the temperature, the nicer the walk* is not valid if the temperature exceeds 40°C.

Moreover, in AI, topoi are often combined in order to form a dependency-graph called *'topical model'*: this is a kind of semantic network where the relations used are topoi (see Appendix A for such models). A valuable application of topical models is to use them in a qualitative simulation for answering questions like 'What-if?' or 'How-to?'. For more details on the use of topoi in AI see References 4 and 5.

2 Hypotheses

In order to test the theoretical models regarding the semio-linguistic and action-based framework on the one hand and the topical model on the other, a number of hypotheses are presented. The main issue here concerns the role of argumentation in the design activity.

Hypothesis 1   *In semio-linguistic terms, every activity can be characterized by analysing its discursive space.*

According to the theoretical framework it is possible to analyse any kind of problem solving activity (for example a design activity or a planning activity, etc.) using the meaning-construction spaces.

From this hypothesis two other hypotheses can be derived related to the different skills of each agent:

Hypothesis 1.1   *Taking the semio-linguistic point of view, the expertise of each agent can be characterized by analysing the discursive space.*

Hypothesis 1.2   *The interaction between the agents is determined by the meaning-construction spaces that they share.*

The next hypothesis is linked to the place of argumentation in problem solving:

Hypothesis 2   *The argumentative activity contributes to the 'anticipatory approach' of the problem solving cycle.*

There are many studies which are concerned with modelling the problem solving process. Lemoigne[11], inspired by Simon, considers problem solving as a decisional process. In cognitive psychology, a problem is defined as '...the representation of a cognitive system built from a task which does not have an already given standard procedure in order to attain a goal'[12].

In order to determine the role of argumentation in a problem solving cycle an anticipatory (or projective) approach of problem solving based on the topical viewpoint is proposed. According to the projectivity principle, each sub-activity integrates the corresponding projective action; this action does not follow after the execution of the sub-activity but is immediately a component part of the sub-activity itself. The access to the real world is then effected through an anticipatory approach. A problem solving process then follows the rule:

Hypothesis 2.1   *Each sub-activity (a component of the global design process) activates by anticipation, and implicitly the sub-activity that is normally its goal-directed component.*

From the two initial hypotheses (2 and 2.1), a third is generated:

Hypothesis 3   *The speech acts of each agent during each step of the design process are linked to the character of each step.*

3 Method

In this study both the team of designers and the individual designer are analysed through the written transcripts of the protocols. A protocol analysis method was elaborated based on the aforementioned theoretical viewpoints. Hence transcripts are used as data referring to the communication between members of the design team and the 'thinking aloud' protocol of the individual designer. When talking about the team members 'Ivan', 'John' and 'Kerry' they are referred to as agents I, J and K, while the individual designer 'Dan' is abbreviated to agent D.

We describe briefly our proposed protocol analysis method based on some analytic tools and corresponding to a hypothesis-testing approach.

In the analysis the main focus is on the relations between objects that each agent explicitly uses in his or her communication during the design process. Examples of objects are bike, backpack, geometry, snap, and pothole. Objects can also have attributes and values: see for example Scheme 17.1.

Scheme 17.1
Example of coding
(object, attribute, value)
uttered by the three
agents

object	attribute	value	space	agent I	agent J	agent K
[BIKE]	cost	high	S	X		price

From the transcripts several empirical data are extracted and an object-oriented modelling approach is used. These data are:

1. A *global dictionary (dic-g)*, containing all (more or less) explicitly uttered objects, attributes and values of these objects; we note O (for 'Object') for any (object, attribute, value). Although some of these objects with their possible attributes and values are repeated during the process, in this global dictionary they occur only once.

2. A *local dictionary (dic-l) for each agent*, containing all (more or less) explicitly uttered objects, attributes and values of these objects for that agent. Between local dictionaries there is some overlap because some objects are used by more than one agent.

3. *Occurrences (OO)* of objects and their attributes/values, which means that every occurrence of any (objects, attributes value) in the transcripts is counted including repeated use of these concepts.

4. *Types of utterances (U) of agents*; an utterance is a set of sentences spoken by an agent in one speech act. The utterances can be distinguished as 'information' (*U1*) and 'support' (*U2*); the latter applies only when an agent supports the others by reformulation, expressing their agreement, etc.

5. *Topoi*. As we have shown before, we extract relations between occurrences and then we identify topoi from the sentences of each agent in order to build his or her topical model (a kind of qualitative semantic network corresponding to his or her viewpoint). This analysis is the most difficult one because topoi are often implicitly used in the argumentative orientation of a sentence.

An example of the way *dictionaries* are determined is presented in Scheme 17.1. We put a capital X in the column of an agent when this agent introduces an object, attribute and/or value for the first time (if it is used previously by another agent we put 'x' between parentheses), and a term if he or she uses a synonym for the attribute or value. Then each attribute of an object has been classified according to their nature, i.e. belonging to the structural or situational space of the design activity: for example, the cost of the bike belongs to the situational space (called S-space).

Occurences and relations between occurrences are coded in the case of a collective activity, i.e. during the interaction, we make the distinction when the agent initiates a discussion by using an (attribute/value of an) object (called 'object 1' in the table), or when the same or another agent mentions an (attribute/value of an) object as a reaction to the first one (called 'object 2'). An example is given in Scheme 17.2.

Scheme 17.2
Coding of initiating an object utternance (object 1) and reacting with an object (object 2) (BP = backpack)

Object 1	time	agent Info (UI)	agent Support (U2)	Object 2
easy-to-attach [BP]	00:22:00	J	I	time-to-remove [user]

3.2 Testing Hypotheses

In order to test hypothesis 1.1 we first identify the importance of the situational and the structural spaces for each agent based on the global and local dictionaries. Next, the spaces for the *topical model* of each agent will be located. This model represents the discursive space and also the size of it. The results of both the team and the individual designer will be used.

Hypothesis 1.2, which is related to the interaction of agents, will be tested by analysing their shared spaces. It means that their *topical model* representing their discursive space will be located. From this, we can explain the interaction between agents: *the less two agents share a specific space, the less they might have interacted within this space.* Here we also focus more on the speech acts, counting the number of utterances and occurrences of the same object, and we try to explain the results according to the different steps.

Looking at the anticipatory approach in designing (hypotheses 2 and 2.1) we analyse (1) the discursive activity of each agent throughout the different steps of the design process, and (2) the contribution of topoi inside the anticipatory approach. For the method used here, a protocol segmentation according to different design steps has to be made first. Then the link between the segmented protocol and the tables from the first analysis is done (see Appendix B for the process steps of the design team).

In testing hypothesis 3 the interaction rules between the agents will be identified for each step of the design process, based on the discursive space shared by the agents.

4 Results

This section presents the main results of our analysis of the Delft protocols of both the team work and the individual work. It is also an illustration of how to apply the topos-based protocol analysis method.

The section is organized into two parts. The first part focuses on the structural aspects of our analysis: what skills are characteristic for each agent and what kind of interaction is observed (hypothesis 1). In the second part the dynamic part of knowledge application is analysed. The assumptions on this so-called 'knowledge in action' can be found in hypotheses 2 and 3.

4.1 Part 1: Structural Analysis

This part focuses on the different skills of each agent in a static way, that is without any link to the different steps of the design process. We present the main results of this analysis guided by hypothesis 1.1 (see the first two subsections) and hypothesis 1.2 (see the third and fourth subsections).

Importance of the situational and structural spaces

The importance of the situational and structural spaces for each agent is analysed first. Next the space in which the discursive skill is performed will be located for each agent.

Based on the local dictionary (dic-l) of each agent, Table 17.1 presents the amount of (attributes/values of) objects in the different spaces and the percentages in each agent's dictionary (dic-l). Also the total numbers of objects, etc., in the global dictionary (Tot (dic-g)) are presented together with the percentages.

Our conclusions are as follows:

- *Team*: According to the absolute amount of information given by each agent we obtain: $O(J) > O(K) > O(I)$. According to the two spaces – the structural one and the situational one (O-s% and O-S%) – we first note that all three agents give their information mainly in the structural space. Secondly, agent J has the largest proportion of situational objects according to O-S%,

Table 17.1 Importance of the structural and situational space. I, J and K are the members of the design team, while D is the individual designer (dic-l = local dictionary per agent; dic-g = global dictionary)

Agent	O	O-s	O-S	O-s%	O-S%
dic-l:					
I	173	143	30	82.7	17.3
J	287	235	52	81.9	18.1
K	204	168	36	82.4	17.6
dic-g:					
Tot	602	491	111	81.6	18.4
D	318	283	35	89.0	11.0

For each agent's dictionary (dic-l) and the total activity (dic-g),
O: number of objects (attributes/values)
O-s: number of structural objects
O-S: number of situational objects
O-s%:   percentage of structural objects
O-S%:   percentage of situational objects

although the differences as to the number of situational objects are small.

- *Individual*: First, in the individual context, we find indeed that agent D utters fewer objects than the agents in the collective activity. Second, agent D uses fewer situational objects (proportionally O-S%) than each member of the team.
- *Comparison between team and individual*: For both the team and the individual work the situational space is much less important than the structural space. This finding may be expected according to the typical content of the design task used in the protocol study. This confirms hypothesis 1.

Location of the discursive space

Based on the topical model of each agent (see Appendix A), the numbers of explicit or implicit attributes of objects (A) and topical relations (R) of different types used by each agent are counted (see Table 17.2).

Our conclusions are as follows:

- *Team*: Firstly, agent K has the largest discursive space, as can be inferred from (1) the total number of attributes (A), (2) the structural space and the situational space (A-s and A-S), and (3)

Table 17.2 Location of the discursive space for each agent according to each agent's topical model. I, J and K are the members of the design team, while D is the individual designer

Agent	A	A-s	A-S	R	R-ss	R-SS	R-sS
I	17	7	10	19	4	9	6
J	25	15	10	25	9	9	7
K	35	17	18	40	7	19	14
Team	62	39	30	66	17	26	23
D	34	24	10	18	6	2	10

A: number of attributes in the topical model of agent Ai
A-s: number of structural objects in the topical model of agent Ai
A-S: number of situational objects in the topical model of agent Ai
R: total number of relations for each agent
R-ss: number of relations between two structural attributes in the topical
         model
R-SS: number of relations between two situational attributes in the topical
         model
R-sS: number of relations between a structural and a situational attribute in
         the topical model

the number of situational relations (R-SS). Looking at the number of R-sS relations, the discursive activity of K is more important than for I and J. Secondly, agent I has a low discursive activity.

• *Comparison team and individual*: It is clear that the discursive space of the team is largest because of the interaction between the agents: A(Team) > A(D) and R(Team) > R(D). The s-space and S-space used for this discursive space are also largest.

The next two subsections concern the interaction between the agents as determined by the meaning-construction spaces they share (hypothesis 1.2). Therefore, only the teamwork is taken into account.

Importance of speech acts during the interaction

Based on the segmented protocol the number of occurrences of (attributes/values of) objects, used by agents during the interaction, are counted. Next, the numbers of utterances by each agent are counted (for a definition of occurrence and utterance, see Section 3, Method). A summary is presented in Table 17.3. We have added the density ratio (D) of the speech of each agent, which is the number of objects divided by the number of utterances.

Table 17.3 Importance of speech acts

Agent	O	O-s	O-S	U	U1	U2	D	D-s	D-S
I	173	143	30	706	374	252	24.5	20.3	4.2
J	287	235	52	733	489	145	39.2	32.1	7.1
K	204	168	36	523	319	145	39.0	32.1	6.9

U:	total number of utterances by each agent
U1:	number of utterances which provide 'information'
U2:	number of utterances which provide 'support'
D:	Density ratio for each agent (O:U)
D-s:	Density ratio for structural space (O-s:U)
D-S:	Density ratio for situational space (O-S:U)

Our conclusions are as follows. Based on the total number of utterances, we can note that U(J) > U(I) > U(K), which means that agent J interacts the most. Although agent I is very near, this is caused by the high number of supporting utterances by I. The same conclusion can be drawn from the density ratio of I compared to that of J and K. For both the structural and situational ratios we find D(J) > D(K) > D(I). In fact agent I has the lowest density ratio in the structural space while agent J has the highest one in the situational space.

Interaction rules based on the location of the discursive space shared by the agents

Another aim of this study is to locate the discursive space shared by the agents and to analyse if it is based on the same space (situational or structural) or on both spaces. We identify two interaction rules, respectively; one is called 'symmetric interaction' and the other 'complementary interaction'.

Looking at the topical models, Table 17.4 shows how many

Table 17.4 Shared discursive knowledge for each agent, according to each agent's topical model

	Shared discursive knowledge						
Interaction	A	A-s	A-S	R	R-ss	R-SS	R-sS
(I,J,K)	6	4	2	5	2	2	1
(I,J)	+2	0	+2	+5	0	+4	+1
(I,K)	+2	+1	+1	+5	0	+3	+2
(J,K)	+5	+2	+3	+3	+1	+2	+0

attributes and relations the three agents share and which meaning-construction space is concerned. For each combination of agents, we analyse the intersection of the topical models according to the number of shared situational/structural attributes and of the shared topoi.

The results show that agent I shares the fewest attributes with both J and K. Agent I shares his discursive skill with the other agents mainly in the structural space. From the data we can check the following interaction rule: *the interaction between the agents depends on the discursive space that they share.* Indeed we can predict that the interaction between the agents (I,J) and (I,K) is more important than for the couple (J,K). A complementary analysis based on the video for example, could confirm this prediction.

4.2 Part 2: Knowledge in Action

In this second part the dynamic part of knowledge application is analysed, while testing hypotheses 2 and 3. It concerns the interaction between the agents throughout the steps and linked to the type of each step.

First, the role of situational aspects related to each step in the design process will be analysed. Second, the postulated anticipatory approach during the design process will be tested. Finally, in order to find out when and why two agents interact, the agents are compared for each step and between steps on their part in information and support giving. In this analysis the process steps are used as defined in Appendix 8.

Importance of the situational space of each design step

The relative importance of the situational space in each step of the process is calculated for the team protocol. These process steps concern both the conceptual and detailed design. Most of the steps are related to the device that has to be designed. In Table 17.5 the proportion of situational space (compared to structural space) for each step in the process is presented, based on the occurrences of the (attributes/values of) objects. A distinction is made between an agent initiating a discussion by using an (attribute/value of an) object (OO_1) and the same or another agent reacting with an (attribute/value of an) object (OO_2): see Section 3.

As an example, the percentage of situational space of OO_1 in step 1 equals 26.8. It means that the structural space equals 73.2%.

Table 17.5 Importance of the situational space for each step: percentages of the total number of situational and structural occurrences of objects (OO)

Steps		OO_1-S%	OO_2-S%	Total
1	Problem presentation	26.8	35.0	30.9
2	Planning	100.0	100.0	100.0
3	Solution generation:			
3.1	Location	21.2	27.8	25.3
3.2	Joining techniques	24.2	20.0	21.4
3.3	Device	26.5	14.0	17.0
3.4	Materials	25.0	6.7	10.5
3.5	Device, general	23.1	32.1	29.3
3.6	Device, specific	0	50.0	38.5
3.7	Joining techniques	0	6.7	5.9
3.8	Device	0	7.7	6.5
3.9	Rack to bike	22.2	17.1	18.0
3.10	Detailing solution	0	3.2	2.6
3.11	Cost estimates	9.1	0	2.4
4	Solution evaluation	100.0	30.0	46.2

Our conclusions are as follows. The character of the design assignment is an important determinant for the use of skills by the agent. Overall the structural space is dominant in this design task except for the planning and evaluation steps. During the solution generation stage (step 3) the situational part is relatively decreasing except in step 3.9. Looking at the last column of Table 17.5 the (sub-)steps in which the situational space plays a relatively important role (> 20%) are 1, 2, 3.1, 3.2, 3.5, 3.6, and 4.

The anticipatory approach In order to test hypothesis 2.1, i.e. that each sub-activity activates by anticipation, and implicitly, the sub-activity that is normally its goal-directed component, only a preliminary analysis is performed. We just want to illustrate here what this hypothesis means with the topical approach of problem solving. It means that we have to analyse the transition from one step to another and to look if there is some topos responsible for this anticipatory approach. Here only two examples are provided from the team protocol:

- Transition between step 3.1 (location) and 3.2 (joining techniques):

  (agent J; 00:35:00) 'yeah well maybe maybe another another ideation tool that might help em is looking at analogous products your other products that mount to bikes other products that clip on to something em'. This transition refers to the following topos: *'The more you look at analogous products, the greater the chance of generating new ideas'*.

- Transition between step 3.2 (location) and 3.3 (joining techniques):

  (agent K; 00:50:00) 'it's harder to get your legs up around the Harley gas tank'.

  (agent J) 'yeah I guess d- like well given the width of the pack you're right you're not going to get around that I mean you're barely with somebody at the knee at the top of their stroke their knee's here on this bike so ... unless there was like y'know up in here'. We can extract the following topos: *'The bigger the width of the backpack, the less acceptable is the middle of the bike as a location for the fastening device'*.

In fact the agents validate the location (step 3.1) for the backpack on the bike only after a preliminary look at joining techniques (step 3.2). Next, at the end of step 3.2 (joining techniques) the discussion on the location on the bike foreshadows the transition to a phase during which ideas for a device are generated.

Interaction rules throughout the design steps

This section concerns only the team and aims to test hypothesis 3. In order to analyse when and why two agents interact, a histogram is presented in Figure 17.2, in which for each step and between steps the agents can be compared on their part in information and support giving (see also Section 3). Based on the segmented protocol the number of utterances for each step in both information and support are coded as follows: [0–4 utterances] = 1, [5–9] = 2, [10–14] = 3, [15–19] = 4, [20–24] = 5, [25–29] = 6, [30–34] = 7, [35–39] = 8, [40–44] = 9, [45–49] = 10.

Our conclusions are as follows:

- Agent I has mainly a supporting role in steps 3.5, 3.6, 3.7 and 3.8. In their interaction J and K give more information than they support the others: an exception for agent K is step 3.7 related to joining techniques.

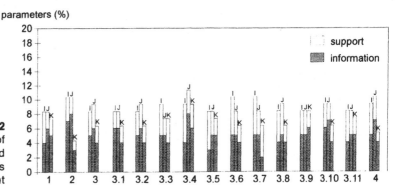

Figure 17.2
Histogram of
informative and
supportive utterances
by an agent

- Characteristic for the interaction of agents I and J is that they give the most information during steps 3.1, 3.2, 3.6, 3.7, 3.8, 3.10 and 4, while agent K has a weak role in supporting during the interaction. The agents I, J and K have a real interaction together during steps 3.3, 3.9 and 3.11 (i.e. during the steps based mainly on the structural space). The characteristic interaction of agents K and J is that they give a lot of information during the steps 3.4 and 3.5, while I has a more supporting role then.

5 Discussion

Critics

Our protocol analysis method is based on the topical viewpoint. However, the analysis is still not complete, because only verbal utterances are taken into account while gestures and movements are not analysed. Moreover, the hypothesis on the anticipatory approach must be demonstrated through a thorough analysis.

Use of an original analysis of design activities

The originality of the theory of this study is due to the focus on argumentation in design. The distinguishing features of this work can be indicated as follows:

- Firstly, there are very few studies on the subject chosen for our study.
- Secondly, our protocol analysis method is in our opinion the first one focusing on the role of argumentation during the problem solving activity.

• Our method addresses both individual and collective activities. It is important to realize that in general there are only a few protocol analysis studies performed on a team.

Main conclusions from our study

This study has partially confirmed the theoretical issues on two topics:

1. The problem solving process can be charactarized by the location of the discursive space used while the importance of the designer's discursive skill is obvious.
2. The approach by designers during their design process can be characterized as an anticipatory process.

Finally, we think that considering design as a discursive activity and extending protocol analyis methods with a semio-linguistic viewpoint are very important in order to have a better understanding of the complexity of the design activity and mainly of the different viewpoints of the designers used during the design process. We think that these results have some significant theoretical implications on 'Design Research' and on 'Design Rationale'[13]. Indeed we can characterize the discursive space used by each agent linked to each sub-step, i.e. according to our topical viewpoint we can define different types of problem solving rationales which occur in the design process.

Acknowledgment

We thank Judith Dijkhuis for her contribution to the first analysis of the Delft Protocol.

References

1   Simon, H.A., The structure of ill-structured problems, *Artificial Intelligence*, **4** (1973) 181–204
2   Goel, V. and P. Pirolli, The structure of design problem spaces, *Cognitive Science*, **16** (1992) 395–429
3   Christiaans, H.H.C.M., *Creativity in Design*. Lemma, Utrecht (1992)
4   Dieng, R. and B. Trousse, 3DKAT, a dependency-driven dynamic-knowledge acquisition tool, *3rd Int. Symp. on Knowledge Engineering*, Madrid (1988)
5   Gobinet, P. and B. Trousse, *Mise en oeuvre de modèles approximatifs dans un architecture a base de tableau*, Report VI.1, contract DRET 90/484 (1992)
6   Trousse, B. and D. Galarreta, The topoi, as a Pragmatic paradigm of knowledge representation, *ERCIM Workshop on Theoretical and Experimental Aspects of Knowledge Representation*, Pisa, Italy (1992)
7   Galarreta, D. and B. Trousse, Cooperation between activities in complex organizations: study directions in the design of space

systems, *Design Science and Technology Journal*, special issue, **2**(1) (1993) 65–86

8   Christiaans, H. and K. Venselaar, Practical implications of a knowledge-based design approach, in N. Cross, K. Dorst and N. Roozenburg (Eds), *Research in Design Thinking*, Delft University Press, Delft (1992)

9   Anscombre, J.C., *et al.*, *Théorie des Topoi*, Editions Kimé, Paris (1995)

10 Bruxelles, S. and P.-Y. Raccah, *Argumentation and Information: How to Express Consequences*, COGNITIVA, Paris (1987)

11 Lemoigne, J.L., *La modélisation des Systèmes Complexes*, Dunod-Afcet Systèmes, Paris (1990)

12 Hoc, J.M., *Psychologie Cognitive de la Planification*, Presses Universitaires de Grenoble (1987)

13 Brown, D.C., Rationale in design, in Workshop Notes on Representing and Using Design Rationale, *Third Int. Conf. on Artificial Intelligence in Design '94* Lausanne, Switzerland (1994)

Appendix A Topical Models of the Team Members, Agents I, J and K

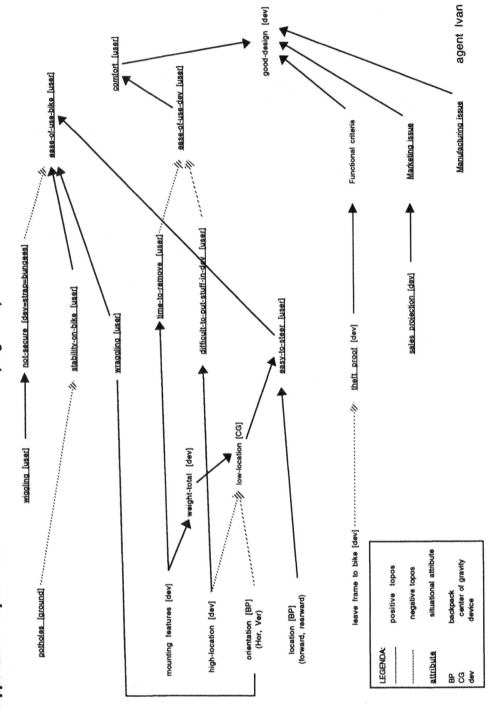

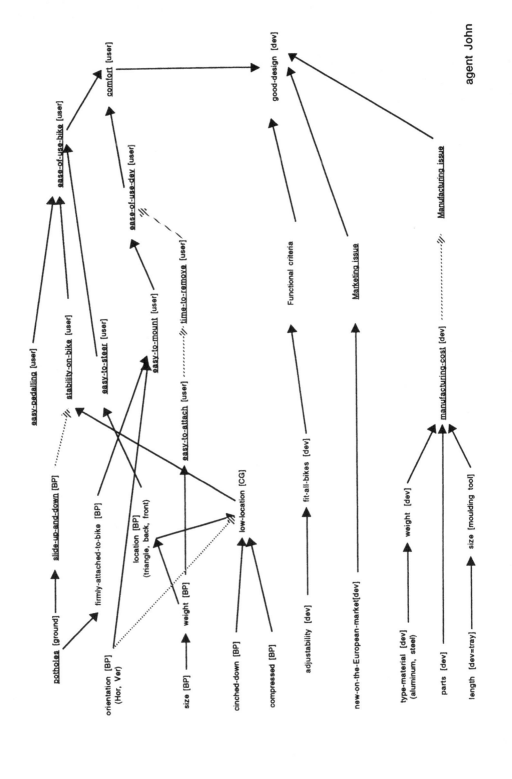

agent John

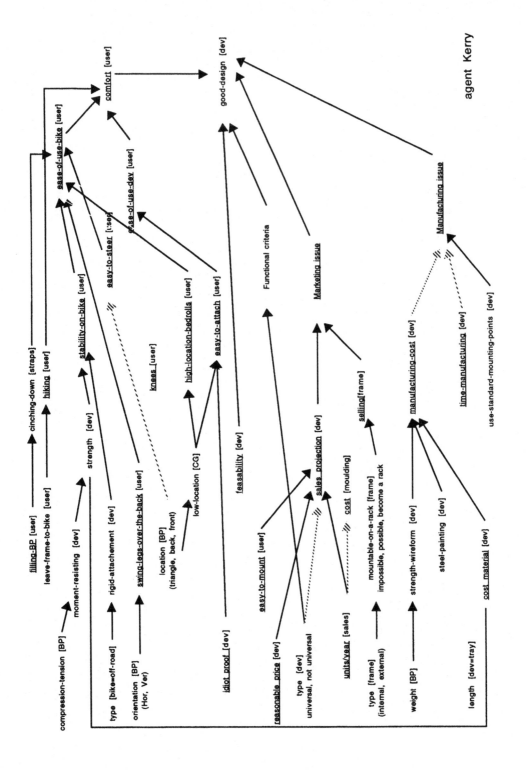

agent Kerry

comfort [user]

good-design [dev]

Manufacturing issue

ease-of-use-bike [user]

ease-of-use-dev [user]

easy-to-steer [user]

Functional criteria

Marketing issue

stability-on-bike [user]

knees [user]

high-location-bedrolls [user]

easy-to-attach [user]

filling-BP [user]

cinching-down [straps]

hiking [user]

leave-frame-to-bike [user]

time-manufacturing [dev]

use-standard-mounting-points [dev]

strength [dev]

compression-tension [BP]

moment-resisting [dev]

low-location [CG]

selling[frame]

manufacturing-cost [dev]

feasability [dev]

sales projection [dev]

cost [moulding]

type [bike=off-road]

rigid-attachement [dev]

swing-legs-over-the-back [user]

location [BP]
(triangle, back, front)

easy-to-mount [user]

mountable-on-a-rack [frame]
impossible, possible, become a rack

strength-wireform [dev]

steel-painting [dev]

orientation [BP]
(Hor, Ver)

reasonable price [dev]

type [dev]
universal, not universal

units/year [sales]

type [frame]
(internal, external)

weight [BP]

idiot proof [dev]

length [dev=tray]

cost material [dev]

18 *Representing Design Workspace Activity*

Maryliza Mazijoglou, Stephen Scrivener and **Sean Clark**
University of Derby, UK

Many writers have recognized the importance of sketching and drawing in design. Goldschmidt[1] has commented on its 'ubiquity' in everyday life. Lansdown[2] claims that drawing is indispensable to designers and is often the only form of external modelling a designer uses. The relationship between drawing and thinking has been noted by Archer[3] who writes that 'Drawing is a very economical way of modelling, it is the fastest and best way of having a quick idea - a visualisation - of what is in your head and thus leads naturally into solid modelling; (see also Fish and Scrivener[4] and Scrivener and Clark[5]). Indeed, Archer[6] believes that 'The capacity for envisaging a non-present reality, analysing it and modelling it externally, is the third great defining characteristic of humankind, along with toolmaking and language use.' Given the recognized value of drawing and sketching it is perhaps surprising that we have so little understanding of design reasoning and how cognitive processes interact with the designer's workspace, i.e. drawing and modelling media, information sources, and so on.

Recently, a number of studies of design have been conducted, often stimulated by a desire to establish requirements for computer systems. Tunnicliffe and Scrivener[7] analysed video recordings and 'self-reports' produced by the graphic designers engaged in a page layout task as part of a knowledge elicitation process. Goldschmidt[1] used similar methods to investigate the reasoning processes supported by sketching. Goldschmidt's analysis focused primarily on the verbal transcripts of utterances produced by designers 'thinking aloud' during design sessions.

Both of these studies used self-reporting but neither provided a detailed analysis of workspace activity or attempted to relate ideas (in the protocol) to design workspace behaviour.

Other studies have investigated the activities taking place during group design. Tang and Leifer[8] proposed and used a framework for investigating workspace activity in terms of 'actions' (listing, drawing, and gesturing) and 'functions' (storing information, conveying and representing ideas, and engaging attention). Bly[9] explored the use of drawing surfaces by employing a similar framework to Tang and Leifer, but extended the settings in which the collaborative activity occurred.

These latter studies are interesting for two reasons. Firstly, the methods overcome some of the criticisms of self-reporting (i.e. that it is unnatural and may lead to distortions) since the utterances occur naturally as part of the communication between participants. Secondly, they recorded and analysed details of drawing surface activities, such as drawing, pointing, and gesturing. However, no attempt was made to represent the drawing surface development over time, or the flow of events within the workspace.

Ishii et al.[10] have drawn a distinction between interpersonal space and shared workspace when discussing work environments. Interpersonal space refers to the communication channels available to participants for interpersonal interaction. Typically, an office space affords opportunities for verbal and non-verbal communication. Shared workspace refers to the tools and media available to all participants, such as whiteboards, pen and paper etc. Here we use the term 'design workspace' to mean both interpersonal space and shared workspace.

In this paper, we focus on the team design session. The design workspace for this task is rich. The designers have at their disposal pen, paper and whiteboard for recording work in progress. The workspace includes information about products, including marketing surveys and drawings, and actual products: a bicycle, a bicycle rack and backpack. The designers are able to move freely between items in the workspace. The resultant design workspace is rich in constituents and the interactions that occur within it (during a design session) are complex in terms of both interpersonal interactions between participants and their interactions with the various workspace artefacts. This complexity is partially recorded in the verbal transcripts, the persistent media and the videotapes, but these records are not easy to scan or reference. As a consequence it is difficult either to develop an understanding of the process or to re-examine the data in the light of experience. In this paper we describe our approach to developing a 'rich picture' of the design workspace session which can be traced back to the raw data: i.e. the transcript, the drawings, and the videotapes.

1 The ROCOCO Project

The work described here builds upon the ROCOCO project which investigated the communication channel usage of product designers as a means of establishing the requirements for a computer system to support geographically separated designers.

The first phase of the project involved a study of face-to-face working[11]. In this study, pairs of product designers took part in design tasks. The designers sat opposite each other across a table. Between them was a pad of A1 plain white paper on which they could both draw using their own pens. Six one-hour, face-to-face design sessions were undertaken during which audio and time-stamped video recordings were made. The aim of Phase I was to identify how design pairs exploited the available communication channels, in a face-to-face situation, when negotiating a resolution to a shared problem. In the context of the whole project, this study acted as a control for comparison with behaviour observed in remote computer-mediated settings.

The communication which occurred (in the recorded audio and video and the actual drawings) was systematically coded to reduce it to a manageable form while minimizing the loss of meaning. Both the verbal (i.e. speech) and non-verbal (i.e. gesture and drawing) communication was coded, yielding a detailed 'communication profile' of each design session. In global terms the coding focused on the verbal exchange, the production and use of drawings, and the non-verbal inter-change. Together these foci led to the production of an extensive system which used in excess of 70 codes. Through observation and communication analysis a wide variety of requirements were formulated regarding the communicational requirements of the subjects.

The second phase of the project investigated these postulated requirements by testing hypotheses concerning the changes in communication activity occurring in conditions where they were not satisfied; typically these conditions involved communication impoverishment and were achieved through manipulation of an electronic workspace and communication environment. In all four conditions were studied: all-on, where pairs could communicate via an audio and video link and a shared drawing surface (designed to emulate the face-to-face workspace and communication environment); Video-off, where the video channel was removed; Speech-off, where the designers were unable to communicate verbally; and Drawing-surface-only, where the shared drawing surface was the only means by which designers could communicate.

In practice it turned out to be extremely difficult to draw conclusions by comparing the conditions because the designers rapidly adapted to each condition in order to complete the task and hence effects due to impoverishment could not be disentangled from those due to adaptation. Neither were attempts to understand changes between conditions assisted by the absence of explicit linking between the coding categories and concepts, issues, problems, etc. explored by the designers. These results have implications for design process and design thinking research. First it suggests, not surprisingly perhaps, that designers adapt their behaviour in quite remarkable ways depending on the situation they find themselves in. We must therefore expect factors such as task, gender, team size, communication media, task time, resources, etc. to influence both design process and design thinking (and hence be cautious about generalizing from single studies). Second, we need to find ways of describing design workspace activity, ideas development, and outcomes in different ways and at different levels of detail.

The potentially profound impact of extraneous factors, such as those identified above, lead us to propose that we should focus our efforts on developing a rich and structured picture of design workspace activity arising in a given study that integrates raw recorded data (such audio and video) and systematically records the relationship between abstractions, such as concepts, and these data. Such a 'rich picture' of structured data and codings (or identifications) will provide a representation that enables researchers to revisit and test the interpretations of others, and to reinterpret the data in the light of their own concerns and interests. In this paper, we describe our first attempts at the production of such a 'rich picture'.

2 Representing Design Workspace Activity

Our aim then is to produce a representation of design workspace activity that coordinates and structures discourse and workspace elements. For the purposes of this paper we have analysed the Delft group design session which we subsequently call the Delft Protocol. Unlike the ROCOCO project we have focused on the data as a whole and have not attempted to explicitly represent individual action or group dynamics. In the following subsections we describe how the design activity, objects of reference, and outcomes, are structured in terms of action flow and artifact development.

2.1 Structuring the Design Workspace Record

In this section we describe ways of structuring records of the design workspace, such as the verbal discourse, drawings, paper notes, whiteboard, and artifacts.

Identifying sub-entities

Aspects of design workspace activity can be captured in static form using different persistent media. In the case of the Delft Protocol, designers used paper and whiteboard for drawing and writing. In the first instance, these records are only partitioned in terms of physical entities, i.e. pages or sheets. On inspection it is clear that these pages can be broken down into sub-entities (*cf.* drawing cluster, Bly[9]). Typically, spatial separation distinguishes one entity from another; alternatively, as for example in a table, headings provide the distinguishing feature. For simplicity, we call such an entity a drawing.

Generally speaking, identifying drawings is not problematic. Our procedure for identifying drawings is to first delimit drawings visually. This first pass yields most of the final drawings. However, there is always some ambiguity: some delimited drawings could in fact be further subdivided while other apparently separated drawings could in fact be a single drawing. The only way to disambiguate such drawings is to refer to the videotape and discourse. This content and time-based information is, in our experience, sufficient for dealing with uncertainties of identification. Consider, for example, the top-right area in Figure 18.1. Here, the bedroll appears visually separated from the backpack. Initially, backpack and bedroll were identified as part of a single drawing. Close inspection of the video revealed that these elements were indeed all produced in a single extended period of visualization activity during which the associated discourse included nothing to indicate that

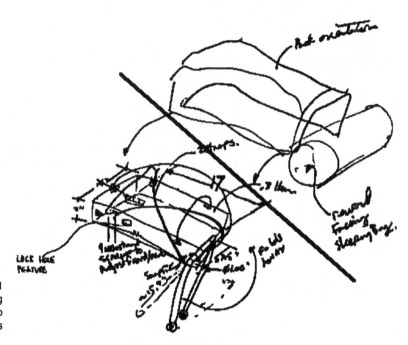

Figure 18.1
An area of drawing
page 24 subdivided into
two drawings

the development of the bedroll was associated with a shift in focus of attention.

Drawing acts

Having subdivided the drawing pad and whiteboard sheets into drawings, the drawings are then further subdivided into drawing acts (this is illustrated in Figure 18.2), a drawing act being defined as 'a continuous sequence of drawing activity terminated by a change in drawing act type or a detectable interruption in drawing'[11]. This is a time-consuming process involving careful scrutiny of the videotape, in conjunction with

Figure 18.2
(a) A drawing, (b) the
same subdivided into
five drawing acts

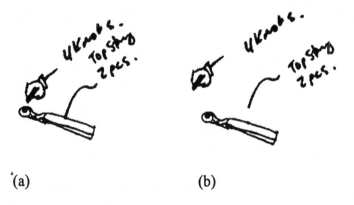

(a)   (b)

the drawing pad and whiteboard sheets, during which the start and end of each drawing act is recorded. Because drawings have already been identified, drawing acts can be associated with particular drawings.

Drawing packets and drawing transition network

Typically, following periods of activity on one drawing the focus of attention moves to another drawing, later to return to the original drawing, and so on. A period of activity on a given drawing is called a drawing packet, defined as being 'a sequence of drawing acts executed by one designer on a given drawing, terminated by a shift in activity to another drawing'[11]. A single drawing may comprise many drawing packets. Because the start and end of a drawing packet are implicitly recorded in the drawing act data it is possible to explicitly record the transition between drawings: this structure we call a drawing transition network. An extract from the drawing transition network associated with drawings on page 24 of the Delft Protocol data is shown in Figure 18.3. The nodes of the network represent (numbered) drawings and the links show the designer's transition between drawings.

Drawing development structure

The drawing transition network represents the flow of movement between drawings during the design sessions. In addition to the temporal relationship between drawings there is also a structural relationship whereby some drawings can be clearly identified as developing from or being refinements of a given drawing. We have also represented this developmental

Figure 18.3
Extract from a drawing surface transition network

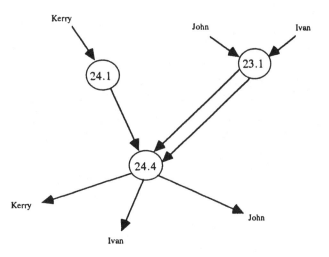

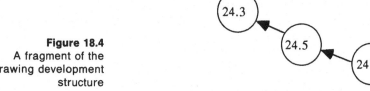

Figure 18.4
A fragment of the
drawing development
structure

aspect of drawing surface in what we call the drawing development structure. Consider for example Figure 18.4. Here Drawing
4 (represented by the 'node' labelled 24.4) is first produced,
recording many of the design ideas to date. As discussion
focuses on the 'top-stay' for connecting the tray to the bicycle
seat post, this is detailed in Drawing 5 (node 24.5). Because the
'top-stay' has to fit into the horizontal base of the tray and
connect to the vertical seat post a twist feature is proposed,
which, although indicated in Drawing 5, is elaborated in
Drawing 3 (node 24.3). The drawing development structure
shows this developmental relationship between Drawings 4, 5,
and 3.

In representing the drawing development structure we only
attempt to identify a relationship where the evidence for it is
strong. Figure 18.5(a) shows drawing 23.1 and Figure 18.5(b)
shows drawing 24.4. There is a lot of similarity between these
drawings, both of which consolidate the design so far, and in
this sense one can be seen as a development of the other.
However, the drawings are separated by time and the flow of
ideas and hence the evidence is less clear that the latter drawing

Figure 18.5
(a) Drawing 23.1; (b)
drawing 24.4

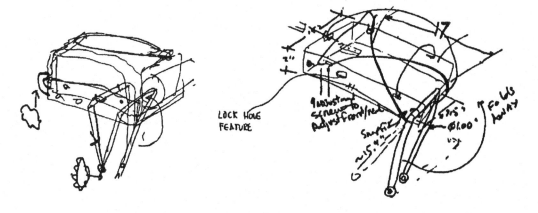

(a)   (b)

Figure 18.6
Turns in a fragment of
the Delft Protocol group
design session

J (turn 1): in it sounds to me that what they're looking for is not they're kinda looking for a an interface a thing that will allow you to carry or fasten an existing backpack to an existing mountain bike (John)

I (turn 2): yeah

J (turn 3): is that how you guys interpret it?

K (turn 4): well also they've got this em Batavus Buster that em

develops from and during discussion of the former drawing. Drawing development is only represented where there appears to be a strong synergy between the development of ideas in the minds of the designers and their partial representation on paper.

Turns in discourse

The discourse as provided is broken down into turns (Figure 18.6), i.e. individual contributions to the discourse, and for the purposes of this paper we have not attempted to break the discourse down further.

Other artefacts in the workspace

The Delft Protocol setting differed from the ROCOCO project in that the design workspace comprised a greater range of artefacts, for example the bicycle and rack, the backpack, the brief, notes, drawings and data on other products. These too could be subdivided and coded, however, we have not attempted this, mainly because of the difficulty of making reliable interpretations from the video.

Deictic references

Certain utterances contain reference to some physical entity other than the participants themselves, by means of a *deictic* word, i.e. one whose interpretation can be determined only in relation to the context in which it is used[12]. Examples of such words include *he, she, it, this, that, here* and *there*. Concrete third-person deictic references (concrete because the reference is to a physical entity and third-person because the reference does not include either the speaker or the listener) relate the discourse to specific physical entities in the workspace. For example, 'so *it* hooks on the front and then you think *that* hooks so *this* snaps in?'.

We first identified such deictic words by scanning the transcript and then identified the artefacts referenced by reviewing the videos. In fact we found that many instances of the words *it* and *its* could not be related to the workspace, instead they seemed to refer to abstract design solutions shared by the designers and held in their minds.

2.2 Coding the Workspace Record

So far we have described how we structure the drawing pad, whiteboard and discourse data in terms of drawings, drawing packets, drawing acts, developmental relationships and temporal transitions between drawings, discourse turns, and deictic references. In this section we present the descriptive coding categories used to characterise drawing act and discourse turn data.

Drawing act categories

In the ROCOCO project, drawing acts were coded as being one of the following types[11]:

1. A non-symbolic act (e.g. a doodle or a squiggle)
2. An other symbolic act (e.g. a line under text or an arrow)
3. An alpha/numeric act (e.g. labels or writing)
4. A pictorial orthographic act (e.g. a plan view of an object)
5. A pictorial perspective act (e.g. a street view receding into the distance).

During the analysis of the ROCOCO data it became clear that further sub-categories within type might be useful. In particular, the other symbolic category included arrows, underlines, etc., which appeared to be used for different functions. Hence, this type (i.e. other symbolic, labelled '2') has been further extended to include:

- Emphasis – such as underlining text or delimiting an area of text or drawing
- Link – usually an arrow connecting text to drawing or drawing to drawing
- Direction/Movement – usually an arrow showing an intended motion or direction of movement
- Abbreviation – for example an acronym
- Other – any other symbolic act which cannot easily be functionally categorized.

In addition, the symbolic alpha/numeric type has also been further extended to enable a functional categorization consistent with the discourse coding; see Appendix A. These extensions to the drawing act type classification reflect our current interest in tools to support collaborative design at a distance; however, we believe that they are generally useful elements of a rich picture of design workspace activity.

Discourse categories

In the ROCOCO project the discourse was analysed in terms of hierarchical structure[13], conversational structure[14–17], discourse progression[11], and communicative acts[18,19]. As explained in Section 3, the aim of the work reported here was to produce a representation of the structure and content of the design workspace activity. With this in mind, and given the constraints of time, we have chosen to code only discourse progression. The ROCOCO data were coded under three categories – problem, solution and progression – and four sub-categories – proposal, development, acceptance and appraisal[8]. We set out to code the Delft Protocol in the same manner but quickly found the three top-level categories too restrictive. As a consequence we enlarged them to include: problem, solution, constraint, requirement, information needs, and process.

Typically, a 'problem' utterance relates to some aspect of a solution being considered, for example, how a backpack rack will attach to a bicycle. The response might be a 'solution' idea of bolting the rack to the seat post. 'Constraints' are taken as 'givens' that the designers have little control over, for example the model of backpack in the Delft Protocol, whereas a 'requirement' is taken to be something over which the designer has some control and which can be traded-off against other requirements, for example the need to quickly release the backpack from the rack. It is perhaps obvious that the decision to categorize an utterance as dealing with a constraint rather than a requirement, and vice versa, can be accompanied by a degree of uncertainty. Furthermore, an aspect of the task may appear as both a problem and a requirement. For example, the designers might regard it as a requirement that the backpack attaches to the bicycle. Because the problem, constraint, and requirement categories are closely related, variation between coders is quite likely to arise. Although we plan to explore this variation, we have not yet had the opportunity to do so. However, we propose to make the data available to others who might wish to test their interpretation against ours. In the meantime, we would recommend that interpretation of these data is treated with care.

Often during the Delft Protocol group design session a need for information arises and designers ask for such information and provide it to each other. This kind of utterance does not naturally fall into any of the other categories, hence the category of 'information needs'. As for the ROCOCO project, utterances about possible designs are coded as 'solution' and about how to work together as a team 'progress'. For all but the 'information needs' category, utterances within category can be coded as

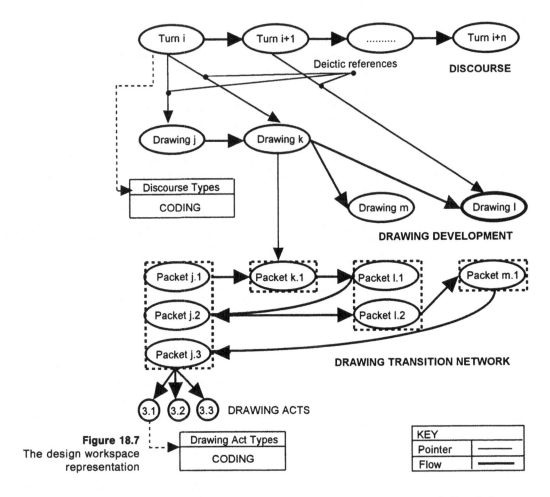

Figure 18.7
The design workspace
representation

proposal, development, acceptance, and appraisal. For 'information needs' the sub-categories request (asking for information),
response (providing the information), acceptance, and appraisal
were used (see Appendix B for a complete list of categories, sub-
categories and codes).

**2.3 An Overall View of
Design Workspace
Representation**

Figure 18.7 illustrates both the structure and the primitive-level
coding of the design workspace representation described in this
section. We believe that the production of this representation
serves a number of useful purposes: first, it helps the design
researcher to develop a rich understanding of the design activity
captured in the videotape and transcript records; second, it
structures the disparate records and artifacts (videotape,
drawings, notes, products etc.) thus assisting the systematic

exploration of the data by the design researcher interested in quickly testing insights and ideas; and thirdly, summary analysis can be derived from the representation itself.

3 Analysis of the Design Workspace Representation

In this section we illustrate and discuss some of the forms of summary analysis that can be derived from the design workspace representation.

3.1 Time Based Analysis

At the most basic level drawing acts and discourse progression primitives are represented in the design workspace representation. Since these records are marked in time they can be clustered and presented against time.

Drawing acts

Figure 18.8
ROCOCO drawing acts over time (five-minute intervals). Types 1 to 5 are 1 = non-symbolic, 2 = other symbolic, 3 = alpha/numeric, 4 = pictorial orthographic, 5 = pictorial perspective

Figure 18.8 (5-minute interval) shows drawing act production over time for the ROCOCO face-to-face study as originally presented[11] whereas Figure 18.9 (10-minute interval) shows the same data presented in terms of 10-minute intervals to conform to Figure 18.10, which shows the comparable results for the Delft Protocol (see Appendix C, Tables 18.C2, 18.C1 and 18.C3 for data).

Figures 18.9 (ROCOCO data) and 18.10 (Delft Protocol data) appear quite different on first appearances. However, a number

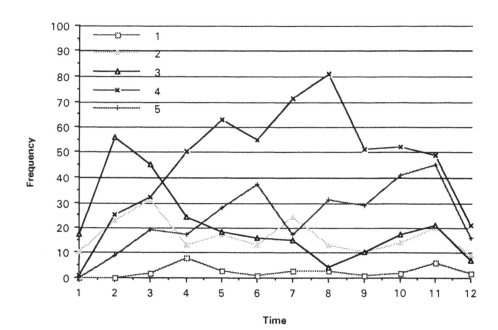

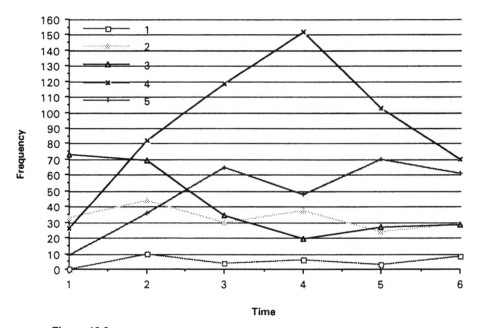

Figure 18.9
ROCOCO drawing acts
over time (10-minute

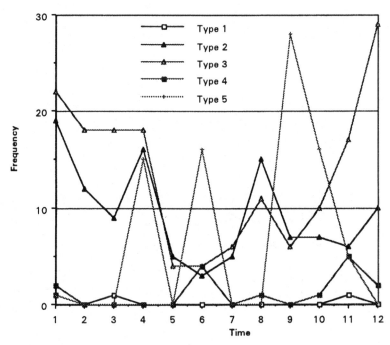

Figure 18.10
Delft Protocol drawing
acts over time (10-
minute intervals)

of factors need to be borne in mind. First, the ROCOCO sessions only lasted for one hour. Second, the ROCOCO data are the summation of six studies. Nevertheless, there are similarities between the data. First, the curves for Type 2 (other symbolic acts) and 3 (alpha/numeric) are similar in each set of data, suggesting that text and other symbolic acts are coupled. Second, increasing drawing activity is usually accompanied by decreasing writing activity, suggesting a change in mental focus of attention. This is particularly marked in the Delft data, where several cycles of writing followed by drawing are obvious. Finally, both studies are characterized by an increase in drawing towards the end of the session. In the case of the Delft Protocol, there is also a great increase in writing at the end of the session. This occurred during the period when the designers were producing a parts list and costs for the design. This requirement was not part of the ROCOCO design brief. One clear difference between the ROCOCO and Delft data is the relative balance between drawing and writing: drawing dominates the ROCOCO data, whereas writing dominates the Delft data. This perhaps reflects different levels of experience (the ROCOCO designers were second-year students) or training.

Whilst the data presented in this way reveal broad cycles of activity, Scrivener and Clark[5] have shown how consideration of the data in finer detail reveals the rapid speed of drawing act production and the rapid and frequent shifts between drawing types.

Discourse

Figure 18.11 shows discourse over time (see also Appendix C, Table 18.C4). Most noticeable here is the dominance of solution- and information-focused discourse over other forms of discourse. Utterances concerned with how the group should proceed as a team are the next most common, with bursts at the beginning and end of the session. This conflicts with the ROCOCO data where there were very few 'process' utterances. Again, this could be due to differences in group experience or group size. An interesting feature of Figure 18.11 is the similarity in shape between the Solution and Inform curves, as if a period of solution-focused working generates the need for information. However, it should be borne in mind that the two peaks in the Inform curve actually reflect different types of information: the first is concerned with information relevant to

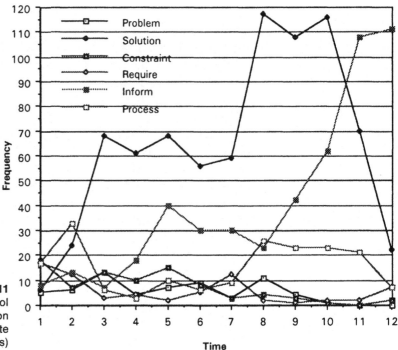

Figure 18.11
Delft Protocol
discourse production
over time (10-minute
intervals)

design; the second with documenting details of the design, e.g. parts and costs.

There are clear correspondences between Figures 18.10 and 18.11. Broadly speaking, increasing solution-focused discourse is associated with increasing visualization, and increasing information-focused discourse is associated with increasing writing activity. This correspondence is particularly strong in the third quarter of the session

Finally, it is interesting to note the similarity between the Solution curve of Figure 18.11 and the Deictic Reference curve of Figure 18.12 (see also Appendix C, Table 18.C5). To confirm this relationship we really need to look for deictic references occurring within solution-focused turns. This reveals the fact that although these summarized views are informative, they almost always lead to further consideration of the data (in this case the transcripts). The intention behind the design workspace representation discussed here is to enable this kind of exploratory investigation to take place.

3.2 Drawing Based

In addition to a time-based analysis, it is possible to present the design workspace analysis in a drawing-based way. Such a

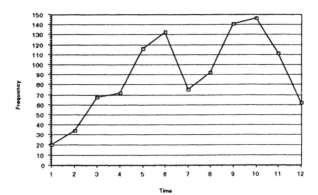

Figure 18.12
Delft Protocol deictic
references over time

representation links all workspace activities to drawings and other artefacts in the design space.

Drawing structure network

Figure 18.13 shows the drawing structure network for page 24 of the Delft Protocol data. The page of drawings was divided into six discrete drawings (see Appendix D). Each drawing was then represented by a numbered 'node' and the structural relationships between drawings (nodes) was derived. In Figure 18.13 it can be seen that drawings 24.2 and 24.5 developed from drawing 24.4 and drawing 24.3 developed from drawing 24.5. Drawings 24.1 and 24.6 were not developed from, or developed into, any other drawings on this page.

Both the drawing act analysis and discourse analysis can be linked to the drawing structure network. Figure 18.14 shows the breakdown of each drawing on page 24 in terms of drawing act type.

Figure 18.13
The drawing structure
network for DPW
94.1.14.B24

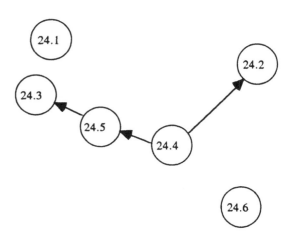

Drawing Number	Drawing Type					Total Acts
	1	2	3	4	5	
24.1	-	-	-	-	4	4
24.2	-	1	1	-	3	5
24.3	-	-	-	-	1	1
24.4	-	11	9	-	19	39
24.5	1	1	2	-	5	9
24.6	-	7	25	1	-	33

Figure 18.14
Delft Protocol drawing
act breakdown for the
drawings on page 24
(Appendix D)

The above data shows that:

- Drawing 24.1 (a small sketch of the backpack) was produced in four type 5 (pictorial perspective) drawing acts.
- The main body of Drawing 24.2 (a large sketch of the backpack) was produced in only three type 5 (pictorial perspective) drawing acts.
- Drawing 24.3 was produced in a single type 5 (pictorial perspective) drawing act.
- Drawing 24.4 (the main drawing on the page) was the focus of the great deal of type 5 activity and contains a high proportion of type 2 (symbolic) and type 3 (text) drawing acts.
- Drawing 24.5 consists of a type 5 drawing (pictorial perspective) together with two type 3 drawing acts (text), a type 2 drawing act and a type 1 (other non-symbolic) act.
- Finally, drawing 24.6 is composed almost entirely of type 3 (text) activity.

Ideally, these data would be presented visually, say by means of pie charts of different sizes depending on the volume of activity, so that one could see at a glance the characteristics of different drawings.

Drawing transition network

A detailed view of the development of each drawing can be obtained through the construction of a drawing transition network for the design workspace. In such a network the production of drawing packets by each designer is traced. Figure 18.15 shows the drawing transition network for the drawing activities associated with drawing 24.4.

The nodes in the above network represent the six drawings on the page (numbered 24.1 to 24.6) and the links show the designers' shift in drawing focus between drawings. For example, Ivan enters the network at drawing node 23.1 (page 23, drawing 1), then adds to drawing 24.4, then adds to 24.6

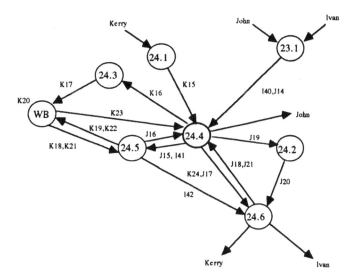

Figure 18.15
Drawing transition
network for drawing
24.4

and then leaves this section of the network. The paths for Kerry and John are more complex. In Kerry's case the path includes a visit to the whiteboard (the node labelled 'WB').

As with the drawing development structure network, it is possible to link the drawing act and discourse analysis to this network. This is illustrated in Figure 18.16 where the visits (i.e. drawing packets) to drawing 24.4 are listed and broken down according to drawing act type. This figure shows that much of the drawing was produced by John in his first visit to the drawing (drawing packet number 24.1.1). Kerry and Ivan then added to the drawing. John then made two further visits, the first of which included the addition of new pictorial marks (type 5, or 'pictorial perspective', drawing acts). The drawing was then completed with a short visit from Kerry and a final addition from John.

Figure 18.16
Breakdown of drawing
24.4 into drawing

Packet	Start	End	Designer	Drawing Act Types				
Number	Time	Time		1	2	3	4	5
24.4.1 (J14)	1:39:39	1:44:46	John	-	3	2	-	10
24.4.2 (K15)	1:42:19	1:42:43	Kerry	-	-	-	-	3
24.4.3 (I40)	1:44:17	1:44:47	Ivan	-	-	-	-	2
24.4.4 (J16)	1:47:31	1:51:45	John	-	5	3	-	3
24.4.5 (J18)	1:56:22	1:57:05	John	-	1	2	-	-
24.4.6 (K23)	1:59:38	1:59:51	Kerry	-	1	1	-	1
24.4.7 (J21)	2:02:27	2:02:29	John	-	1	1	-	-
Totals				0	11	9	0	19

4 Computer Assisted Design Workspace Representation Development

This paper has described an approach to representing design workspace activity which aims to provide a rich and structured picture of the workspace activity occurring in a given study. This 'rich picture' can then be used as the basis for further study. Our goal in this research is to ensure that such a picture integrates recorded data (such as audio and video) with codings, allowing researchers to revisit and test the interpretations of others and reinterpret the data from their own perspective.

Although it is certainly possible to produce such a description through the analysis of pen-and-paper drawings and recordings of a design session, it is an extremely time-consuming process. We believe that this process would be greatly enhanced if computers could be used to capture the design workspace data and software tools were used to assist its analysis.

Clark and Scrivener[20] describe a system in which a computer sketching tool known as the ROCOCO Sketchpad can be used to capture drawing data prior to computer assisted analysis. The ROCOCO Sketchpad allows up to eight designers to collaborate via a 'virtual' drawing surface. The drawing surface takes the form of a large window which is replicated on each designer's computer screen. A stylus can be used to 'draw' in the window a variety of colours and pen thicknesses. It can also be used as a telepointer to 'point' to areas of the drawing surface.

An important feature of the ROCOCO Sketchpad is that it is able to log all drawing surface activity occurring in a computer-mediated design session. This computer log can then be used in design workspace analysis, rather than having to use pen-and-paper drawings. Clark and Scrivener found that it is possible to *automatically* derive the drawing acts occurring in a design session from such a log and have developed a computer tool which allows the user to encode each drawing act in terms of the categorization described earlier[5,20]. The next stage in this research is to develop software tools which allow drawing acts to be grouped into drawings and drawings to be linked to discourse. Finally, it is planned that the playback of the ROCOCO Sketchpad log can be synchronized to the playback of a video recording of the design session, resulting in a system that supports the rapid analysis and review of design workspaces.

5 Conclusions

We recognize that the proposed design workspace representation is by no means complete (we have, for example, not explicitly represented individual contributions of group dynamics); much more needs to be added. The same comment applies to the codings and identifications derived from the data. In addition to enlargement, pruning of the representation may be required; some coding, for example, may turn out to be unnecessary. Experience and the results of others will help us to decide what stays, what goes and what is added. Indeed, we expect to incorporate descriptions provided by other researchers in the workshop.

We believe that the production of this representation serves a number of useful purposes: first, it helps the design researcher to develop a rich understanding of the design activity captured in the videotape and transcript records; second, it structures the disparate records and artefacts (videotape, drawings, notes, products etc.) thus assisting the systematic exploration of the data by the design researcher interested in quickly testing insights and ideas; and thirdly, summary analysis of the DWR can be derived directly from it as a basis for and in support of interpretation.

It may be time-consuming to produce the Design Workspace Representation (DWR), but at this stage of research into design thinking short-cuts should be avoided. Computer methods can come to our aid in two ways: first, a hypermedia representation of the DWR would make it more accessible to the researcher; second, artificial intelligence methods could be employed to assist in the generation of the DWR. All of this would be greatly assisted if design studies were actually computer-based.

References

1   Goldschmidt, G., The dialectics of sketching, *Creativity Research Journal*, **4**(2) (1991) 123–143
2   Lansdown, J., Computers and visualisation of design ideas: possibilities and promises, in T. Maker and H. Wagter (Eds) *CAAD'87 Futures*, Elsevier (1987) pp. 77–80
3   Archer, B., The three Rs, *Lecture delivered at Manchester Regional Centre for Science and Technology*, 7 May 1976
4   Fish, J. and S.A.R. Scrivener, Amplifying the mind's eye: sketching and visual cognition, *Leonardo*, **23**(1) (1990) 117–126
5   Scrivener, S.A.R. and S.M. Clark, Sketching in collaborative design, in L. MacDonald and J. Vince (Eds), *Interacting with Virtual Environments*, Wiley, Chichester (1994) pp. 95–118
6   Archer, B. The nature of research into design and design education, *Proc. DATER'91*, Loughborough (September 1991) pp. 1–11
7   Tunnicliffe, A.J. and S.A.R. Scrivener, Knowledge elicitation in design, *Design Studies*, **12**(2) (1991) 73–80

8   Tang, J. and L. Leifer, A framework for understanding the work-space activity of design teams, *Proc. CSCW'88*, Portland (September 1988) pp. 244–249

9   Bly, S., A use of drawing surfaces in different collaborative settings, *Proc. CSCW'88*, Portland (September 1988) pp. 250–256

10 Ishii, H., M. Kobayashi and J. Grudin, Integration of inter-personal space and shared workspace: clearboard design and experiments, *Proc. CSCW '92*, ACM, Toronto (1–4 November 1992) pp. 33–42

11 Scrivener, S.A.R., S.M. Clark, A. Clarke, J. Connolly, S. Garner, H. Palmen, S. Schappo, and M.G. Smyth, *ROCOCO: Phase 1 Report*, LUTCHI Research Centre, Loughborough University (1992)

12 Lyons, J., *Semantics*, vol. 2, Cambridge University Press, Cambridge (1977)

13 Coulthard, M., *An Introduction to Discourse Analysis*. Longman, London (1977)

14 Atkinson, J.M. and P. Drew, *Order in Court: the Organisation of Verbal Interaction in Judicial Settings*, Macmillan, London (1979)

15 Dore, J., Conversation and preschool language development, in P. Fletcher and M. Garman (Eds), *Studies in First Language Development*, Cambridge University Press, Cambridge (1979) pp. 337–375

16 Levinson, S.C., *Pragmatics*, Cambridge University Press, Cambridge (1983)

17 Schegloff, E. and H. Sacks, Opening up closings, in R. Turner, (Ed.), *Ethnomethodology: Selected Readings*, Penguin, Harmondsworth (1974). pp. 233–264

18 Hancher, M., The classification of cooperative illocutionary acts, *Language in Society*, 8 (1979) 1–14

19 Searle, J., The classification of illocutionary acts, *Language in Society*, 5 (1976) 1–24

20 Clark, S.M. and S.A.R. Scrivener, Using computers to capture and structure drawing activity, *Revue Sciences et Techniques de la Conception*, 3(1) (1994) 51–59

**Appendix A
Drawing Act
Types and Sub-
types**

TYPE	SUB-TYPE	CODE
Non-Symbolic	–	1
Other Symbolic	Emphasis	2a
	Link	2b
	Direction/Movement	2c
	Abbreviation	2d
	Other	2e
Alpha/Numeric	Label	3a
	Problem	3b
	Solution	3c
	Constraint	3d
	Requirement	3c
	Information	3f
	Process	3g
Pictorial Orthographic	–	4
Pictorial Perspective	–	5

**Appendix B
Discourse
Categories, Sub-
categories and
Code**

CATEGORY	SUB-CATEGORY	CODE
Problem	Proposal	P1
	Development	P2
	Acceptance	P3
	Appraisal	P4
Solution	Proposal	S1
	Development	S2
	Acceptance	S3
	Appraisal	S4
Constraint (issue)	Proposal	C1
	Development	C2
	Acceptance	C3
	Appraisal	C4
Requirement	Proposal	R1
	Development	R2
	Acceptance	R3
	Appraisal	R4
Information needs	Request	I1
	Response	I2
	Acceptance	I3
	Appraisal	I4
Process	Proposal	H1
	Development	H2
	Acceptance	H3
	Appraisal	H4

Appendix C
Drawing Act Data
for ROCOCO and
Delft Protocol

Table 18.C1 ROCOCO: frequency of drawing act types over time (10-minute intervals)

Type	Interval					
	t1	t2	t3	t4	t5	t6
1	0	10	4	6	3	8
2	33	44	30	37	24	29
3	73	69	34	19	27	28
4	26	82	118	152	103	70
5	9	36	65	48	70	61

Table 18.C2 ROCOCO: frequency of drawing act types over time (5-minute intervals)

Type	Interval											
	t1	t2	t3	t4	t5	t6	t7	t8	t9	t10	t11	t12
1	0	0	2	8	3	1	3	3	1	2	6	2
2	10	23	31	13	17	13	24	13	10	14	20	9
3	17	56	45	24	18	16	15	4	10	17	21	7
4	1	25	32	50	63	55	71	81	51	52	49	21
5	0	9	19	17	28	37	17	31	29	41	45	16

Table 18.C3 Delft Protocol: frequency of drawing act types over time (10-minute intervals)

Type	Interval											
	t1	t2	t3	t4	t5	t6	t7	t8	t9	t10	t11	t12
1	1	0	1	0	0	0	0	0	0	0	1	0
2	19	12	9	16	5	3	5	16	7	7	6	10
3	22	18	18	18	4	4	6	11	6	10	17	29
4	1	0	0	0	0	5	0	1	0	1	5	2
5	2	0	0	15	0	16	0	0	28	16	5	0

Appendix D Page
DPW.94.1.14.2.24
Segmented into
Six Drawings

Table 18.C4   Delft Protocol: frequency of discourse types over time (10-minute intervals)

Type	Interval											
	t1	t2	t3	t4	t5	t6	t7	t8	t9	t10	t11	t12
P	5	6	13	4	7	9	3	11	4	1	0	0
S	6	24	68	61	68	56	59	117	108	116	70	22
C	18	7	13	10	15	8	3	4	3	1	0	2
R	17	12	3	4	2	5	12	2	1	2	2	7
I	8	13	7	18	40	30	30	23	42	62	108	111
Pr	17	33	6	3	10	6	9	26	23	23	21	7

Table 18.C5   Delft Protocol: frequency of diectic reference over time (10-minute intervals)

Type	Interval											
	t1	t2	t3	t4	t5	t6	t7	t8	t9	t10	t11	t12
D	20	34	67	72	116	132	75	92	141	146	111	62

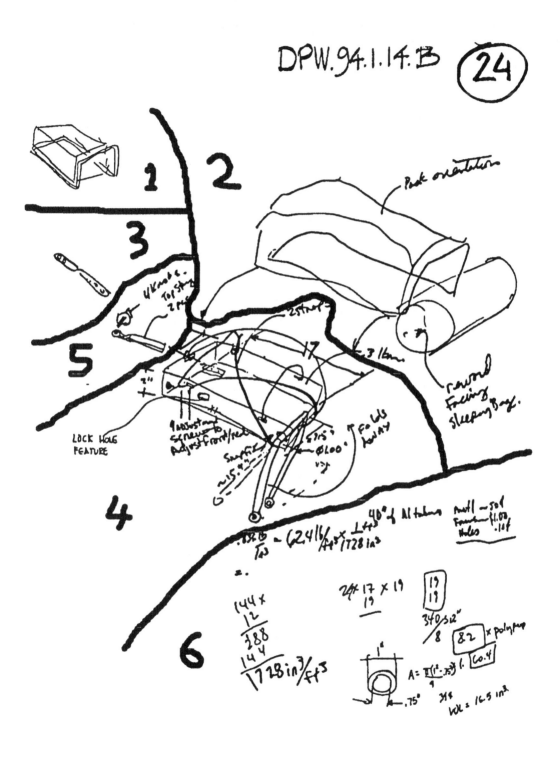

19 A Bike in Hand: a Study of 3-D Objects in Design

Steve Harrison and **Scott Minneman**
Xerox PARC, Palo Alto, CA, USA

Shortly after reading the instructions, one of the designers in the group exercise picks up the backpack and puts it on (at approximately 00:26:30, 'Kerry' puts the backpack on while 'John' gets on the bike and 'Ivan' is at the whiteboard). The designers have stuffed the bag with some of their possessions and it is now weighted and filled out in rough approximation of its projected use. Doing this, the designers have produced a set of questions (some voiced, some not) and a few answers about the pack: where is the centre of gravity, how much does it weigh, how is the weight distributed when carried, what are reasons for not just wearing the backpack instead of carrying it on a bike carrier, what size and shape is it when filled, and should alterations to the pack be part of the final design? They learn that the straps dangle and tangle and must be accounted for, that the pack is rounded when filled, and is a kind of floppy amorphous shape when not. The designers learn this almost all at once; while the one wearing it learns more of this, the others (who see this) express some understanding gleaned from this experience.

So why do we care about objects in design? Our interest in the study of design is to build new tools that support the process of design. To do that, we must take seriously the social aspects of design. Looking at both the detailed moment-to-moment activity of designers as well as the larger processes, we see that their work results in the establishment and maintenance of shared understandings among the participants. These are obtained through imagination, conversation, experience, power relations, personal dynamics, persuasion, and personal insight. We believe that to support design work practices, be it with technologies or methods, one would do well to understand the detailed workings of this communicative activity, this building of understanding.

The communications in realistic design projects interact with and reflect a lot about the situation, including the existing and emergent artifacts involved in the effort, analyses and abstractions of those artifacts, the processes (both moment-to-moment organization and more formalized plans) employed (or not) in making headway (or not), and the relations that exist and develop among groups and individuals. These communications are seldom amenable to an information theoretic approach; design communications are subtle, mutually developed among parties having special (often conflicting) interests, and highly contextual. Although admittedly complex, these communications are available to us for study (note that a researcher need not ask the participants in a group design exercise to think aloud); furthermore, understanding design communications is an important step in appreciating the phenomena of designing[1].

One striking aspect of some design communications is the inclusion of physical objects; i.e., talk about artifacts often includes the use of artifacts. While much has been made of the role of sketching in design activity[2,3], little has been done to shed light on the myriad ways that solid objects play into design activity. How do artifacts function in the generative, performative process of taking ideas to realities? What purposes do the artifacts serve? When do designers turn to the physical world?

Thus, we begin our study with an eye towards uncovering the intricacies of using tangible things, in this case focusing primarily on the bicycle and backpack, in service of designing other things, in this case the design of a mounting system for transporting a backpack on a bicycle.

Our other eye is focused on our motivation: we are specifically interested in how multimedia communications technologies can contribute to freeing designers from the 'tyranny' of physical co-location. This perspective both colours and is coloured by our view of designing as a social activity. Since we believe the telephone companies are going to drive these technologies into the workplace, our interest is in understanding how design can be poised to take best advantage of this eventuality[4,5].

So, the work described in this paper is an initial attempt to serve that two-pronged agenda – to arrive at a sound descriptive account of how solid objects are used in design settings, from which we may design systems that adequately support the full range of interactions that occur around and with solid objects.

1 Design Objects

There are many kinds of objects that are present in design situations: our current study is of the objects that are the *subjects* of the design exercise. That is, we are primarily concerned with the interactions with the backpack and the bicycle, *not* with the chairs, tables, pens, books, etc. For convenience, the former will be referred to as the design objects.

By the nature of the project under study, most of the design objects could be animated by manipulation; few of the objects were too large or too heavy to be articulated or displayed directly. Also by the nature of the activity on this particular project (revealing its status as an exercise), the design objects were rarely, if ever, changed. Basic alteration of the physical properties of the objects under consideration had a secondary role to qualities of position and relation to one another and to the designers. That said, we believe that aspects of this design exercise provide a realistic first look at how objects might function in design settings.

2 How we Looked

We reviewed the videotapes of both individual and group design sessions. Contrary to the other papers in this book we studied all three design teams and two individual designers which were involved in the experiments at Xerox PARC. When we observed someone having an interaction that engaged in some way with objects, it was noted. We brought very few pre-conceived notions of what we would see; analysis proceeded by a method of continual refinement[1,6]. Having obtained a reasonable cut at what we were looking for, we then extended this observation to the other three sessions that were undertaken as part of the experiments. We first noted and tried to categorize interactions at the minute-to-minute scale (Figure 19.1).

Following a general review of the observations and categories, we returned to segments of the video where the interactions

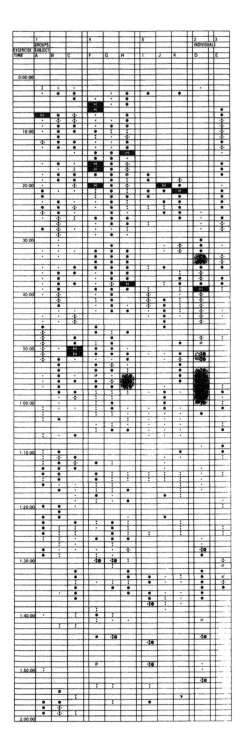

Figure 19.1
Exercise overview

were more complex and quickly changing, looking for resources
the participants rely upon and for fine-grained regularities in
their activities. The detailed kind of analysis we employed is
highly derivative of the ethnographical method known as
'interaction analysis'[7-9]. (There are other forms of ethnographic
methods that have been brought to bear on design process,
most notably Bucciarelli[10] and Cuff[11].) We applied it to both the
group and individual session data.

3 What we Saw

This was an early design effort, concerned with hardware for
mounting one object to another (backpack to a bicycle); the
manipulations were part of interacting over and with those
existing objects, putting them into particular relationships. The
designers were often concerning themselves with the spaces
between.

Figure 19.1 shows a timeline of the minute-by-minute observations of all five sessions side-by-side. It is coded to show the
engagement each designer had with the design objects occurring
at any time during that minute. So, consecutive minutes with
the same notation do not necessarily mean that a particular
form of interaction was continuous in that period – for example,
an object might have been put down and picked up again.
Furthermore, if more than one kind of action with respect
to the objects took place during the minute, the more engaged
activity is indicated; the coding scheme reflects and generally
follows an early version of the categorization detailed later in
this paper. In this first round of observation, the uses occurred
in different kinds of encounters and they are presented without
definitive ties to specific kinds of encounters or communicative
functions.

The first observation is that the design objects were *frequently*
part of the activity. The general pattern for all designers was to
engage with objects soon after reading the instructions. (The
group and individual selected as the primary subjects (group 5
and individual 2) take longer to become engaged with any of
the design objects than the other groups and individual
designer, the individual only doing so after telephoning the
Blackburn Company seeking information. Curiously, he learns
that heel and thigh clearance are key issues in placing the rack
on the back of the bike. The group learns the same thing in
about the same amount of time by getting on the bike and

trying various configurations.) The frequency and duration of engagement with objects diminishes as the designers move into documenting their designs; however, the kinds of interactions with the objects (except for staring at them) tends to remain the same. The two individual designers do less gesturing over the objects than do the designers working in groups, but they still do gesture over and with objects. Another salient phenomenon is that the objects are often used as part of a conversation (in the case of the groups) or as part of an explanation (in the case of individuals).

A common thread that runs through all of the ways that objects were used is the relation of hand and eye: on one hand, one set of hands could animate a situation or set of relations involving the design objects for both the presenter and the viewers. On the other hand, the experience of manipulating the object is not equally available to both the viewers and enactor. How does this play with our notions of shared experience? Does everybody have to do everything? What does the viewer experience?

Related to this is the observation that the objects were often stand-ins for other objects (e.g., the tape measure was used as a surrogate for a strut). Often, it would be difficult to tell if the designer was using a specific object (such as the backpack) as a generic. The spaces between objects or over them became the location for imaginary objects that would be acted out, acted upon, or pointed at. These stand-in uses and imaginary objects might be alternative representations, so the relationship of the objects to other forms of representation also becomes an issue. Since drawing and sketching are outside of the scope of this project, we have chosen to leave out representation in these other forms from this study, while acknowledging it as an important issue worthy of separate study.

Furthermore, all of these would be occurring simultaneously (in the case of groups) and in rapid succession, each flowing into another, in a kind of improvised 'choreography'. These exercise-scale observations suggested more detailed study of the activity. To do this, we looked at representative segments in both the individual and group exercises, exploring how conversation, manipulation, and design development worked across a few minutes' activity. Looking at the activity as a whole, the interrelatedness of these constituents was a choreography of object, action, and communication.

4 The Dimensions of Objects

There are a number of characterizations of visible involvement between the designers and the objects: kinds of orientations (e.g. 'What is the person doing with the object?'), kinds of gesticulation (e.g. 'How does the person move with regard to the object?'), and as a component of the language (e.g. 'Do the spoken utterances depend on the gesture or the object?').

As a coarse approximation, these categories roughly correspond to eye, hand, and mouth. The gaze of the eye is an indication of attention to an object; the hand is the primary means of working with an object; and speaking is the central means of communicating about an object. More accurately, the first category is object-centred, the second is body-centred, and the third is communication-centred. But this is misleading: just as the act of seeing requires the coordination of not just the eyes, but the entire body, descriptions in each category overlap the other. Describing the relation of designer to design object requires components in each. In fact, some of the subcategories in kinds of orientations to objects and kinds of gesticulations are the same, only from a different perspective. Once these major categories have been characterized, the significance of them in relation to one another and the experience of design will be laid out.

4.1 Kinds of Orientations to Objects

To what extent does the activity rely on the design object?
What is the person doing with the object?
Are they looking at it?
Are they touching it?
Are they using the object?

Across the various settings in this project, we have seen a range of orientations to design objects: attending, moving in relation to, touching, manipulating in relation to another, and acting. This is a continuum (from the most distant or most 'outside-looking-in' to the closest, most enveloping 'inside-looking-out') with fuzzy boundaries between them. Another way to look at this continuum is that the designers' eyes are augmented by varying amounts of bodily activity and contact with the object(s).

Attending to an object

While it is not possible to know if anyone is actually paying attention, it is readily clear when they are looking directly at something. At times, people would gather around the bicycle or

Figure 19.2
Attending –
DPW.94.1.13.2 00:42:44
'...or or sit on top of
the wheel...'

would glance across the table at the backpack lying on it (Figure 19.2). Another indication of attention to an object is the way it is talked about, such as referring to it with a pronoun or special language, such as calling a carrier mounted on the top tube 'motorcycle gas tank'.

Moving in relation to an object

Moving their hands over it, designers could do more than just look at a design object, they could use it in relation to their body. They would point at it or to features of it or would animate imaginary parts (Figure 19.3). In fact, moving in relation to objects was one of the most common engagements with objects, since it was often part of a communicative gesture (see 'Component of Language', below).

Touching an object

Frequently, designers would pick up the pack, turn it over, gesture with it, or even just lean on it while talking (Figure 19.4). Being larger and clumsier, the bicycle would be held or leaned on.

Manipulating one object in relation to another

Not only would they work with one object, almost every designer took the pack and held it in various positions around the bicycle (Figure 19.5). Sometimes explicit purposes were ascribed to this (like 'can the rider's thigh clear the pack in this position?'); sometimes it went unremarked.

Figure 19.3
Moving in relation –
DPW.94.1.14.5 00:50:12
'...I mean you're
barely with somebody
at the knee at top of
their stroke...'

Figure 19.4
Touching –
DPW.94.1.13.2 00:46:57
'...I'm going to ask
one question while I'm
at it here em can you
show me some
literature on the Hi
range...'

Acting

Finally, there is using the object in action (Figure 19.6). The most obvious examples are riding or mounting the bicycle, wearing the backpack, or measuring with the tape measure. The measuring tape was one of the most frequently used objects manipulated in relation to other objects. Obviously, it was used to understand distances, but the property of physical extension

Figure 19.5
Manipulating –
DPW.94.1.13.2 00:53.52
'...he points out is that
er when you get up
there your er thighs em
bump in...'

Figure 19.6
Acting –
DPW.94.1.14.5 00:26:38
'...I mean between
the in the triangle...'

allowed it to be used as a stand-in for struts or other parts of the rack. (Many reference gestures, such as appearing to attach the pack to the rack with a bungee cord, involve motions similar to their corresponding 'actions', but without the actual object.)

4.2 Kinds of Gesticulation	*To what extent does the activity rely on the designer's body?* *How does the person move with regard to the object?* *What does the hand do independent of the eye?*

The bodily motions the person employed in the engagement are phenomena of this category: moving in relation to, touching, and acting. Most often, gestures were made with hands, such as pointing to locations, but a few like glancing, were not. Just as holding an object informs seeing, seeing can inform holding.

Moving in relation to an object

Moving their hands over it, designers could do more than just look at a design object; they could highlight aspects of it, drawing focus on a location or property. The video is rife with examples of situations where gestures provide demonstration and elaboration around the object. This activity may take various forms, ranging from the use of the body as a surrogate for the object, to using the gesture to indicate or suggest an alternative form for the object. Occasionally they would animate with their hands how a strap might fasten the pack to the rack or how vibration might dislodge the contents from the pack (Figure 19.3). Again, moving in relation to objects was one of the most common engagements with objects, since it was often part of a communicative gesture (see Section 4.3, 'A Component of Language', below).

Touching an object

Getting hold of an object provides the opportunity to manipulate it, to see it in a more comprehensive way, to experience it. Picking up the pack, turning it over, gesturing with it, exposing its parts, or even just leaning on it while talking would provide the designer with its weight, its surface texture, the ease of access, etc. (Being larger and clumsier, the bicycle would be held or leaned on.)

Acting

Finally, there is using the object in action The most obvious examples are riding or mounting the bicycle, and wearing the backpack. These provided some sense of the object in use. (Many reference gestures, such as appearing to attach the pack to the rack with a bungee cord, involve motions similar to their corresponding 'actions', but without the actual object.)

4.3 A Component of Language

To what extent do the spoken utterances depend upon a gesture or the object to be understood?
Is the utterance amplified by properties of the object?

Often, gestures may be an integral part of the communications. They might clarify 'which one?' or 'what?' when 'this' or 'that' is used; they might make the phrase 'the bag will shake' more specific by showing the various motions of a shaken backpack.

> '... gestures are the person's memories and thoughts rendered visible. Gestures are like thoughts themselves. They belong, not to the outside world, but to the inside one of memory, thought, and mental images. Gesture images are complex, intricately interconnected, and not at all like photographs. Gestures open up a wholly new way of regarding thought processes, language, and the interactions of people.' – David McNeil[12]

The psycho-linguist McNeil[12] uses the following taxonomy:

- *Iconics*   These gestures illustrate an act or exhibit an object that is referred to in a concurrent utterance. They tend to be parallel expressions of meaning, often revealing different aspects of some event or thing. They tend to have three phases: preparation, stroke, and retraction.
- *Metaphorics*   These are like iconics in their pictorial or representational nature, but their content is of an abstraction rather than a concrete object or event. The example given is that of the speaker offering a gesture of holding a container while referring to the genre of cartoons.
- *Beats*   Beats are two phase (in/out, up/down) gestures – so called because the producer appears to be beating out musical time. They index a particular word or phrase as having some discourse-pragmatic import (summing up, introducing new topics or characters, etc.).
- *Cohesives*   These gestures are essentially the converse of beats, emphasizing the continuities in a string of discourse rather than highlighting discontinuities. Their form can be of almost any sort, including iconic, metaphoric, pointing, and even beats. Repetition plays a role here – for example, continuity is indicated by returning to a gestural stance (say a particular hand position) after an interruption (one that utilized the hands for another gesture).
- *Deictics*   This is the use of pointing to resolve unbound referents. While most clearly used in instances where a concrete object or event is available, there need not be anything present to point at.

K	why not see like you mount shoulders back **here**
J	yeah yeah just maybe maybe you just mount a child seat back **there** and you give them a child (laugh) and make him wear the backpack
I	or a manikin
K	a manikin
I	with the top towards the
K	Harry the backpack holder
J	a backpack a a manikin with with clamps coming off of it (laugh) hold on
K	now we're kind of assuming that there's some rack to attach **this** to but what if the rack was really um something that attaches to **this** and just flips down so maybe you hook it on to a bracket up **here** but you just flip down and it clips in **here**

Figure 19.7
Conversational props – DPW.94.1.14.5 00:42:48– 00:43:22, just where is 'here' or 'there'? What are the properties of 'here' that make it a candidate location for a bracket? What is the first 'this' referring to? Is the second 'this' the same thing or place?

While McNeil is describing objectless, free-handed gestures, the quotation suggests the powerful communicative role they play. Besides being part of the content of the communications, gesture is also used to develop, sustain, and coordinate interaction[13]. At its most basic (in the case of the group exercises), picking up the backpack or moving the bicycle provided a non-verbal means to indicate focus of activity to others. While this use of gesture has been well-studied in conversation analysis – and even in relation to drawing and sketching in design[14,2] – there is precious little that addresses the general ways in which objects are conversational 'props' (see Figure 19.7).

Deictics: resolving references

Hands would be visible around and on parts in coordination with pronoun references like 'this' and 'that'. Reference to the object acted in concert with words to convey meaning. The shared understanding of this diectic reference at once depended on the worlds and the image of the concrete object under discussion. There are a number of gestures around objects that, in concert with talk, were used to identify and clarify subjects: pointing, holding, and framing with fingers or hands.

Illustration

Objects were also used explicitly as illustration. In fact, they were often introduced in conversation for the express purpose of illustrating a particular quality that could not be addressed as directly by talk or sketching. Sensory characteristics were used to identify specific instances as well as convey properties. The object (more than the conversation) conveys its appearance, its material, and how it might fit with the other parts.

- *Appearance* What the surface looks like; this is easily conveyed in many communicative exchanges (but being shown how or what to see remains important; we'll come back to this). While it appears that little additional commentary would be needed, often discussion about recently introduced objects would begin by describing elements of appearance.
- *Form* The shape of an object and its fit with other things. This was conveyed by manipulating the objects (usually the smaller, more manoeuvrable one, the pack) to show their relationship to each other and with the world.
- *Feel* The ways in which an object might respond to touch or other interaction, such as pushing, sliding, scraping, or rolling. This is only partially conveyed by showing – viewers have to infer feel from the way it reacted to touch, from explicit explanation, or from repeating the touching themselves.
- *Material* Although appearance, form and feel can give some clue as to the material qualities, it is not a comprehensive evaluation. Material questions were often direct, 'What is it made of?' – 'Tubular steel, I think'.
- *Performance* These are interactions with the fundamentally 'mechanical' properties of objects: for example, showing the way the crank turns, or the shape the backpack deforms to when loaded.

Communicating to whom?

Various aspects of the situation offer the communicant opportunities to carefully construct a view for the other; so hands could be visible touching parts. Sometimes the parts would be held in view or simply touched while conversation about them proceeded. That said, there is no way to distinguish the degree to which a communicative gesture was informing to either the sender or the receiver, but there were a number of interactions where the gesture or activity was tailored for a particular audience. In the example of Figure 19.8, 'Kerry' is positioning the pack on the handlebars for 'John', the rider, to understand that the radical increase in moment of inertia related to steering would be unacceptable. This use of interaction as a form of monitoring and assessing the degree of mutual understanding is an important area for further study.

Figure 19.8
Feeling – DPW.94.1.14.5
00:27:35, by placing the
backpack on the
handlebars, 'Kerry' is
using it as part of her
communications to
'John' that this location
will make steering
difficult

When is a movement not a gesture? This seems difficult to answer if a functional, motivating description is sought, especially in cases of acting with the object such as measuring with the tape. By this definition, though, these are also examples of what we mean by 'gesture'. The following category will address the communications phenomena so as to clarify some of the ambiguity.

5 Activities Before, During and After

What was the context in which the use of objects took place? What happened prior to, during and after objects were actively engaged with? To get some idea of the sorts of triggers that bring about interaction with objects and, in turn, the sort of activities that objects trigger, let us look at just the segments where objects were handled.

One simple hypothesis could be that most engagements with objects are triggered by an explicit seeking of information, either of the concrete, closed-end form, 'How long is the backpack?, or more open-ended, 'Which configuration works best?' This hypothesis doesn't hold water though; while there are a few clear-cut instances of information-seeking missions, there are also quite a number of 'spontaneous' engagements. Furthermore, there appear to be other equally compelling explanations that account for the change from an activity without to one with objects: to control the dynamics of a conversation, to change topics, to ground gestures, and to confirm or to recalibrate imaginary objects.

Both during and after, the use of pronouns increased as the body of referents increased. During subsequent activity, it was possible to make reference back to events and objects during the engagement; engagement had reset context.

While we did see these recurring phenomena, no causal explanation accounts for most, much less all of them. It is important to consider the cumulative effect of making these purposeful both in the small (such as to measure a distance) and in the large (such as to build a more robust understanding).

6 The Experience of Design

What does this framework say about design activity? First, that objects are more than a source of information; they are constituents of the activity. Second, that they are constituents of and frames for the communications. Third, they alter the dynamics of interaction, especially in multi-designer settings.

Let us consider an example. Returning to the early part of the group exercise, 'Kerry' puts on, then takes off, the backpack while 'John' sits astride the bike. 'Ivan' is at the whiteboard, but is looking at both of them. 'Kerry' carries the pack to the back of the bicycle, holding it over the rear wheel. She holds it in a number of positions, including the upright one of the earlier version mentioned in the reference materials. She then takes it to the front of the bike, placing it first on the handlebars, then on the cross-tube between the handlebars and the seat (Figure 19.9).

How can we understand this in terms of the categories and their significance as a description of the experience of designing?

First, consider how the activities *rely on the presence of the objects*. The bike and the pack are both subjects of attention on the part of all the designers. The physical properties (the weight and bulkiness of the pack and the height and instability of the stationary bicycle) restrict movement and inform the designers about their behaviour in use and their utility. These properties are available in relation to one another.

An example of the next category – the activities that *rely on the bodily interactions with objects* – the bicycle (through being ridden) and the pack (through being worn) provide questions and some answers about their properties. At the end of the

(K: puts on backpack)
(J: is 'riding' bike)
J   so I I keep thinking that there's all this weight in this area
             between the seat but that could be used but I wonder if that would
             really work when you're like pumping really hard pedalling up a
             hill it sounds like
K   between the seat where?
J   I mean between the in the triangle
K   well your knees work (inaudible)
J   that was a pull on the brake-lever (laugh)
I   can can you un
J   un-?
I   there's a way to un-.. un it
J   oh to unhook it right there so you can
I   I guess we don't need to (inaudible)
J   that's OK so
00:27:00
K   so we want to put it in there but I let's see if you got (in-
             audible)
J   yeah you'd never really be able to
K   you wouldn't be able to get your knees pedalling
(K: puts pack over rear wheel, horizontally)
K   OK now what about
(K: puts pack over rear wheel, vertically)
K   They've gotta a prototype that kinda has it this way
J   is a facing forew
I   yeah that's right facing forward OK
J   would it be too funky to have on the like projecting from the
             front wheel?
I   handlebars? yeah try that
(K: carries pack to front of bike)
J   or off this handlebar stem even because that's fixed but if
             it's off the handlebars you know it's like an old bike basket that
             way like the Wizard of Oz (laugh)
(K: puts pack over handlebars)
K   OK now steer tends to
J   you could turn it long ways
I   or if you could get it down low where the
K   (inaudible)
I   centre of gravity is low
(K: puts pack between handlebars and seat)
J   oh yeah like a motorcycle gas tank
I   oh yeah the gas tank there we go
J   you know like on your on your Harley (laugh)
K   except on a Harley you don't pedal so you don't get your
             knees cranking
(J: moves legs in reverse pedal, hits pack with knees)
J   maybe they're gonna have to redesign the top of their
             backpack

Figure 19.9
Example DPW.94.1.14.5
00:26:30–00:28:00

sequence, 'John' pedals the bicycle, immediately 'seeing' the interference with the backpack.

Lastly, the objects are *part of the communicative activities.* 'Kerry' puts the pack upright behind 'John' and says that the prototype 'kinda has it this way'; this is an example of dyxsis (resolving a reference). The next interchange also has deictic value because what is meant by 'facing forward' is established. A few moments later, the pack becomes the 'gas tank' for a 'Harley', fleshing out an abstraction. And, at a slightly different level, 'Kerry' may have placed the pack on the handlebars to say to 'John' that this location will not work, communicating without speech (shown previously in Figure 19.8).

These phenomena are part of the complex fabric of the experience of design. There are two primary ways in which object engagements constitute experience: as a rich source of information and as a major shaping of the 'unwitting' learning aspects of design.

One component of the experience of design is the learning process: the act of designing changes the designer (rendering it impossible for designers to do truly the same design twice). The engagement with objects produces a particular kind of learning process which can be demonstrated by one of the aspects of object-centred phenomena: apparently repetitive interaction with objects. One obvious example of this: starting fairly early on, and repeated throughout each design session, designers would hold the pack in various positions over the bike. We see designers, after settling on a particular geometry, hold the pack over the rear wheel, put the pack down and a few minutes later repeat the process. While this might be thought of as an inefficiency in the process, each encounter produces two results: information and additional experience. The experience with the object teaches two things: how to interact with it, and the scope of possible information it can supply.

We have seen how the communications are more robust because of the introduction of objects and references to them; we can infer from this that some qualities of the information are more robust as well. Furthermore, touching, moving, riding, wearing, etc. produce (for the toucher, mover, rider, wearer) a kind of information that is not well described in external forms. And this is not restricted to first-hand experience, either: the close-at-hand action of the actor is available to the others, either through directed viewing or through legitimate peripheral participation[15]. Being both action in relation to an object and a communicative act, the engagements of others are themselves part of the fabric of the design process.

7 Conclusions

This study suggests a few new important questions in the attempt to understand how design is carried out.

We have seen how the objects at hand were used and used pervasively. The significance is not that they provide a rich source of information for the designer (which they do) or that they are superior to abstract forms of information (which they may or may not be), but rather that the processes of interaction with objects have communicative value and alter the dynamics in multi-designer settings.

Where to go from here? The relationship of objects to representations has come up in a number of interesting ways in the current study, but has not been well pursued. We have seen the objects used and referred to as themselves or as stand-ins for other objects; we have seen the unfilled area between the backpack and the bicycle represent the backpack carrier being designed; we have seen the tape measure be a gnomon of distance, a variable length prototype support and a device for extending gestural reach. As a starting point, a follow-on study might look at these questions: How do embodied representations (using hands to animate or tape measures for struts) compare with drawings? What is their relationship? Is this the same kind of externalization?

Acknowledgments

The authors wish to thank the other members of the Xerox-PARC Design Studies Group (David Bell, Margot Brereton, David Cannon, Catherine Marshall, Susan Newman, Lucy Suchman, and Randy Trigg) for reviewing the tapes with us and who worked with us on developing the dimensions of engagement and methods of observing designers at work. We also would like to thank Deborah Tatar for suggesting alternative interpretive schemes and sources of related research, and the Xerox Corporation and Delft University for supporting this work.

References

1   Minneman, S.L., *The social construction of a technical reality: empirical studies of group engineering design practice*, Xerox PARC Technical Report (Doctoral Dissertation, Stanford University) (1991)

2   Tang, J., *Listing, drawing, and gesturing in design; a study of the use of shared workspaces by design teams*, Xerox PARC Technical Report SSL-89-3 (Doctoral Dissertation, Stanford University) (1989)

3   Rowe, P.G., *Design Thinking*, MIT Press, Cambridge, MA (1987)

4   Harrison, S.R. Computing and the social nature of design, *ACADIA Quarterly*, **12**(1) (1993) 10–18

5   Harrison, S.R. and S.L. Minneman, Design tools for the communications age, *Stanford Design EXPEerience* (March 1994)

6   Strauss, A.L., *Qualitative Analysis for Social Scientists*, Cambridge University Press, Cambridge (1987)

7   Irwin, S.L., *Technology, talk and the social world: a study of video-mediated interaction*, PhD dissertation, Michigan State University (1991)

8   Goodwin, C. and M.H. Goodwin, *Formulating planes: seeing as a situated activity*, XeroxPARC Technical Report, Palo Alto (January 1991)

9   Goodwin, C. and M.H. Goodwin, Context, activity and participation, in P. Auer and di Luzo (Eds), *The Contextualization of Language*, (Eds), Benjamins, Amsterdam (1989)

10   Bucciarelli, L.L., An ethnographic perspective on engineering design, *Design Studies*, **9**(3) (1988) 159–168

11   Cuff, D., *Architecture: The Story of Practice*, MIT Press, Cambridge (1991)

12   McNeil, D., *Hand and Mind: What Gestures Reveal about Thought*, University of Chicago Press, Chicago (1992)

13   Sacks, H., E. Schegloff, and G. Jefferson, A simplest semantics for the organization of turn-taking in conversation, *Language*, **50**(4) (1974) 696–735

14   Allen, C., *Situated designing*, MS thesis, Carnegie-Mellon University, Pittsburgh, PA (1988)

15   Lave, J. and E. Wenger, *Situated Learning: Legitimate Peripheral Participation*. Cambridge University Press, Cambridge (1991)

20 Can Concurrent Verbalization Reveal Design Cognition?

Peter Lloyd, **Bryan Lawson** and **Peter Scott**
Cranfield University, Bedford, UK and University of Sheffield, UK

A newspaper article about the architect Richard Rogers' dyslexia observed that 'his [spoken] sentences meld into a peculiar mush that can be hard to follow.'[1] Although dyslexia is usually thought of as a difficulty in writing, clearly there is also some difficulty in speech, or assigning words to thoughts. It is tempting to imagine how Richard Rogers would fare in an experiment using protocol analysis.

The theory behind classical protocol analysis[2] is that by asking a person simultaneously to perform a certain task and to 'think' aloud one can gain direct access to that person's thought and hence the patterns and sequences of the thought required for the task. Words 'thought' aloud in this way are thus considered to be unconstrained by the need to communicate. It is not clear how a disability such as dyslexia, primarily a disorder of language, would affect an analysis dependent on language ability.

The example may be atypical but suffices to highlight the problem of using concurrent verbal reports as a means of obtaining cognitive information. If dyslexia could be considered as a filter between thought and speech, then we might hypothesize that such a filter, to a greater or lesser extent, exists in everyone. That is to say, thought is always mediated, and sometimes offset, by channels of communication formed by constitution and experience. A central problem for protocol analysis as a research tool is to determine just how much these channels of communication affect the thought preceding the communication. Generally the paradigms of information processing, and particularly the psychological research area of problem solving, assume that the effect of communication channels on problem solving thought is minimal. This may well be true for 'conventional' problem solving tasks but the relatively recent application of protocol analysis to design tasks has again brought the assumptions that underlie protocol analysis under scrutiny.

It is our contention that designing is a way of thinking incorporating many separate modes of thought in much the same way as talking is a way of thinking incorporating many modes of thought[3]. Sometimes a mode of 'design' thinking and a mode of 'talking' thinking may be concomitant, in which case one may elicit designing behaviour using concurrent verbalization. At other times, though, 'design' thinking and 'talking' thinking may be unrelated to one another, in which case prompting a subject to 'think aloud' will result in verbalization that is not a reflection of designing behaviour and additionally the verbalization will affect the task.

1 The Elicitation of Design Thinking

Gilbert Ryle[4], in discussing a class of knowledge he terms 'knowing how', has questioned whether knowledge, in the form of propositions, precedes the act of doing something, that is whether people plan their actions before executing them or whether they simply execute them. His assertion is that people simply execute actions without conscious or propositional planning. Concurrent verbalization seems in some way to be attempting to elicit fragments of propositional knowledge which may, in actual fact, not exist. It is a common notion amongst designers (and artists) that if they could say what they were attempting to do they wouldn't have to design/draw/compose it. A recent interview with Richard MacCormac, a well-known architect, touched on the problem of talking while designing:

> 'Architecture is a medium of thought which is very powerful and that in the same way, as say, mathematics and music are media of thinking, we have our medium of thinking and the difficulty with it of course is, like music, that it is a medium that's extremely difficult to talk about.'[5]

What, though, does concurrent verbalization actually represent if not design activity? It is thought there may also be times when propositional knowledge does precede action and this is thought to be when a person is learning a skill. In such a case it may well be possible to elicit concurrent verbalization that is a fair representation of thought. If we consider that designing is an ability that is constantly being learned then we will always have some degree of propositional and hence verbal thought to reflect design behaviour.

In this paper we present our interpretation of the individual designing protocol of 'Dan' and show how concurrent verbalization effectively reveals the close interaction between design problem and design solution; how, for example, problems are worked out through the conjecture of solutions. We then go on to look at aspects of design thinking that are not effectively revealed by concurrent verbalization in Dan's protocol. These types of thinking fall into two categories: (1) whether words 'thought aloud' are an adequate representation of thought, and (2) whether an experimental design situation actually affects the design thinking it seeks to analyse.

We conclude that although concurrent verbal reports can reveal some aspects of design thinking there are many types of design thinking that remain impervious to concurrent verbalization requiring different methodologies for analysis. We argue against design as a unitary concept, and propose a view of designing as consisting of many interlocking and overlapping processes.

2 Protocol Analysis

Our usual experimental methodology[6,7] is slightly different from the present study, in that it attempts to preserve a 'real' design process. The design brief is usually given to the designer at least a week before the experimental design session. During the design session the roles of the experimenter (client) and the designer (consultant) are clearly defined. The designer is not required to think aloud; instead a detailed later review attempts to elicit design thinking during, and previous to, the design session. The designer is also free to do anything they wish (within reason) during the course of the experiment.

Our analysis usually proceeds by summarizing the design session into a number of phases. A subjective interpretation of these phases is then arrived at and these are incorporated into a global model of the design session. Interpretations, though subjective, are based on models from the design literature[7,8]. Our studies usually involve a number of designers (up to five), varying in experience so that we can observe how theoretical designing mechanisms might change as experience of design is gained.

2.1 Analysis of Dan's Protocol

The present methodology is thus somewhat different from our own. Before we go on to show how concurrent verbalization *does not* adequately reveal aspects of Dan's design thinking, we analyse the protocol according to the procedure outlined above to show what, in our view, it *does* reveal about Dan's design activity. A summary of the design activity is provided in Table 20.1. The protocol is divided into 11 phases and for each phase a brief interpretation is given.

Dan dwells for some time on the issues that the problem provides without any concrete commitment to a particular solution form. Indeed the solution form arises from these design issues (phase 6). In phases 1–3 Dan first tries to explore the generally accepted issues of carrier design by generalizing from both other market products and a phone call to a manufacturing company. He then goes on to sort out what he personally feels are important issues to concentrate on (phase 4). Phase 5 is where he reconciles these two views and from phase 6 onwards a solution begins to develop. A solution does not, however, develop independent of the issues Dan has uncovered. Phases 6 and 7 are more a means of exploring the issues, finding out what they mean in the terms of a physical solution, behaviour that has been termed in the literature as 'conjecture–analysis'[9].

In phase 8 Dan goes on to sub-problems of fixing the carrier to the frame of the bike even though the form of a solution is unresolved. The final phases of the protocol (10–12) are extensions of this form of behaviour where Dan looks at sub-problems of fixing types, materials, mounting brackets, etc. At the end of the protocol the sketches give a good idea of the low level problems involved in the design although the complete form of the design remains vague. The final phase could thus be approximated as synthesis: grouping the low level problems into a higher level solution. Overall Dan's design process can loosely be interpreted as a process of analysis–synthesis, the traditional model of the design process[10,11].

2.2 The Interaction of Design Problem and Design Solution

The protocol of Dan illustrates many different forms of behaviour. Overall is an analysis–synthesis type mode of behaviour, individual episodes reveal conjecture–analysis type behaviour.

There are also illustrations of the way the initial problem is structured into sub-problems and the protocol shows the close interaction between problem and solution, sub-problem and sub-solution. There are times when Dan seems methodical in his designing (phases 1 to 3, for example) and conversely times when he skips between seemingly unrelated pieces of information (phase 10, for example). A particularly interesting episode comes in phase 10 which ends in Dan 'suddenly' discovering the two bushings on the bike specifically for mounting a carrier.

After coming up with a tentative 'triangular' solution form (extending to a point) Dan needs to find a way of attaching the carrier to the frame of the bike just below the saddle. A first solution is a sheet metal bracket which turns out to be 'not removable enough'. After the insight of a 'pop-in' bracket, he explores the possibility of having a 'windsurfing boom' type attachment. Proclaiming that the carrier is 'not an after market product' (01:28:03) he feels a better solution would be to have the pop-in type attachment directly brazed on to the frame of the bike. The Batavus company is not willing to do this which leads to Dan 're-understanding' the nature of the product as a 'retrofit' (01:29:40). Silent for a while, Dan suddenly notices the 'two bushings' already brazed on to the bike. It turns out these are specifically for mounting rear carriers.

In any field apart from design this information would be extremely disheartening; after all Dan has just spent a concentrated ten minutes on working out a type of bracket only to be told that one already exists. The episode is not wasted, though, as Dan has found out valuable information about the motivations of the Batavus company. The episode is one of discovery, Dan finding many things about how far the company is prepared to go in backing the new product. What is interesting is that this information was obtained by a very indirect means – Dan was originally focused on the attachment sub-problem. A simple question such as 'how far are you prepared to go in backing this product?' would have resulted in a suitably reassuring and vague reply. By working out the answer to this question through solutions Dan has (perhaps unintentionally) understood a problem.

The protocol clearly illustrates how Dan's short-term focus explores the links between problem and solution. There are, however, many aspects of Dan's design thinking that remain unverbalized or restricted in some way. In the remainder of the paper we come back to the two issues we highlighted earlier.

Table 20.1 Summary and interpretation of protocol

Phase	Description	Analysis
1.	Reads criticism of prototype: – Doesn't take advantage of frame – Centre of gravity high Dan – 'Keep backpack low, don't reinvent the wheel'.	*Dan familiarizes himself with the task, trying to understand the issues that the deficiencies in the initial prototype raise. He formulates an abstract strategy.*
2.	Looks at pros and cons of alternatives already on the market.	*Trying to get more general information he asks for comparable market products.*
3.	Thinks what to do in order to 'get up to speed'. Phones company who manufacture carriers. Finds out a number of things: (1) Low mounting on the front of the bike is OK (2) High up is bad (3) If mounting on the back then the main factor is heel and thigh clearance (4) Keep fastening as low as possible (5) Side mounted could be OK.	*Needing more information he thinks about the best way to find out the key factors.* *Phase 1–3 are all problem analysis. Dan is sorting out issues that are generally thought of as important for cycle carriers.*
4.	Measures the HiStar backpack for length, width and weight. In the background is where to put the backpack (and hence carrier). Looks at bike, gets on bike, measures thigh clearance. Checks all locations.	*Continuing to analyse the problem, Dan tries to identify design factors first hand.*
5.	Back to criticisms of prototype seeking another perspective and some form of consensus between personal issues (phase 4) and general issues (phases 1–3). Rigidity and centre of gravity emerge as key issues.	*With first and second hand knowledge of the problem phase 5 follows as a phase when the two views are integrated.*
6.	Makes a decision to go for rear mounting: 'Hey the place to put it is back here.'	*A solution then begins to evolve with positive and concrete generation.*

Table 20.1 (*continued*)

Phase	Description	Analysis
7.	Need 'parallelogram' for stiffness. Different perspectives of the bike lead to a 'triangular' form. Stiffness → structure → solution of triangular point ('our proprietary feature').	*With stiffness a key factor Dan generates a structural solution which also has the added benefit of having a saleable feature.*
8.	Attachment to frame. Top attachment is to be detachable. Uses windsurfing bracket as a possible solution.	*There is now a positive sense of direction in the design process. The design problem now takes on a structure to it, being divided up into a number of sub-problems.*
9.	Considers cost.	*Concrete constraints are not forgotten.*
10.	Finds lugs on sample bike specifically for top attachment. Looks at their specification. Returns to triangular point feature, and how it will connect to the frame so as to be able to fit any mountain bike. Working through, finds that the triangular form might not actually work. 'This [point] is not going to come together.'	*In a phase of productive enquiry Dan 'discovers' several problems of his conjectured 'triangular point' solution. The problems are to do with attaching the carrier. Working out an attachment leads Dan to find out the extremely valuable information of just how much the Batavus company are prepared to support the product.*
11.	Reminded of the time has to 'go for a solution'. Decides to go in one direction and take the consequences. Encounters problems of materials, mounting, brackets, etc.	*Attempting to solve as many sub-problems as he can Dan begins to piece together a solution (synthesis). He achieves what seems a temporary resolution of the overall design problem.*

Firstly, are Dan's verbalizations really expressing what he is actually thinking for most of the protocol? Secondly, how 'real' is Dan's design process?

3 The Mapping of Words to Thoughts

The problem of what words and language actually represent is a problem that occupies a central position in twentieth century philosophy, the interaction between knowledge and language being a central question within this problem. We therefore feel cautious about accepting a direct link between words and thoughts in a commentary provided by concurrent verbalization during designing.

One of the central tenets of a protocol analysis methodology is that the problem situation can be manufactured to such an extent that a verbal report provides an accurate representation of the thoughts of the participant. The situation is contrived so the words of the verbal report can have no other function than to express thought. Researchers such as Nisbett and Wilson[12], investigating the validity of verbal reporting as an experimental method, concentrate on obtaining information in a number of different forms (for example physiological, behavioural, observational) to test the consistency of the verbal information. In all such studies the underlying assumption is that the verbal report either confirms or denies other forms of evidence; it is either correct or incorrect to use such a methodology.

The standard text of protocol analysis[2] summarizes its application by saying that there are strict limits for reports on thought processes, particularly the perceptual and retrieval processes that lead to any given thought being attended. Presumably it could be argued that the larger the element of perception and long term memory retrieval required in a given task, the less complete a verbal report becomes. Nisbett and Wilson echo this point, citing several 'perceptual' experiments, saying that subjects do not report the higher mental processes of subliminal perception and perception without remembering.

Concurrent verbal reports, then, do seem a fair reflection of a certain type of thought – thought appearing in short term memory (our previous analysis concentrated on the short term focus of Dan between problem and solution). It thus seems valid to use protocol analysis to elicit short term thought processes. Design, though, is a task that involves many other types of thinking, and it is these other mechanisms that are not amenable to concurrent verbalization. As an example the interview with the architect Richard MacCormac revealed how using different drawing instruments mediates different ways of thinking:

'There are different frames of mind which involve different instruments actually for producing [and] representing what you are doing.'[5]

It is these 'frames of mind' which concurrent verbalization does not adequately convey.

Additionally, empirical studies of the design process[8] suggest that the function of words often seems to change according to the specific situation; that words are intimately linked to the process of design rather than being a simple commentary on design thinking. Some designers suggest that words often make them 'over committed' to a particular thought. Others stress that words provide sufficient externalization for them to 'find out what they mean' or 'explore what I just said'.

To discuss the issue of when concurrent verbalization can and cannot reveal 'design thinking' we look at four types of situation arising in Dan's protocol. Each situation serves to highlight a separate psychological issue. These are non-verbal thought, verbal thought, perception, and insight. The list is not intended to be either mutually exclusive or exhaustive.

3.1 Non-verbal Thought in Design Tasks

An extremely pervasive view in the psychology and creativity literature, extending right back to Freud, is that the creative process consists of two distinct modes of thinking. This idea has been also applied to design[13,14]. Such a view is indeed supported by neurological evidence suggesting that the left and right hemispheres of the brain provide different aspects to human thought. The left is concerned with temporal activities including verbal memory and speech articulation; the right is concerned with perceptual and spatial activity.

Such a theory might suggest that activities in design such as sketching and understanding a brief would be difficult to verbalize and there is indeed evidence to support this view[15]. Unfortunately it is difficult to isolate an episode of purely 'understanding the brief' but there are many points in Dan's protocol where his verbalizations stop for more than a few seconds. Analysing the video of the protocol it seems obvious that in these short periods of silence Dan is thinking deeply about something. The fact that he stops talking while thinking seems to point to some incompatibility of these particular thoughts with simultaneous verbalization. Table 20.2 gives some examples of times when Dan stops verbalizing. These seem to be points of value where decisions are made but we have no idea of the thinking behind these decisions. We focus on one such episode.

After the two-hour mark in Dan's protocol we reach a point where Dan realizes he has to begin to commit himself to a solution. Prior to this (1:59:00–2:02:00) he has been talking about

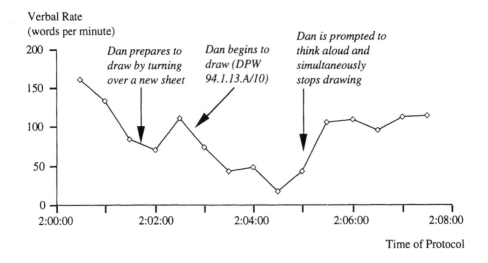

Figure 20.1
Different types of thinking begin to impinge on Dan's rate of verbalization. The third comment shows how a prompt from the experimenter seems to influence Dan's 'mode' of thinking

the issues arising from his other drawings. When he is reminded that he has only 15 minutes remaining he prepares to draw.

Figure 20.1 shows how Dan's verbal rate decreases as he prepares and starts to draw. The verbal rate is calculated by counting the number of words in the 30-second period directly preceding the point on the graph added to the number of words in the 30-second period directly following the point on the graph. The number of words in each 30-second period is thus used twice and averaged out. A 30-second period was found sufficient to illustrate our point. We need only note the general trend of down to up and the position of several key points in the eight minute episode. The general pattern is similar for the other examples of Table 20.2.

Figure 20.1 seems to imply that a non-verbal form of thought is contributing to the designing. The fact that this episode is accompanied by Dan sketching – traditionally thought of as a spatial activity – suggests that Dan is engaged in an abstract activity that he cannot verbalize. We do not suggest what form this thought might take, merely pointing out that concurrent verbalization, almost by definition, cannot provide a commentary on non-verbal thought. In this episode we do not know how and why Dan generates the drawings he does. This would not be a problem if the activity didn't seem to be as fundamental as it does when studying the video.

At its very loosest interpretation Figure 20.1 illustrates how the verbal rate of a subject can vary within a short

Table 20.2 Episodes of 'silent thinking'

Protocol time	Episode
00:23:30	'Interestingly enough ...' [15 secs silence]
00:31:52	7 secs silence before question 'do you [have any material]'
01:30:08	'see what the issues are here...' [12 secs silence]
01:02:02	'em then that's gonna be er, er, and this is um...' [15 secs silence]
01:30:02	'the erm brake guide to em ... [13 secs silence] let me just check'
1:44:39	'they want it to be quick connect em [28 secs silence] and so em em OK'

period of time. It seems logical to suggest that a possible indicator of a mismatch between designing and talking at the same time might be a decrease in the verbal rate. Conversely a verbal rate that did not vary significantly might indicate that concurrent talking and designing were in some way complementary.

3.2 Verbal Thought in Design Tasks

If concurrent verbalization does not seem an accurate reflection of a subject's thoughts, are there any times that concurrent verbalization does seem to reflect a subject's thoughts? Using our own criterion we might expect such behaviour to be accompanied by a relatively consistent verbal rate.

A whole theory of 'conventional' problem solving is built on the rubric of 'information processing', Newell and Simon being perhaps its greatest exponents[16-18]. Such experiments centre around the validity of the subject's verbalizations but also use the very constrained (state-space) nature of the problem in subsequent analysis. In short there are limits to what a subject can do in solving such problems. Design, being inherently 'ill-defined', seems a qualitatively different type of problem, although according to Simon ill-defined problems can be broken down into related well-defined problems having more in common with conventional problems[19].

The solution process of conventional problems, being a relatively short term task, is extremely directed and deductive. We expect a full analysis of the problem to take the subject a long way towards a potential solution. This may or may not be the case with design problems but the designer must spend time 'analysing' the problem, deducing implications from the brief. It is in these periods when the designer may unconsciously begin to structure the design problem into more well-defined problems. In our protocol such behaviour can be identified when•Dan comes to measure up and generally analyse the backpack and mountain bike. It is to this period we look to find whether concurrent verbalization seems a better representation of thought.

Figure 20.2 shows Dan's verbal rate, calculated as in the previous section, during the four-minute period in which he analyses the backpack and bicycle. During this period the verbal rate stays between 111 and 160. The pattern is quite different from the previous graph of Figure 20.1.

In this episode Dan is clearly focused on the sub-problem facing him which could be loosely defined as information gathering and more specifically as problem analysis. Each utterance is quick, understandable, and often deductive in nature. We get a clear idea of the information Dan is collecting and although the behaviour is difficult to predict at the micro level – after all we don't have a clear indication of Dan's motivation physically to collect information – the episode is coherent and Dan gets the information he sought.

The structure to this episode is simple: Dan has a need to understand the reality of the problem facing him and fulfils this

Figure 20.2
The graph shows a relatively stable verbal rate illustrating when verbalization might be a good representation of thought

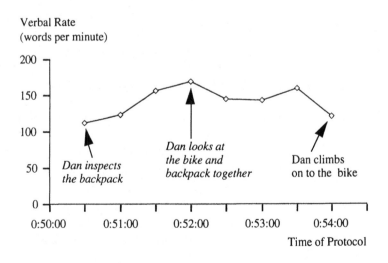

Verbal Rate
(words per minute)

need by getting out of his chair and physically measuring the problem artifacts – information that could equally well be obtained from technical drawings. During the episode we have a clear idea of the focus of Dan from both his body language in the video and his verbalization. The behaviour makes sense given his general aim – the solution of the sub-problem he set himself. During this period his verbal rate also stays fairly constant.

It seems that in this episode Dan's verbalizations are relatively close to his thoughts. Regular directed, deductive behaviour at the utterance level indicates Dan has a well-defined problem to solve. His intentions and motivations remain less obvious; these are more easily constructed by relating the episode into the context of the whole design process.

3.3 Perceiving the Design Problem

Often in design, how a designer produces a solution depends on the way that they 'see' the problem. For this reason it can be of tremendous help to know how a designer is perceiving the problem facing them. This can be expressed in quite abstract ways. Schön[20] focuses on just such aspects of design when he contrives situations in which designers reflect on their design process and review how they see the problem (discussing '*the problem* of the problem'). Designers also discuss the reasons for, and intentions arising from, a particular way of viewing a design situation. Other research shows that perceptions often occur with the first contact a designer has with a problem and can set the tone for subsequent solving behaviour[21].

In Dan's protocol there is little intimation of the way he actually understands the problem, the way he fits it into his framework of knowledge. One possible reason for this is that by asking Dan to concurrently verbalize his thoughts one is demanding that he always be speaking. This speech can, however, operate as a superficial mask on design thinking. As an example of what we mean by this we quote the passage when Dan looks over the problem brief after first having read it out almost verbatim. The timed silences include any paralinguistic activity (ums, errs, false starts etc.).

'OK so I am going to ← 3 secs → make a concept design for the device and this is a, I will write this down so that I will not forget it, carrying fastening ← 6 secs → device ← 5.5 secs → and it is to attach to a bicycle, a

mountain bike and to me that makes it different. ← 3.5 secs
→ mountain bike. OK let me just verify that ←——— 12
secs ————→ the HiStar backpack is a framed backpack,
it's an external framepack external ←— 6 secs ——→ OK
so we know that it's actually an existing frame backpack it's,
an existing bike and we're making a device that is going to
attach one to the other. Em and I have to focus on ease of use
←—— 4 secs ——→ good looks ←— 6 secs ——→ techni-
cal issues ←— 4 secs ——→ price..' (00:19:14–00:21:00)

There is clearly more to this passage than there seems. The
silences indicate that each new 'chunk' of the problem is being
assimilated in some way. We are given no clue as to the way
this new information is being understood. Could it be that by
reading the brief aloud Dan has hit on a mechanism for
masking his actual thoughts of his initial perceptions of the
problem? There is a reversal in the function of the verbal commentary: rather than commentating (externalizing) on his
(internal) thoughts, Dan reads aloud, and thus internalizes, the
(external) brief.

*3.4 Brief Moments of
Insight*

Any form of problem solving demands that the solver have
information about the situation facing them. The information
may be of many types but in solving the problem the solver
somehow has to relate this new situation to their existing
knowledge. Depending on the complexity of the situation the
solution may be found quickly (in a flash) or more methodically
as the problem situation is slowly understood. In relating
existing knowledge to the problem situation solvers often report
a 'moment of inspiration' or a 'time when everything fitted
together'. Gestalt psychology and in particular the work of Max
Wertheimer tend to call this point of a solving process
'insight'[22]. It has also been described as the 'ah ha' phenomenon[23]. A designer in one of our recent studies reviewing his
design process describes insight:

> 'All of a sudden I just twigged off. I remembered that
> drawing [an old "block" diagram] and I thought "ah just the
> right thing. Just the right thing"'[7]

In our protocol Dan describes the key points as being 'the telephone conversation' and the decision to go for 'tubular aluminium'. These are large general descriptions of the process but in

the protocol there are several lesser moments when something 'clicks'. These little insights, although often not fundamental to the design solution more often than not being ignored later in the process, occur regularly. Table 20.3 lists five examples in Dan's protocols.

The trouble with including these insights into a description of the design process is that they happen quickly and unexpectedly. Clearly there is a buildup of information before, a period often termed 'incubation'[24], but there seem to be no general rules as to when insights occur. The cognitive process seems so deep, abstract, fundamental, and above all to do with retrieval and long term memory, that it seems almost unfeasible that a designer could commentate on such moments. Indeed if it were possible then design problems would be much easier!

In Dan's protocol the process of insight has a regular form. He will be methodically working on an aspect of the problem and then, when insight occurs, his tone of voice will change from that of a commentary to that signalling recognition (the bold type in Table 20.3). What seems to happen is that for one moment part of the solution suddenly 'drops' into place. Dan seems to happen across these solutions; they don't emerge systematically. It is the process of recognition that alerts him to

Table 20.3 Episodes involving insight (bold type indicates change in voice tone)

Protocol time	Episode
00:51:50	Holding the backpack up to the bike – 'My first thought is, **hey the place to put it is back here**'
01:24:20	Talking about quick release clamps – 'What would happen is this thing would plug in and there would be er, [pause] **alright we got it**'
01:25:49	Putting the bracket on different sizes of frame – 'The catch is that those tubes have different height **AH good idea** ...'
01:28:44	A permanently fixed bracket – 'So if this bracket is on all the time **AH HA ... Better ... They're going to build it into the bicycle**'
01:30:28	Finding the two lugs already brazed on to the bike – 'I had another thought and that is to take advantage of that em space between the em er this lug and the em brake guide to em er to em ... let me just check one more thing here **oh no there is something funny about this thing see** erm ...'

a 'good fitting' solution and this whole event triggers a change in voice tone and conversational style.

In the minutes and seconds before an insight occurs concurrent verbalization may or may not be able to capture the designer's thoughts. If insight occurs in an episode of what we describe as 'non-verbal' thought then we really have no idea of what triggered the insight. Alternatively if the insight occurs in an episode of what we describe as 'verbal' thought then we may have a better idea of what triggered the insight. In each case it seems clear that a background cognitive process is operating, a process of fitting and connecting new information with existing knowledge. Although the information itself may occupy the subject's short term processing, the background recognition operation ending in the insight remains hidden to verbalization, the external pieces of the problem merely offering cues to a deeper form engrained in the mind of the designer.

4 Capturing a 'Real' Design Episode

So far we have discussed whether Dan's verbalizations during the experiment represent his thoughts, and by implication whether we can piece together these thoughts to give a causal description of the design process. What we have assumed in doing this is that Dan was actually *designing* and not performing some other form of complex problem solving. For this to be true the experimental situation should intrude on the subject's behaviour as little as possible and certainly not influence the subject in any way.

In Dan's protocol there is some evidence to suggest that the methodology is interfering with the design process. Several times Dan explicitly tries to second-guess what the experimenter 'wants to hear' and often seems engaged in activity that he feels he should do rather than what he would normally do. For example the instinct to get up and look at the bike and backpack seems a strong one which Dan seems to quell for some time, while 'forcing himself to go through other types of information'. The design situation is artificial not least because it imposes a short time limit on designing which does not allow any change in the intensity of the situation. We expect (for example) that somehow all behaviour will be evenly directed towards achieving a satisfactory solution; we don't suspect any alternative motivation(s).

We go on to explore just how artificial this situation is by

relating Dan's protocol to other theories and accounts of designing that talk of design as a social activity, a solitary activity, and
an activity continually varying in intensity. We discuss four
aspects of design that we feel necessarily contribute to the design
process: negotiation between designer and client, discussion
between a designer and their peers, variations in the intensity of
the design activity, and design synthesis as a considered process.
For each aspect we draw on our own data for general examples,
and also specific examples from Dan's designing. Again these
aspects are representative and not exhaustive.

*4.1 Negotiating with a
Client*

In a design project it is the client who sets the borders for what
the designer can do. It is rare for a designer to receive *'carte
blanche'*. The problem as it appears in the brief, though, is far
from definitive. It is more a collection of variables, and even
then an incomplete collection of variables. One of the jobs of the
designer is to find out how the client understands these variables and the range of allowable values for the variables. The
designer investigates where the client will draw the line and say
'no, that's not what I want'. When the designer has a clear view
of how the client sees the problem they can start to work on a
solution. During this process, however, additional variables may
be 'uncovered' and again these need to be negotiated with the
client. The relationship between the solution development of the
designer and client negotiation is thus a close one. There are
models of the design process that stress this relationship[3]. An
interview with Eva Jiricna, a Czechoslovakian architect, revealed
how designers typically describe client–designer negotiation:

> 'I try to express in words what they (the client) want, and in
> the same way – in words – I try to twist it into a different
> statement and make them accept a verbal statement and only
> then draw it.'[5]

This illustrates how a designer can very subtly explore what the
client means in their statement of the design problem. The
designer can then take advantage of this meaning by exploiting
the ambiguity of words in a design situation. Between them
they negotiate the limits of the design brief. Design problems
vary in the amount of involvement a client will have in the
solution process but there will always be a 'client factor'. In
Dan's protocol, although there is a design brief, there is no
client. This situation is artificial because it portrays the design

problem as a static, definitive entity and not the dynamic, malleable entity we know it to be. One might even argue that a definitive statement of the design problem cannot exist, only the designer's and client's interpretation of that brief.

In Dan's protocol there are several times when he needs to negotiate with a client. Examples of this need come when Dan wishes to discuss whether brackets can be brazed onto the bike frame (he receives a short answer); a second example comes when he wishes to find out if the 'two bushings' are in line (he refuses to believe the reply he gets). The experimenter fulfils part of the client's role but the interactions are not what we would normally expect to be negotiation. An example highlighting the absence of a client comes when Dan attempts to work out how much weight the backpack will hold. Initially, after quite a detailed calculation Dan arrives at a maximum value of 12 kg:

> 'I will assume that er you've got it filled up with something that has a specific gravity, let's just see just try I dunno what specific gravity now point em three er so forty it'll be ... em that's er forty kilograms so you probably can get a specific gravity of em point three I'm going to assume that er three four twelve kilograms er max' (00:50:00)

Reflecting on this amount he reduces this calculated value to one he intuitively feels fits better (9 kg):

> 'I would say probably in the order of er twenty pounds alright approximately nine kilograms I think that that's more like it em of what a person might have OK' (00:51:40)

When he returns later in the protocol to the backpack weight issue, he reduces the value further (7 kg):

> 'This pack can't possibly hold that much weight so I'm going to lower my weight estimate to er no more than fifteen pounds in a pack like this so...' (01:53:00)

The example serves to illustrate how a value that one would normally assume a client would have a firm idea of, varies considerably when the designer has no one to negotiate with. Dan is free to fit the value to his design in this example. Without the client in the design process, the designer possesses a freedom that they would not normally be allowed if the designing were 'real'.

4.2 Discussing Ideas

Distinct from negotiations with a client are other social interactions in the design process involving dialogue, typically with peers. Such interactions have also been incorporated into models of designing[25]. This form of communication, rather than allowing the designer to understand the client's view of the situation, allows designers formally to externalize their own ideas about the problem and gain objective criticism. Such discussion commonly takes place between colleagues and often hinges around a shared language[26]. As an example of such an interaction we quote from an interview describing a scene in which a group of architects were focusing on the design problems of a residential housing centre:

> 'I can't quite remember what happened and somebody, either Dorian or I, said "it's a wall!" and I said "yes it's a wall, it's not just a lot of little houses, it's a great wall 200m long and three storeys high, but we'll make it a bit higher because we've got some plant, all the service circulation will get it up, we'll get a high wall and then we'll punch the residential elements through that wall as a series of glazed bays which come through and stand on legs".'[5]

The 'wall' is being used in a metaphorical sense, but it is a sense that all the participants in the conversation understand, and a sense from which they can discuss the progress of the design. This emphasis on design as a dialogue is not so much on group design, but on how the individual designer feels comfortable in expressing their own ideas; how they extract information from others about their own project.

This form of interaction is frustratingly denied Dan in the protocol. Several times it appears Dan wishes to engage in discussion; each time he receives a short reply. These are 'scenes' that social psychologists might find unusual, in which rapport and body language attempt to deny the existence of a second party. This is inevitable given the role of the experimenter in protocol analysis as being in some way 'outside' the situation so as not to interfere with the task. The opposite seems to happen, however.

A good example comes in the episode that we described in detail earlier in the paper (phase 10 in Table 20.1). This episode concerns the mounting brackets for the carrier. In the period 01:23:00–01:31:30 Dan looks for suitable mounting for the rear carrier he has in mind. After being told the Batavus company is not prepared to braze on additional components to the bicycle's frame, Dan's solving loses some momentum. He notices two bushings, however, and asks the experimenter what these are for:

Dan	I mean I've these two little bushings here I don't know what they are [for], and you can tell me what they are right, these two little bushings? What use are those little bushings?
Experimenter	They are mounting points for rear carriers
Dan	They are mounting points for rear carriers?
Experimenter	Yep.

The episode is odd because it appears the experimenter has withheld this information from Dan. He has unwittingly reinforced the experimental situation to Dan. The episode appears alien because such a situation would never arise in a 'real' design situation. Any extra person present in a real design situation would help along the design process, pointing out useful aspects of the situation to the designer rather than leaving them to fend for themselves. The artificiality has arisen because the experimenter has assumed the role of experimenter and not that of a colleague. At a stroke the experimental situation succeeds in alienating the designer, unnecessarily affecting the design process.

The way to incorporate both negotiation with a client and discussion with peers into a design experiment is to have well-defined roles for all people taking part in the experiment. The role of the client, then, is to have a definite agenda for the design product. The role of the peer is to provide support for the designer and give a different perspective on the design process. If there is only one experimenter then he or she might have to switch between these, and other possible, roles. The basic point is that these interactions are vital to a 'real' design process and should therefore be encouraged by the use of role play.

4.3 Displacement Activity

Many texts in psychology, and other disciplines where creativity is a factor in solving problems, stress a need for the solver to think in different ways about a problem (de Bono[27,28] for example). It is suggested that it is best for the solver to 'do something else for the time being' or 'forget about the problem for a while'. The process is known as displacement and the manner in which this behaviour can 'unconsciously' help the solver is termed 'incubation'. Such incubation may lead to insight, discussed earlier. Design, again, is no exception and it is well known to designers that changes in the environment and intensity of solving behaviour can often provide solutions to nagging problems. Richard MacCormac, summarizing how he sees the nature of the designing process, describes the variation in intensity of problem solving:

> 'It's a process that is both fast and slow, I find, because there are these very hot moments as it were when things happen very quickly but there are also periods of reflecting and criticism.'[5]

Anderson[29] explains the reason for the effectiveness of such behaviour as a 'change in mental set', or a period where the short term memory processes information not ostensibly to do with the problem solving. Obviously one doesn't gain insight into a problem purely by doing something different; clearly there are other factors involved, but if we are to understand anything of such processes in design activity then the experimental methodology should be flexible enough to support such behaviour.

In Dan's protocol there are unplanned moments of displacement that occur. For four minutes (00:33:00–00:37:00) there is preparation for a telephone call, for example. However, it is the moments when Dan seems to realize that there is a need for a change in the intensity of problem solving that are of interest. Several times during the protocol Dan outlines what he might have done at an appropriate point in the designing. Table 20.4 lists three such episodes and a later comment from the post-session review.

Within the experimental procedure of protocol analysis Dan is not able to carry out his wishes and clearly he is able to design without such excursions. It is not clear, however, how such excursions *might* have affected his design. What we get is not a proper representation of design. At an extreme we might even consider the rest of the design activity as in some way invalid, for what if Dan had called up his friend and suddenly proclaimed half-way through the conversation that he'd just had a brilliant idea for a solution? In such a situation the design process may have been very different.

One might argue that any change in environment (or stimulus) is a change in mental set, right down to asking for

Table 20.4 Dan's expressions of a need to change the intensity of problem solving. Note how in the first example Dan seems to want to go on talking about something not to do with the design. He really only need say one sentence.

Protocol time	Episode
00:45:00	'I had considered calling another friend em, I'm not gonna do that right now because, in the interests of time here, this is a friend and I'll end up having to talk to him about other things [...] if I were normally designing this thing, rather than calling him during the day I might call in the evening and have a chat with him – but then I'm gonna have to ask about his wife and his kids and and all these other things and also how the bicycle business is...'
00:58:58	'I might actually at this stage of the game actually do some trials of my own, go out and see if I found people who had bike racks and do a little more field testing myself, but since I don't have much time...'
01:25:24	'actually what I would do is run down to the local windsurfing store and buy a boom extension, and I would mock it up with a boom extension or something like that...'
Review	'normally I'd sort of go sit down and think about it for a while [...] I'm not one who's strong into em, you know I could have asked for a profile on the company...'

some new information about something, and indeed there are moments when the intensity of designing does vary. The beginning of the protocol, for example, is very much more relaxed than the end. Underlying protocol analysis, however, there seems to be a requirement for the solver to remain at one level of intensity, thinking aloud at a regular pace, and methodically working towards a solution. Such assumptions render the design process artificial.

The constraint of time within a design process is an important one. Too much time and a designer has the chance to change their mind over and over again. Too little time and mistakes are made and ill-considered solutions are 'thrown' together. These of course are extreme views, and usually there is a fairly large time range that a design can be completed in. Indeed it is often noted how design activity increases to fit the time allowed.

In a description of a 'normal' design process we would expect a design brief to arrive some time before a designer might begin to work on that brief. Many design studios and companies employ designers to work on a number of projects and so any new project must wait its turn – that is unless the content proves irresistible for the designer or the project is a 'rush job'.

Time as a background constraint, however, is at the very least in the order of days, and is usually weeks or months. The point is that, although there may be inspirational moments, overall designing is a considered process. Client and designer together would like to think that they are getting the best possible solution for a particular problem. In achieving this end there must be a period where a designer weighs up alternatives, critiques their solution, or simply just leaves the problem solved for a few days before 'officially' submitting it.

We are of course talking here about a complete design process, from brief to detail solution. If we were to perhaps isolate the 'conceptual' design process this might be finished early on, though the amount of time between receiving the design brief and identifying a design concept will still be quite large. (One never really knows if an initial concept will remain through the whole design process.)

With regard to Dan's protocol there is obviously an increasing sense of desperation in the last half an hour or so of the protocol. We have already noted the conflict between Dan's commitment to think aloud and his inclination to think in other ways but additionally he seems to cut corners and artificially hurry towards his final goal which is not, as it turns out, a design of sufficient detail to manufacture. Concurrent verbalization is forcing Dan to change the nature of his design process:

'OK, now I'm ready to tell, I have a half hour left, I haven't even gotten to some of the calculations but I'm not gonna worry about them right now, em, I have to essentially come

up with a thing so lets just draw something out here alright
we have a em … er …' (01:54:27)

Dan is beginning to miss out things he might normally do, for
the sake of the experimental methodology. At the end of the
two hours Dan frantically tries to draw everything together to
meet his deadline:

'I'm gonna actually want to go and weld those things there
so it's gonna be a detail, what's this thing? Let's do that em,
this is the top view, I'm going to draw it like that … OK so
you're going to stop me in about ten seconds' (2:13:30)

The foregoing examples illustrate that the methodology
protocol analysis prescribed means that the design task must be
presented as unseen by the designer (to capture *every single
thought* a designer might have about the problem). This again
renders the designing artificial as this simply does not
happen. As an analogy we wouldn't think of the essay a student
writes sitting an hour-long English exam as being a fair representation of one she might compose in her own time. Why then
do we expect a designer to be properly designing in such a
situation?

The time when a designer sits down and focuses on a design
task takes place some time after they have received a brief.
There is still much valuable information to be gained at such a
point, initial thoughts still need to be worked out, the design is
not 'over and done with'. A far more accurate experimental
methodology would be one that did not attempt to give the
designer a new task 'cold' but allowed a period of time for the
designer to digest the task – though not to focus on it; in other
words a methodology that didn't force the designer into
working any faster than they felt happy doing.

5 General Discussion

Dan's protocol contains many different episodes and more
importantly many different senses in which Dan uses language
– he is clearly not thinking aloud for much of the protocol.
Dan's commentary consists of direct questions, rationalizations,
well-formed sentences, hesitation, verbatim reading, retrospection, statements of strategy, phone conversations, pet
phrases ('off the bat' is used at least four times), avoidance strategies ('OK, yeah, I'm thinking aloud, alright…') and, we might

conclude, some thinking aloud. Language is being used in many ways; all words in the protocol are not equal in function.

What we hope to have illustrated is that protocol analysis, and the constraint it brings both theoretically and methodologically, *interferes* with designing. This much we feel is beyond doubt. Our initial analysis attempted to show something of what aspects of design a concurrent verbal methodology does reveal; our subsequent critique showed what it cannot, or at least does not seem to reveal. What we feel would be helpful for the use of classical protocol analysis in design is to define the characteristics of failure. When, for example, does protocol analysis *not work* as a research method? This issue, to our knowledge, has not been addressed.

A misconception that remains implicit in using protocol analysis in design is what we term 'the unitary notion of design', that is the idea that designing is one 'thing'. The premise of protocol analysis is that by putting a designer into a laboratory and asking them to design while thinking aloud we are able to capture much information about this 'thing'. What we hoped to have shown is that design is not a unitary thing – it would be absurd, for example, to put a golfer in a laboratory and hope to capture everything about golf – but a collection and pattern of many things. The time to put a designer into a laboratory is when an aspect of design can be isolated and studied either separately or in context.

Allied with a view of design as a unitary 'thing' is the idea that every designer must be following a valid design process and that the study of any designer, no matter how inexperienced, must tell us something about designing. Clearly the incarnations of design vary considerably and this seems to imply that perhaps better questions to ask are what it is that distinguishes a good designer from a bad one, a novice from an expert, an architect from an engineer.

There are many methods of investigating the design process and each has salient features. In the present paper we have used examples and theories arising from many types of enquiry. Interviews[5], retrospective reports[30], concurrent reports[31], teaching[32], and introspection[33] all have something to contribute to an empirical understanding of the design process. What one should then look for is *consistency* in the results that each method proffers. If we cannot use protocol analysis to elicit the design thinking of Richard Rogers, then we must use other methods of analysis. After all Richard Rogers is probably too busy for a demanding session of protocol analysis!

References

1   Pilkington, E., Ghetto blaster, *The Guardian* (5 May 1994)
2   Ericsson, K.A. and H.A. Simon, *Protocol Analysis: Verbal Reports as Data*, MIT Press, Cambridge, MA (1993)
3   Lawson, B., Parallel lines of thought, *Languages of Design*, **1** (1993) 357–366
4   Ryle, G., *The Concept of Mind* Penguin, Harmondsworth (1949)
5   Lawson, B., *Design in Mind*, Butterworth, Oxford (1994)
6   Lloyd, P.A. and P.J. Scott, Difference in similarity: interpreting the architectural design process, *Environmental and Planning: Planning and Design*, **22** (1995) 383–406
7   Lloyd, P.A., *Psychological investigations of the conceptual design process*, PhD thesis, University of Sheffield (1994)
8   Lloyd, P.A. and P.J. Scott, Discovering the design problem, *Design Studies*. **15** (1994) 125–140
9   Hillier, B., J. Musgrove, and P. O'Sullivan, Knowledge and design, in W.J. Mitchell (Ed.), *Environmental Design: Research and Practice*, University of California, Los Angeles (1972)
10 Jones, J.C., A method of systematic design, in N. Cross (Ed.), *Developments in Design Methodology*, Wiley, Chichester (1963) pp. 9–31
11 Pahl, G. and W. Beitz, *Engineering Design*, The Design Council, London (1984)
12 Nisbett, R.E. and T.D. Wilson, Telling more than we can know: verbal reports on mental processes, *Psychological Review*, **84** (1977) 231–259
13 Tovey, M., Thinking styles and modelling systems, *Design Studies*, **7** (1986) 20–30
14 Tovey, M., Intuitive and objective processes in automotive design, *Design Studies*, **13** (1992) 23–43
15 Daley, J., Design creativity and the understanding of objects, *Design Studies*, **3** (1982) 133–137
16 Newell, A. and H.A. Simon, *Human Problem Solving*, Prentice Hall, Englewood Cliffs, NJ (1972)
17 Newell, A., On the analysis of human problem solving protocols, in P.N. Johnson-Laird and P.C. Wason (Eds), *Thinking*, Cambridge University Press, Cambridge, (1977) 46–61
18 Simon, H.A., A. Newell, and J.C. Shaw, The processes of creative thinking, in H.E. Gruber, G. Terrell and M. Wertheimer (Eds), *Contemporary Approaches to Creative Thinking*, Atherton Press, London (1962), pp. 63–119
19 Simon, H.A., The structure of ill-structured problems, *Artificial Intelligence*, **4** (1973) 181–201
20 Schön, D.A., *The Reflective Practitioner*, Temple Smith, London (1983)
21 Darke, J., The primary generator and the design process, *Design Studies*, **1** (1979) 36–44
22 Wertheimer, M., *Productive Thinking*, Harper, New York (1978)
23 Köhler, W., *Gestalt Psychology*, London (1947)
24 Hadamard, J., *The Psychology of Invention in the Mathematical Field*, New York (1945)

25   Lawson, B., *How Designers Think*, Butterworth, London (1990)
26   Medway, P., Building with words: discourse in an architect's office, in *Carleton Papers in Applied Language Studies*, Vol. IX (1992)
27   de Bono, E., *Teaching Thinking*, London (1976)
28   de Bono, E., *Lateral Thinking*, London (1977)
29   Anderson, J.R., *Cognitive Psychology and its Implications*, W.H.Freeman, New York (1990)
30   Guindon, R., Designing the design process: exploiting opportunistic thoughts, *Human–Computer Interaction*, 5 (1990) 305–344
31   Akin, Ö., An exploration of the design process, *Design Methods and Theories*, 13 (1979) 115–119
32   Schön, D.A., *Educating the Reflective Practitioner*, Jossey-Bass, San Francisco (1987)
33   Galle, P. and L.B. Kovàcs, Introspective observations of sketch design, *Design Studies*, 13 (1992) 229–272

Lightning Source UK Ltd.
Milton Keynes UK
UKOW07n0600251116
288471UK00002B/6/P